This book is dedicated to the victims of environmental disasters.
A part of the royalty from this book will go to help such people.

Unfortunately, trees have to be felled to produce any book.
Another part of the royalty will help grow more trees.

Preface

To the student

'Oh, no! Not one more subject and one more book to be mugged up!' As it is, the load is heavy enough with so many tough and boring things to study. How can I manage the new course? Why should I bother about the environment?'

This could well be your response when told about the new compulsory course on Environmental Studies. As you read this book, however, you will discover that this subject is different from many others: It is about you, and the future of your family, community, humanity, and this fragile planet.

About this book

This book is meant for undergraduates of any discipline and it assumes no background in mathematics or science beyond the tenth standard. It is based on the new University Grants Commission (UGC) syllabus that came in the wake of the landmark judgement of the Supreme Court of India in 2003, to prescribe a course on the environment for colleges and to consider the feasibility of making it a compulsory subject at every level in college education.

A short extract of the Supreme Court judgement is given below:

> … for more than a century there was a growing realisation that mankind had to live in tune with nature, if life was to be peaceful, happy, and satisfied. In the name of scientific development, man started distancing himself from nature and even developed an urge to conquer nature. Our ancestors had known that nature was not subduable and, therefore, had made it an obligation for man to surrender to nature and live in tune with it.

Even as you cover the entire UGC syllabus, learn the concepts, and try the short-answer and essay questions, you will:

- learn about the major environmental problems, such as runaway growth, imperilled ecosystems, disappearing forests, endangered species, dwindling natural resources, escalating pollution, growing population, dangerous toxins, green laws, etc.;
- find out what is being done about these problems;
- discover how YOU can make a difference to the state of the environment;
- savour reading over one hundred short environment-related stories about crises, solutions, successes, failures, interconnections, and inspirational individuals; and
- reflect on the prologues, quotations, poems, and deeper issues.

In addition, this book will aid you in:

- finding ideas and guidance for meaningful field work;
- surfing the Internet for more information on environmental issues which interest you;
- locating books and magazines on the subject;

- choosing a satisfying and rewarding career in environment-related areas; and
- joining an environmental organization or forming one.

To make things easier for you, there is a glossary of terms.

This is in fact more than a textbook. It is about your life and what is in store for you. It is about the present and future of the Earth. The text and the stories are meant to evoke surprise and shock, despair and hope, resolve and action. You will surely return to it even after the course. It may in fact change your life!

A note on the references

The material in this book has come from a variety of sources: books, textbooks, journal papers, magazine articles, news reports, documentary films, etc. You will find at the end of the book an extensive list of references. Most of the websites cited were accessed during August–December 2004.

Feedback is welcome!

Many a topic covered in this book could fit in more than one chapter. In this matter, I have followed the UGC syllabus except for a few minor adjustments. The publishers and I will be delighted to receive feedback and suggestions from you on any aspect of the book such as:

- an evaluation of the content, style of presentation, production, other features, etc.;
- any errors, inconsistencies, wrong figures, unclear illustrations, difficult terms, etc.;
- topics to be included in the next edition;
- an update on the stories and any additional information you may come across;
- new stories of environmental degradation, conservation efforts, individual and group attempts, successes and failures;
- other websites, stories, poems, quotations, etc.; and
- your experiences in environmental conservation.

Your feedback will help us improve the book. All feedback can be sent to *rrgopalan2005@gmail.com.* Thank you.

On a personal note

The writing of this book absorbed me totally for about a year. I went through various moods—curious, interested, mildly hopeful, extremely depressed. The horror stories were many, the hopeful ones were few. At times I wanted to give up, weighed down by the thought of the coming apocalypse. But I pushed on thinking of the one student whom the book may influence and convert into an environmentalist. The book is really for that unknown student. I hope, of course, that there will be many more than one!

February 2005 **R. Rajagopalan**

Acknowledgements

I do not know what I would have done without the help of Alan and Ramjee. Apart from providing detailed comments and corrections on every chapter, they kept my spirits up, whenever I was discouraged. They were also my sounding board on many issues concerning the book.

I would not have embarked on this project, had I not been involved for the last five years in conducting environmental workshops at Auroville. I must thank Tency Baetens for inviting me to join the organizing team. It has been a great experience working with Alan, Bhavana, Claude, Harini, Lata, Prashanth, Tency, Tineke, Silvano, and others at Auroville.

It was my entry ten years ago into the International Ocean Institute (IOI) that made me think deeper on environmental issues. The Late Professor Elisabeth Mann Borgese, founder of IOI, was a great source of inspiration. I thank all my colleagues in IOI for their appreciation of my work with coastal communities. In particular, I am grateful to Krishan Saigal and Masako Otsuka for their constant support.

My friends in Navadarshanam—Ananthu, Atma, Jyoti, Om, Pratapji, Pushpaji, and Sudeshji—encouraged me in the task of writing this book and I thank them.

I would like to thank Francois Gamache for making my stay in Vancouver so comfortable that I could spend almost all my time on reference work in the libraries of that great city.

I am grateful to my friends who have always been a source of strength—Ahana, Chitra, Dilip, Muraleedharan, Narendran, Sekhar, Subramanian, Vishwanath, and many others.

I am very grateful to the authors of several excellent American textbooks on Environmental Science. In particular, I have drawn heavily from Botkin and Keller (1995), Chiras (2001), Cunningham and Saigo (1995), Enger and Smith (2004), Miller (2004), Nebel and Wright (1998), and Raven and Berg (2004). Their canvas was much wider and the books were aimed at undergraduates who had chosen Environmental Science as a subject. However, I learnt a great deal from reading their books and I have used many of their ideas. If this book is welcomed in Indian academia, they will have a share in its success.

My thanks for reprint permissions:

- Excerpts from pages 85 and 87–88 are from Eknath Easwaran, *The Compassionate Universe*, Penguin India edition, copyright 1989, reprinted with the permission of Nilgiri Press, P.O. Box 256, Tomales, CA 94971, *www.nilgiri.org*.
- Excerpts from the essay by Lois M. Gibbs, 'We Have Been Asking the Wrong Questions', in G. Tyler Miller, *Living in the Environment* 13e, copyright 2004, pp. 530–31, are reprinted with the permission of Lois M.Gibbs.

I have requested for reprint permission in the following cases:

- Excerpts from the Rio Summit speech of Severn Suzuki from David Suzuki, *Time To Change* (1994, Stoddart, Toronto), pp. 227–30.
- Excerpts from, 'How Can One Sell the Air?', from the book of the same name published by The Book Publishing Company, Summertown, USA.

I have taken many poems from websites and I record my appreciation for the creative work of the authors.

If the environmental lawyer M.C. Mehta had not filed the case in the Supreme Court in 1991 and again in 2003, this book would never have been written. He does not know me, but I must thank him not only for taking initiative in the matter of environmental education, but also for all the legal battles he has been fighting in the cause of India's environment.

I thank Oxford University Press India for supporting the idea of writing this textbook and for extending invaluable editorial help to me.

Finally, I thank Malathi, Kumar, and Anand for being patient with me all through the months I spent sitting before the computer screen.

R. Rajagopalan

Contents

MODULE 1

The Global Environmental Crisis

CHAPTER 1

The Global Environmental Crisis

Only after the last tree has been cut down,
only after the last river has been poisoned,
only after the last fish has been caught,
only then will you find that money cannot be eaten.

Native American Prophecy

What will I learn?

After studying this chapter, the reader should be able to:

- appreciate the fact that the world is facing an environmental crisis;
- describe the global environmental crisis through some examples and statistics;
- understand that the crises of security, development, and environment are interrelated;
- describe the carbon dioxide, consumption, population, and species extinction spikes and their implications;
- trace the origins of the environmental crisis to the Idea of Progress and the change in our attitude towards nature;
- explain the concept of ecological footprint through examples;
- recall the major international meetings and agreements on the issues of environment and development;
- define terms like environment, ecology, and environmental studies; and
- appreciate the interdisciplinary nature of environmental studies.

Box 1.1 Crisis: Desperate district

The monsoon has rarely failed this area. It gets an average annual rainfall of 1250 mm, more than what Punjab receives. The water table in some places is very high. Yet, Kalahandi in western Orissa is frequently in the news for its extreme poverty and deprivation. Often it is under drought and sometimes it has floods. The people migrate to all parts of India looking for work and survival.

Kalahandi, however, was not always a place of hunger and deprivation. Nineteenth century travellers have talked about a mass of jungles and hills in the region. Just a few decades ago, it was all green here. The forests provided a livelihood for six months and agriculture for the next six months. With a large diversity of crops, it was one of the richest areas in eastern India. How did things change?

In the past, there was a network of about 30,000 traditional water-harvesting structures in the area. They included ponds, lakes, check-dams, and even tanks within paddy fields. The whole system was designed to suit the topography of the land so that no part of the rainwater went waste. What is more, the system was under the control of the community, which ensured proper maintenance and water-sharing through *jal-sabhas* (water councils).

Many tanks were built with the free labour given by the people. There was also a social system to protect the forests in the catchment areas. About 50 per cent of the cultivated land was irrigated using the water bodies. Failure of rainfall did occur, but there was never any scarcity of water. Came Independence and the troubles started. First, the government took over many of the structures, but did not maintain them properly. In other cases, fearing a takeover, the landowners converted the tanks into croplands. The focus shifted to large irrigation projects like the Hirakud Dam, though the canal system often did not suit the local topography.

Concurrently, the forests were being cut down for timber and the resulting erosion of the soil dumped a lot of silt on the catchment areas. River-beds also silted up, leading to floods downstream. A large amount of water was being lost as run-off. From then on, even a slight shortfall in rain brought water scarcity and a large-scale crop failure. Agriculture became a difficult proposition and people started migrating in large numbers, affecting community life and preventing any revival of the water-harvesting system. Kalahandi thus slid into a vicious circle, from which it has never recovered.

Source: Mahapatra and Panda (2001)

Is there a global crisis?

Kalahandi is a classic case of environmental degradation leading to poverty and deprivation. Is Kalahandi a rare case or is it a typical one? Is there really a global environmental crisis? If there is a crisis, what is the cure? These are the questions addressed in this book.

What is the state of the world today? The United Nations Environment Programme (UNEP) periodically produces a comprehensive global state of the environment report, called the *Global Environment Outlook (GEO)*. The latest was *GEO-3*, published in 2002. The Worldwatch Institute, an independent research organization based in Washington, D.C. in the US, also publishes an annual *State of the World* report listing the significant environment-related events of the previous year. The Centre for Science and Environment (CSE), New Delhi, a public interest research and advocacy organization, periodically issues reports on the state of India's environment. The latest one, the *Citizens' Fifth Report*, came out in 1999.

Based on *GEO-3*, the *State of the World 2004*, the *Citizens' Fifth Report*, and data from several other sources, here are some indicators of the current global and Indian situation:

Water

- One fourth of the world's population, or about 1.5 billion people, have no access to safe drinking water, and half the world's population lacks sanitation facilities. Some 80 countries suffer serious water shortages.
- In India, more than 60,000 villages do not have a source of drinking water. Diarrhoea, caused by drinking contaminated water, claims the lives of 1.3 million children in India every year. Further, 45 million people in the world are affected annually by poor quality water.

Biodiversity

- Worldwide, 24 per cent of mammals, 12 per cent of birds, 25 per cent of reptiles, and 30 per cent of fish species are threatened or endangered, that is, 100 to 1000 times the rate at which species disappear naturally.
- More than 10 per cent of India's recorded wild flora and fauna are threatened and many are on the verge of extinction.

Forests

- The net loss in global forest area during the 1990s was about 94 million hectares. At present tropical forests are being cleared at the rate of one hectare per second.
- India's forest cover declined from 40 per cent a century ago to 22 per cent in 1951 and further to 19 per cent in 1997. The quantitative decline in forest cover is supposed to have been arrested since 1991, but the qualitative decline persists.

Land

- Each year, six million hectares of agricultural land are lost due to desertification and soil degradation. This process affects about 250 million people around the world.
- India has nearly 130 million hectares of wasteland as compared to 305 million hectares of biomass producing area.

Pollution

- At least one billion people around the world breathe unhealthy air, and three million die annually from the effects of air pollution.
- The World Health Organization (WHO) consistently rates New Delhi and Kolkata as being among the most polluted megacities in the world.

Traffic jams and air pollution in an Indian metro

Coastal and marine areas

- Worldwide, 50 per cent of the mangrove forests and wetlands that perform vital ecological functions have been destroyed.

- Over the past 40 years, India too has lost more than 50 per cent of its mangrove forests. The absence of the mangrove forests once found along the Orissa coast accentuated the damage to life and property when the supercyclone struck in 1999.

Disasters

- Across the world, the numbers of people affected by natural disasters have risen from an average of 147 million a year in the 1980s to 211 million a year in the 1990s. Floods, droughts, and windstorms accounted for more than 90 per cent of the people killed in natural disasters.
- India is the most disaster-prone country in the South Asian region. Drought, floods, earthquakes, and cyclones occur with grim regularity. Ten thousand people died in the Orissa supercyclone of 1999, 16,000 in the Gujarat earthquake in 2001, and at least 30,000 in the tsunami of 2004.

Energy

- Two billion people around the world go without adequate energy supplies.
- India imports more than 50 per cent of its oil needs, primarily for the transportation sector.

Global warming

- There are clear signs of global warming. The period September 2003 to November 2003 was the warmest quarter in recorded history. The permafrost and glaciers in the polar and other regions are melting, and the resultant rise in the sea level is threatening the very existence of many small islands.
- In May 2002, the summer temperatures in Andhra Pradesh rose to 49°C, resulting in the highest one-week death toll on record. Glaciers in the Himalayas are retreating at an average rate of 15 m per year, which is consistent with rapid global warming.

Urbanization

- About half the world's population now lives in urban areas, as compared to little more than one-third in 1972. About one-quarter of the urban population lives below the poverty line.
- About 23 per cent of the population in India's metros lives in slums. Dharavi in Mumbai, the largest slum in Asia, is home to 500,000 people in a small area of 170 hectares.

Overcrowding in an urban area—slums, apartments, crowded streets, traffic jams, etc.

We can continue quoting statistics, but the picture is clear—severe environmental degradation is occurring throughout the world. What is the reason for this depressing state of affairs? A large number of scientists, environmentalists, politicians, social workers, and thinking citizens have come to a single conclusion:

We are consuming natural resources at a rate much higher than that at which nature can regenerate them. We are polluting the environment at a rate greater than its ability to absorb the pollution. This is an unsustainable way of living and it can only lead to an environmental and social catastrophe.

Should we not worry more about armed conflict than about the environment?

It is true that we have a security crisis. Armed conflicts did not end with World War II and global security remains a problem. The second half of the twentieth century was marked by many wars—proxy wars, civil wars, ethnic wars, secessionist wars, and resource wars. At the same time, weapons became more powerful, sophisticated, and destructive. By the year 2000, money spent on armies globally exceeded US$ one trillion per year and the number of refugees crossed 20 million. New wars have begun in this century and the uncertainties continue. Resources that could go into development are instead used for destruction.

Clearly, wars of any kind destroy people and their livelihoods, particularly in the poorer countries. They also degrade the environment through the planting of landmines, destruction of irrigation systems and water resources, interference in planting and harvesting crops, etc. There is also the real danger of future wars being fought over scarce resources like oil and water.

What about the hunger, poverty, and deprivation faced by millions of people?

The crisis of development arises from the fact that 20 per cent of the world's population remains desperately poor, even after several decades of international development aid. Poor countries have become enormously indebted and incoming aid is often used only to service old debts. Many countries have cut their social welfare programmes leading to further deprivation of the poor. The disparities between the rich and the poor have only widened. Again, the indigenous people, and the children and women among the poor are under particular stress in many developing countries.

Wars have turned large populations into refugees, without food, water, and shelter. In addition, there are millions of 'environmental refugees', displaced from their lands to make way for large developmental projects like dams. Again, as in the case of Kalahandi, environmental degradation also turns people into refugees.

What is the crisis of environment?

The world is heading for an environmental disaster, and like the three wise monkeys, we do not want to see, hear, or talk about it. What is even worse, we sometimes do talk about it, and then go about our business as if the environment and our lives were quite separate.

Many environmental problems arise from the deliberate or inadvertent abuse, misuse, and overuse of natural resources by human beings. Land, water, energy resources, air, and space have all been adversely affected by human intervention.

Surely nature will take care of these problems over time? The earth has existed for over five billion years, humanity for three to five million years and civilization for about 10,000 years. They have survived many crises and cataclysms. Looking back over the centuries, nature seems to have always absorbed disturbances and maintained a balance. Thousands of species have survived over a long period of time and, consequently, may be expected to continue to exist forever. Will we not survive the current environmental crisis too?

We may be correct in expecting the Earth and many species to survive, but we may be wrong in assuming that humanity will also continue to exist along with nature in the same way. The reason is that there is something different happening now.

What is different about the current scene?

In the past, changes were always slow, but this is no longer true. Human activity has drastically increased the pace at which changes in the environment are taking place. What is happening now can be explained with the help of a simple mathematical curve or graph showing the *exponential growth* of a quantity with time. This curve is relatively flat at the beginning, but it becomes steeper with time.

Box 1.2 explains exponential growth, while Figure 1.1 contains a graphical representation of a curve showing exponential growth. Box 1.3 recounts a story that illustrates how powerful exponential growth can be.

Box 1.2 Concept: Exponential growth

Suppose you invest Rs 1000 in a bank, which gives you an annual interest of 10 per cent. You ask the bank to reinvest the interest earned every year (we call this compound interest). At the end of one year, your account will have grown to Rs 1100. At the end of the second year, the accumulated balance will be Rs 1210 (i.e., Rs 1100 + 110). Thus, every year, an increasing amount of interest gets added to the principal. What will be the accumulated amount at the end of 50 years? Guess the amount before reading further!

In 50 years, the amount would have grown to more than Rs 117,000! If you did not know the power of compound interest, you would have surely underestimated the value. This kind of increase is called exponential growth.

Compare this growth with the case of simple interest. The initial deposit of Rs 1000 will earn a

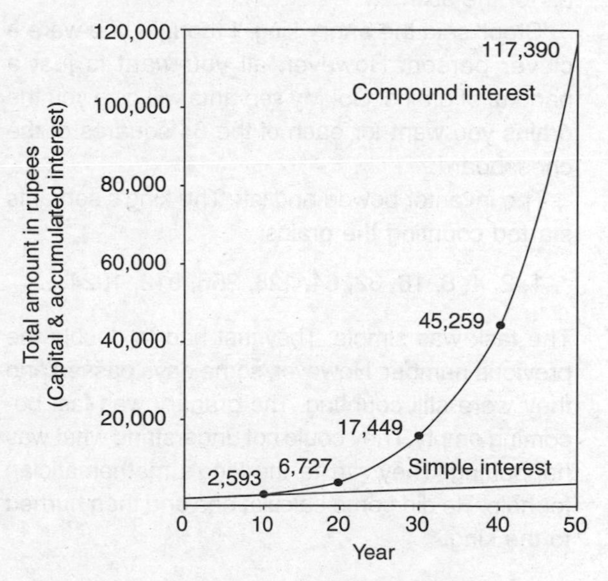

Figure 1.1 Curve showing exponential growth

constant amount of interest (Rs 100) every year. At the end of 50 years, the total amount will be only Rs 6000 (original deposit of Rs 1000 + interest of Rs 5000). You can see for yourself how compound interest makes so much difference. This is the power of exponential growth.

Mathematically, any growth is exponential if the increase is at a constant rate per time period, rather than a constant amount. If you show exponential growth as a graph, the shape of the curve will be like the letter 'J'. Figure 1.1 shows the growth of your investment of Rs 1000 at a compound interest rate of 10 per cent. You can also compare the growth with simple interest on the same value.

Many natural phenomena exhibit exponential growth. For example, the population of the world has been increasing exponentially. With every case of exponential growth we associate 'a doubling time'. This is roughly the time it takes for the quantity to double in value. You can get the doubling time by dividing the number 70 by the growth rate of the quantity. This formula is derived from the basic mathematics of exponential growth. World population doubles every 45 years or so.

Box 1.3 The power of exponential growth: The king and the inventor of chess

There was once a king and in his kingdom lived the man who had invented the game of chess. The king was so happy playing chess that he wanted to reward its inventor.

'Tell me what you want,' said the king to the inventor, 'and I will give it to you.'

The inventor thought for a while and said, 'You are very kind, Great King. I would like to have one grain of wheat for the first square of the chessboard.'

'Is that all?' asked the surprised king.

'No, sir. Give me two grains for the second square, four for the third, eight for the fourth, 16 for the fifth, 32 for the sixth, ...'

'Stop!' said the angry king. 'I thought you were a clever person. However, all you want is just a handful of grains. Go! My servants will give you the grains you want for each of the 64 squares of the chessboard.'

The inventor bowed and left. The king's servants started counting the grains:

1, 2, 4, 8, 16, 32, 64, 128, 256, 512, 1024, ...

The task was simple. They just had to double the previous number. However, some days passed and they were still counting. The granary was fast becoming empty. They could not understand what was happening. They ran to the king's mathematician for help. He did some calculations and then hurried to the king.

'Sir, we cannot give the inventor the grains he wants.'

The king was now very angry. 'You are a bigger fool than the inventor himself,' he told the mathematician. 'Why can't we give him the grains?'

'Sir, I have calculated the number of grains we need. That much of wheat is not available in the entire world.'

The king was shocked. 'How many grains are needed?' he asked.

'**18,446,744,073,709,551,615**.'

The story goes that the king, who was anyway weak in mathematics, collapsed.

You might still wonder what was the meaning of that big number? After all, the inventor wanted only that many grains. Why should the granaries become empty?

The king had to give more than 18 million million million grains. We can assume that one cubic metre of wheat contains about 15,000,000 grains. This means that the inventor should be given 12,000,000,000,000 cubic metres or 12,000 cubic kilometres of wheat. If we build a granary 8 metres high and 10 metres wide to store this much of wheat, its length will be about 150 million kilometres, which is the distance between the Sun and the Earth. Such is the power of exponential growth!

Four mega phenomena or spikes have been occurring, with profound implications for life on earth. The four quantities that have been growing exponentially are:

- Concentrations of carbon dioxide gas in the atmosphere
- The number of biological species becoming extinct every year
- Production and consumption of goods and services
- The size of the human population

A remarkable fact is that, in each case, the curve was flat over centuries until the spike began in recent times. What makes matters worse is that the spikes in the four quantities are interconnected, each amplifying the other three. Let us take a closer look at these phenomena.

Are carbon dioxide emissions excessive?

Concentrations of carbon dioxide gas in the atmosphere started increasing exponentially from 1800 onward, and the trend continues unabated. The main source of these emissions is the burning of fossil fuels like coal and oil. Excess carbon dioxide reduces the earth's capacity to radiate heat and leads to higher temperatures.

The Intergovernmental Panel on Climate Change (IPCC), a body of scientists from 100 countries, concluded in 1995 that human-induced global warming was occurring rapidly, which would lead to many natural disasters. It advocated immediate cutbacks in carbon dioxide emissions. The scientists could only propose policy changes; it was up to various governments to implement them. However, many industrialized countries are against any such action that will affect economic growth and involve changes in lifestyle. We will discuss this issue in Chapters 2, 8, and 19.

Are we losing species too fast?

There have been mass extinctions on Earth caused by natural changes, but they had occurred over hundreds of years. For millions of years, the number of extinctions per year seems to have been very low. The rapid slaughter of species probably started in the sixteenth century, with the hunting of millions of animals by the Europeans. The massive clearing of tropical forests in the last four centuries and the large-scale planting of single species have decimated plant and animal species. Many biologists believe that we have now entered a phase with the fastest mass extinction rate in the Earth's history.

Like the silent rise in carbon dioxide levels, the extinction of species is also invisible. We do not realize that the wide variety of species is our natural capital and that the process of evolution depends on this biodiversity. The loss of this diversity will affect our ability to stay healthy and to adapt to a changing environment. It also affects our immediate future, including our food security. Variety in food species is necessary if they are to develop a natural resistance to pests and diseases. We will return to the topic of biodiversity in Module 3 (Chapters 5 and 6).

Are we consuming too much?

The amount of natural resources used up every year began spiking around 1900. Steady economic growth since then has led to extraordinary levels of unsustainable consumption. The rise in consumption is visible, but its effects are only partially noticeable. The increasing number of cars is clearly seen, but the fact that world oil resources will soon run out does not sink in. In the mad race to consume more, we are using up the Earth's finite resources, such as the topsoil, water, and forests far faster than natural processes can regenerate them.

Are there too many people on Earth?

The exponential growth in human population began in 1650 and now, every three days, the size of the world's population increases by as much as it did in a whole century throughout most of human existence. The equivalent of seven Kolkatas (about 80 million people) are added to the world's population every year. Of this 80 million, most are poor and need food, water, shelter, education, medical facilities, and so on. This task looks even more formidable in the light of the impact of the first three spikes. Chapter 15 deals with the topic of population growth.

Why should these quantities spike over the last two or three centuries?

These spikes are due to a fundamental change in our relationship with nature. This change began with the Scientific and Industrial Revolutions that occurred in Europe in the sixteenth and seventeenth centuries. The new attitude towards nature came with the Idea of Progress that advocated the superior role of humans as masters of Nature. We could indefinitely exploit Nature and our progress towards a better life would be linear and continuous. Science and technology would help us in this quest for ceaseless progress and development.

First through colonialism and later by the export of western education and culture, the Idea of Progress was conveyed to large parts of the world. Today, most countries swear by this notion of growth and development through science, technology, and industrial expansion.

Can we blame all our ills on the Scientific and Industrial Revolutions?

Clearly, these two revolutions marked the beginning of the exponential rate of consumption of natural resources by humans. There is also the view that the problems began much earlier, when humans shifted from being nomadic hunter-gatherers to settled agriculturists. A discussion of such arguments is, however, outside the scope of this book.

The current problem is that we are living beyond our means; our *ecological footprint* is growing larger and larger.

What is meant by ecological footprint?

Assume that you live in a small house in a city. Let us say that the house is surrounded by a small garden and there is a wall around the compound that marks your plot of land. Can you isolate yourself in your home and continue to live indefinitely? You cannot, since you need many things from outside: food, water, and various other items. Your garden may give you some vegetables, but it cannot provide you all the food items you need. There may be a well on your land, but the water may not be potable and you may have to depend on outside sources for drinking water.

Suppose we ask the question: How large a piece of land would you need, if you are to sustain yourself completely? That area is your ecological footprint.

Let us conduct a mental experiment. Take the physical area of a city like Chennai and cover it with a huge glass hemisphere. We let in sunlight, but we do not allow anything else to enter or leave the enclosed space. How long will the city survive? Not for many days!

The city cannot produce enough food for all the people. There will be severe water scarcity, because the tankers from the surrounding villages will stop entering the city. The enormous amount of solid waste generated every day cannot be sent away or dumped into the sea. The air trapped in the hemisphere will soon become so polluted that people will find it difficult to breathe. The 'carrying capacity' of the city area is not sufficient to sustain the lives of the population.

Suppose we are able to expand the size of the glass hemisphere to take in more and more of the surrounding area. Assume also that this area has diverse natural resources like a mini-Earth. We can now ask another question: How large should the area covered be, if we want the city to survive indefinitely on the land, water, and energy resources available within the hemisphere? That area is the ecological footprint of the city.

The millions of people living in Chennai have huge needs and they draw their requirements from a very large area surrounding the city. The ecological footprint of Chennai is many, many times larger than its actual area. Similarly, we can talk about the ecological footprint of a country, or we can compare the ecological footprint of a citizen of the US with that of an Indian citizen.

What has the world done about the environmental crisis?

Dissatisfaction with the Idea of Progress began surfacing in the 1960s, when the adverse environmental impact of unbridled growth became clear. Books like Rachel Carson's *Silent Spring* (Carson 1962) set the tone for an environmental movement (see Module 3: Prologue, for an a description of this book).

The United Nations Conference on Human Environment in 1972 at Stockholm was the first international initiative to discuss environmental problems. Later, the World Commission on Environment and Development (WCED) was set up in 1983 with Gro Harlem Brundtland of Norway as its Chairperson. The WCED report, *Our Common Future,* emphasized the need for an integration of economic and ecological systems. The commission supported the concept of *sustainable development* and defined it as '*development that meets the needs of the present without compromising the ability of future generations to meet their own needs.*'

The idea of sustainable development has been criticized for being interpretable in many ways. Yet, it has caught the attention of many people, since it seems to imply that development and environmental conservation can go together.

Besides Stockholm, the other major effort was the United Nations Conference on Environment and Development (UNCED) held in 1992 in Rio de Janeiro. Attended by more than a hundred heads of state and 30,000 participants, UNCED came up with several documents including:

- 'The Rio Declaration on Environment and Development', listing 27 principles of sustainable development,
- 'Agenda 21', a detailed action plan for sustainable development in the twenty first century, and
- 'The Convention on Biological Diversity'.

The next conference, popularly known as Rio+10, was held in Johannesburg, South Africa, in 2002. This conference recognized that the implementation of the Rio agreements had been poor. It marked a shift from agreements in principle to more modest but concrete plans of action.

Why is there a general lack of public awareness about environmental issues?

There are many factors that limit our awareness of environmental issues. To the millions of poor people, the problems of daily existence are more important than environmental degradation. The more prosperous are afraid that, in the name of the environment, their comforts may be taken away. In any case, most of us do not pay the real costs of exploiting nature.

There is also false anti-environment propaganda by vested interests like the large corporations, for whom constant growth is vital. At the political level, parties are more interested in short-term gains and will not adopt unpopular measures to conserve the environment. Further, in this age of extreme specialization very few can look at the larger picture of what is happening to the world. There is also a paucity of reliable and clear information on the environment. To make matters worse, indicators like Gross Domestic Product (GDP) give us a misleading picture of what is desirable.

What is the role of environment-related studies? If subjects like environmental studies, environmental science, or ecology are taught to everyone at the school and college level, there will be an increase in awareness levels. This was the thinking behind the Supreme Court order of 2003 that resulted in the course that you are now doing.

Could we define environmental studies and related terms?

Generally, there are two ways in which we use the term *environment*. In one, the term refers to what surrounds an entity. Any entity, say a person, a living organism, a citizen, a company, etc., has its environment. We thus talk of the home environment, the business environment, the political environment, and, of course, the natural environment.

In the second way, we use the word environment by itself. Here we mean the natural environment: the air, water, soil, living beings, plants, trees, mountains, oceans, etc.

An entity interacts with its environment, that is, it influences and is influenced by its environment, either positively or negatively. For example, the natural environment affects human beings. We in turn have an impact (often a negative impact) on the natural environment.

The Oxford Advanced Learner's Dictionary defines 'environment' as 'the natural world in which people, animals, and plants live.'

Ecology is a closely related term that we must make friends with. Ecology is the science that studies the relationships between living things and their environment. It is often considered a discipline of biology.

Environmental Science is the systematic and scientific study of our environment and our role in it.

What then is *Environmental Studies,* which is the subject of this book? The two terms 'environmental studies' and 'environmental science' are often used interchangeably, but we could make a distinction.

Environmental Studies can be defined as the branch of study concerned with environmental issues. It has a broader canvas than environmental science and includes the social aspects of the environment. It does deal with science where necessary, but at a level understandable to the non-scientist. That is why the subject made compulsory for all undergraduates in India is called 'Environmental Studies'.

Why do we say that these subjects are interdisciplinary?

The subject of environment is inherently interdisciplinary. We study the complex relationships that exist in our natural environment among people, animals, other organisms, water, soil, air, trees, the ocean, and so on. The interconnections are numerous and involve many different disciplines. We need inputs from biology, botany, zoology, soil science, technology, oceanography, atmospheric science, economics, sociology, anthropology, ethics, and so on. A simple rule to remember is that everything in this world is connected (read the two stories in Box 1.4 and Box 1.5).

Box 1.4 Connections: Drink coffee in the US and make the songbird vanish in South America!

How can drinking a cup of coffee in a chain shop in the US lead to the disappearance of songbirds in South America?

In recent years, chains of coffee shops have proliferated in the US and many other western countries. The coffee comes from central and South America and the growers now face a heavy demand. Traditionally coffee is grown under the shade of the rainforest (without using chemicals) and the ripe berries are handpicked. In order to increase productivity and meet the new demand, the growers have been shifting to unshaded plantations treated with fertilizers and pesticides. Over the past 25 years, about 50 per cent of the shade trees have been replaced with unshaded plantations.

The shade plantations have always provided shelter for a large number of birds, vertebrates, and insects. Many migratory songbirds spend the winter in the rainforests. With a decrease in shade trees, there has been an alarming decline in the population of birds such as warblers and orchard orioles.

Conservation groups like the Rain Forest Alliance and the American Birding Association have initiated programmes to certify coffee as 'shade-grown' and encourage consumers to ask for such varieties. In fact, one brand already available is called Songbird Coffee!

There are thousands of such connections in nature. Even a simple act of ours could have widespread ramifications.

Source: Raven and Berg (2001)

Box 1.5 Connections: Get rid of malaria, but invite the plague!

In the mid-twentieth century, malaria was rampant in the Indonesian island of Sabah, earlier known as North Borneo. In 1955, the WHO began spraying the island with dieldrin (a chemical related to DDT) to kill mosquitoes. The attempt was successful and malaria was almost eradicated.

Dieldrin, however, did other things too. It killed many other insects including flies and cockroaches. The lizards ate these insects and they died too. So did the cats that ate the lizards. Once the cats declined, rats proliferated in huge numbers and there was the threat of plague. The WHO then dropped healthy cats on the island by parachute.

Dieldrin had also killed wasps and other insects that consumed a particular caterpillar, which was somehow not affected by the chemical. The caterpillars flourished and ate away all the leaves in the thatched roofs of the houses and the roofs started caving in.

Ultimately the situation was brought under control. However, the unexpected chain of events showed the importance of asking at every stage the question: 'And then what?'

Source: Miller (2004)

Does everyone agree that we have an environmental crisis?

While there are many voices and hundreds of books that speak of an environmental crisis, there are also opposing views. We have to discount the versions of lobbyists for large corporations or industry associations, who say that there is nothing seriously wrong with the environment and that the scare stories are exaggerated.

There are, however, experts or academics, who have hard questions to ask of the environmentalists. They often question the scientific basis of doomsday assertions. We have to take their arguments seriously and come to a conclusion. There is also the view of developing countries that environmental issues should be debated in the context of poverty and deprivation.

Ultimately, we can only believe what we see and experience for ourselves. Over the last decade, for example, the water situation in India has steadily worsened. Many rivers and dams have less water than ever before, most cities have water scarcities, and large sections of the population are buying water for their daily needs.

The congestion in cities is increasing, traffic jams are worsening, and the air quality is deteriorating. Power problems are common, the various state electricity boards are all in the red, and the cost of power is steadily increasing. There are many cases of industrial pollution affecting local populations.

In the face of all this, how can anyone say that the environment is improving?

Is there any hope for the future?

While this book describes the environmental crisis, it also contains many stories of hope—of successful efforts by individuals, voluntary groups, international organizations, and even governments.

However, positive action here and there is not enough. We must act individually and collectively to save the planet and the human species through sustainable development.

To begin with, each of us should view the environmental crisis as a real one, which needs urgent action from all of us. We have to move from linear to systemic or circular thinking that recognises complex interactions and natural cycles in the world.

We should examine beliefs, such as the following, that we have imbibed from our education and culture:

- The world works today, will work tomorrow
- Economic growth is good
- Happiness is consumption

Ultimately it is a question of changing one's mindset. If that happens to a sufficiently large number of people, we may yet begin managing our resources in a wise and sustainable way.

A quick recap

- The world is facing a global environmental crisis in addition to the crises of security and development, and the three are interconnected.
- Our unsustainable way of consuming natural resources at a rate greater than that at which nature can regenerate them and the rapidity with which we are polluting the planet, can only lead to a catastrophe.
- The crisis of today is rooted in our attitude of domination and exploitation of Nature, based on the Idea of Progress.
- Many phenomena like population and consumption have been growing exponentially over the past 100–200 years.
- The concept of ecological footprint expresses the amount of land needed to sustain the lifestyle of an entity—a person, a city, a country, etc.
- Several international meetings have deliberated the issues of environment and development and have come up with agreements. The accepted approach is that of sustainable development.
- Any study of the environment has to be an interdisciplinary one.

What important terms did I learn?

Ecology
Ecological footprint
Environment
Environmental studies
Exponential growth
Sustainable development

What have I learned?

Essay questions

1. Explain the root cause of the current environmental crisis through the four spikes.
2. Give a non-mathematical explanation of exponential growth and compare it with linear growth. Give one or more examples.
3. Why do we say that any study of the environment becomes an interdisciplinary one?
4. Give an update of the indicators of the state of the world and of India given at the beginning of this chapter. Have any indicators changed for the better?

Short-answer questions

1. Briefly describe the Idea of Progress and our attitude towards Nature.
2. Study the definitions of the following terms, given in the chapter, and suggest modifications: ecology, environment, environmental science, and environmental studies.
2. Explain the concept of ecological footprint through an example.
3. List the three major international conferences held on environment and development. What were the major outcomes in each case?

How can I learn by doing?

Study your neighbourhood, your town, or your ancestral village. What are the key environmental issues faced by the place? What are the conflicting interests in each case?

What can I do to make a difference?

Practical ways of helping to save the environment are given in most of the succeeding chapters. To begin with, you could try the following:

1. Join a local environmental group or voluntary organization and work with them at the local level.
2. Try to reduce your ecological footprint by examining the source of everything that you consume. Can you grow vegetables and fruits in your backyard or terrace? Can you collect the rainwater that falls on your roof?

Where are the related topics?

- Chapter 2 describes the carbon cycle.
- See Module 3: Prologue, for an excerpt from the book *Silent Spring* (Carson 1962).
- Chapters 5 and 6 cover biodiversity.
- Chapters 8 and 19 refer to global warming.
- Chapter 15 deals with population growth.

What are the deeper issues that go beyond the book?

1. There is a view that all our problems and our domination over nature started when we started practising agriculture about 10,000 years ago. Examine this viewpoint and give arguments in favour and against.
2. In your opinion, what should be the difference in the way the two subjects, environmental science and environmental studies, are taught to undergraduates?
3. Why are many environmental problems so difficult to solve? Is it because they have often many dimensions and many stakeholders with conflicting interests?

Where can I find more information?

- Mahapatra and Panda (2001) and Sainath (1996) for stories on Kalahandi.
- UNEP (2003), Starke (2004), Agarwal et al. (1999), and Brown et al. (2002) for global and India indicators.
- Ayres (1999) and Mason (2003) on the four spikes.
- Wackernagel and Rees (1996) and CSE (2000) for more on ecological footprint.
- Carson (1962), the book that started off the environmental movement.
- WCED (1987) for the concept of sustainable development.
- Appendix A2 of this book for United Nations websites on the Rio and Johannesburg summits and Appendix A6 for web links to the Rio Principles and other such documents.
- Appendix A5 of this book for a timeline of the major world events concerning the environment including international conferences and agreements.

What does the Muse say?

Here are the famous first lines of the poem *Auguries of Innocence* by the English poet William Blake (1757–1827):

To see a World in a grain of sand
And a Heaven in a Wild Flower,
Hold Infinity in the Palm of your hand
And Eternity in an Hour.

Prologue

The man who planted trees

JEAN GIONO

I was then young and fond of travelling. Once I was walking in an area I had never visited. The land was dry and hot. I walked on for three days and I did not meet anyone. I did come across some villages, but these had been abandoned. The houses and even the temples were crumbling.

I realized I was in an area where the people used to make charcoal from wood. They went on cutting trees until there was nothing left. The land became barren and dry. The wells dried up and there was no water. The people had to leave the villages and go away.

As I walked on the hot and dusty land, I felt the howling wind that made things even worse. My water bottle was empty and I was very thirsty. Just then, I saw something like a tree at a distance. I walked towards it and, to my great relief, I saw a shepherd. He had some sheep and a dog.

The shepherd did not show any surprise on seeing me. He gave me some water and took me to his hut. The stone hut was very neat and tidy. I noticed that the man was also neatly dressed. He did not talk much. He just made me welcome. It was clear he expected me to spend the night as his guest.

He made some simple food and shared it with me. As I rested after the long walk, he was busy. He brought a bag full of seeds to the table. He examined carefully each seed and put aside the good ones.

When he had selected a hundred seeds, he put them in a separate pile. We then went to sleep.

In the morning, he let his sheep out for grazing, with his dog to guard them. He set off with a stick and the seeds and I followed him. He would make a hole on the ground with his stick, place a seed in the hole and close it. He continued doing so until he had planted all the hundred seeds.

I found out that his name was Elzeard Bouffier. When his wife and son died, he came to live here. Seeing the dry and barren land, he decided to plant trees. Over three years, he had planted 100,000 seeds and expected at least 10,000 trees to grow and survive. The land was not his and he did not know who owned it. He did not really care. All he did was to plant the seeds, a hundred of them every day.

I took leave of him and later I even forgot the incident. A few years later, I returned to the area. Suddenly I remembered Elzeard and went looking for him. On the way, I saw some of the trees planted by him. He was now far away from where I had met him. However, he was still doing exactly the same thing: planting a hundred seeds every day.

I started visiting Elzeard once in two or three years. As the trees rose tall and as the forest grew, the birds came, followed by many animals. The howling wind became a gentle breeze. What was even more wonderful, the rains came and the wells had water again. Slowly, people returned to the villages. Houses and temples were rebuilt. There was once again life and laughter.

The Forest Department noticed the change and thought that a natural forest had come up by itself. One day, a forest ranger told Elzeard that he could not light a fire in his hut, because the natural forest had to be protected! Elzeard then moved further away, but continued his work.

A forest officer was my friend and I took him to meet Elzeard. My friend was also greatly impressed with Elzeard's dedication. He told me, 'Elzeard has discovered a wonderful way to be happy!'

My friend gave strict instructions to the rangers that no tree in the forest was to be cut. Thus, the forest continued to grow even as Elzeard continued to plant new seeds.

Elzeard Bouffier died quietly at the age of 75.

When you see a tree being cut or you read about disappearing forests, perhaps you tell yourself, 'What can an individual like me do?' This moving story shows that it is possible for one person to make a big change—if the person is selfless and persistent.

This is an abridged version of the story written by the French author, Jean Giono (1895–1970). It was first published in the magazine *Vogue* in 1954 under the title, 'The Man Who Planted Hope and Grew Happiness'. Giono said that his purpose in creating Bouffier was, 'to make people love trees, or more precisely, to make them love planting trees.'

The story has been translated into many languages. It became a cult classic during the environmental movement of the 1960s and it retains its appeal even today. To any individual who wants to save the environment, it gives hope and inspiration.

As you will see later in this book, there are real Elzeards in India and elsewhere, whose only purpose in life is to plant or save trees.

What awaits you in Module 2?

This module is about ecological systems, that is, nature in all its variety. You will find out what sustains life on this Earth and examine the complex web of relationships between plants, animals, and their environment. You will learn about specific types of ecosystems: forests, grasslands, deserts, ponds, lakes, rivers, wetlands, and, of course, the great ocean. Another topic will be the ways in which human activities are adversely affecting these ecosystems.

CHAPTER 2

Ecosystems: Basic Concepts

One cannot but be in awe,
when one contemplates the mysteries of eternity,
of life, of the marvellous structure of reality.
It is enough if one tries to merely comprehend
a little of this mystery each day.
Never lose a holy curiosity.

Albert Einstein

What will I learn?

After studying this chapter, the reader should be able to:

- explain what is meant by an ecosystem, and define terms like species, population, and community;
- describe the major components of an ecosystem, the biotic community, and abiotic conditions;
- appreciate the edge effect that comes into play in the transitional zone between ecosystems;
- recall the First and Second Laws of Thermodynamics and their relevance to ecosystems;
- understand the meaning and role of producers, consumers, and decomposers in an ecosystem;
- trace the flow of matter and energy in ecosystems and explain concepts like food chains, trophic levels, and ecological pyramid;
- explain the concepts of ecological succession, habitat, and ecological niche;
- describe the water and carbon cycles;
- appreciate the importance of the Sun as the sustainer of all life; and
- list the services provided by ecosystems, and appreciate the economic value of such services.

Box 2.1 Crisis: Where the earth meets the sky

'If there is a paradise on Earth, it is here, it is here, it is here.' So said the Mughal emperor Jehangir while visiting Kashmir in the early seventeenth century. The beauty of the great Himalayas is timeless.

The majestic Himalayas are not merely a range of mountains. They are considered to be the cradle of Indian civilization and have exerted great influence over Indian thought, culture, and of course, the environment. They are not just the 'abode of snow' (as their name suggests), but the kingdom of the gods.

If the Himalayas were not there, the rain clouds from the Indian Ocean would pass over the subcontinent, leaving India a desert. Many rivers that nourish the land, including the great Indus, Ganga, and Brahmaputra, originate in the Himalayas. It is on the banks of the Indus that the Harappan Civilization flourished.

The Himalayas are spread over 600,000 sq km, in a broad arc 2500 km long between the Indus and the Brahmaputra. The forest cover is about 40 per cent and there is immense biodiversity, with flora and fauna varying extensively from one region to the other. One-tenth of the world's known species of higher altitude plants and animals occur in the Himalayas. What is not common knowledge, however, is that the region also has deposits of boron, lead, lithium, coal, chromium, iron, copper, tungsten, zinc, limestone, dolomite, and marble. The Himalayan belt is inhabited by 51 million people.

Today, however, this paradise is in danger and has already suffered irreparable ecological damage. Human activities like the clearing of forests, road construction, and mountaineering have taken a heavy toll, leading to soil erosion, landslides, and floods. In addition, natural events like earthquakes, avalanches, and the melting of glaciers adversely affect the natural and human environment.

With the growth of towns in the region, the commercial felling of trees has increased to unsustainable levels. Deforestation continues in spite of resistance by the local people (the Chipko Movement and the efforts of activists like Sunderlal Bahuguna and Chandi Prasad Bhat, for example), the creation of protected areas, and a government order banning the felling of trees.

Mountaineering and trekking teams leave behind enormous amounts of garbage. They consume large amounts of local vegetation as fuelwood and fodder, and their movements disturb the wildlife in the mountains.

A dense network of roads has been built in the Himalayas. While they improve access for the local people, they also make it easy for outsiders to come in and exploit the forest resources of the region. The very act of road construction, using methods like blasting, disturbs the ecosystem and leads to landslides and loss of vegetation. The debris also damages agricultural fields and human settlements. The scale of quarrying of stone, to construct roads and to build houses for the growing population, has been increasing. This causes a loss of vegetation and topsoil, lowers the water table, disturbs wildlife, and increases air pollution. The hydropower potential of the region has led to projects like the Tehri Dam, which could turn out to be a social and ecological disaster.

In spite of their scale and grandeur, the Himalayas constitute a fragile ecosystem in delicate balance. That balance may have already been upset, with unpredictable consequences for the entire subcontinent.

Environmentalists, activists, international agencies, and the governments of the region are now working to save the Himalayan environment (read Box 2.2 that describes the inspiring efforts of one individual).

Source: www.icimod.org/ and *www.mtnforum.org/*

Box 2.2 Inspiration: A gentle warrior fights for the Himalayas

'Himalaya is a land of penance. Nothing in the world can be achieved without penance. I am doing this on behalf of all who are striving to save our dying planet. Why should a river, a mountain, a forest, or the ocean be killed, while we cling to life?'

This is what Sunderlal Bahuguna wrote to his friends when he undertook a second long fast in 1996 on the banks of the Bhagirathi River demanding a review of the Tehri Dam Project. He broke the fast on the seventy-fourth day on an assurance given by the then Prime Minister that the project would be reviewed.

A two-minute conversation (when he was thirteen) with Dev Suman, a Gandhian and a freedom fighter, completely changed the life of Sunderlal. He joined the non-violence movement, organized students, and was sent to jail even as a young boy. After Independence he followed Gandhi's injunction that all the freedom fighters should go and work in the villages.

Bahuguna, and his equally committed wife Vimla, have been living in the Himalayas, working for the welfare of dalits, lobbying against deforestation, encouraging forest-based small-scale industry, and campaigning to save the Himalayan ecology. Bahuguna became well known when he led the Chipko Movement against the commercial felling of trees in the Himalayas (see Chapter 7).

In 1981, Bahuguna walked 4800 km across the Himalayas, from Kashmir to Kohima, carrying the message: 'Save the tree, save the future.' On the way, his group held meetings in countless villages—on the environment and development issues, talked to children, and learnt of the problems faced by the people.

For the last two decades, his struggle had been against the building of the huge Tehri Dam, set to displace 25,000 people and devastate one of the most sacred and beautiful landscapes. He was (and still is) convinced that the dam would destroy the Garhwal-Kumaon Himalayas and create a perpetual threat of floods. In the true Gandhian tradition, he has gone on fast several times on the Tehri issue.

In spite of all the efforts of Bahuguna and others, the Tehri Dam is nearing completion and the old Tehri town is under water. Bahuguna, however, is undeterred. 'My fight is to save the Himalayas, and it will continue', he says.

My movement has not gone to waste because it will give strength and direction to the fight ahead. Sooner or later my voice will be heard. Truth never dies, it ultimately prevails, no matter what. Our goal is to have a comprehensive Himalaya policy to save the hills. The Himalayas are a mountain of emotions, not rocks and boulders, and we must preserve and nurture it to save our culture, to save our souls.

Where do we start in our journey through environmental issues?

In this book, we will see how from the Himalayas to Kanyakumari, and in most other parts of the world, human activity is causing severe degradation of the environment. First, however, we should learn about nature itself—that is, about ecological systems including forests, deserts, ocean, rivers, lakes, etc. That is the task of the next three chapters.

What is an ecosystem?

When you hear the term 'forest ecosystem', what comes to your mind? Surely you think immediately of trees, animals, birds, butterflies, and so on. The mention of the term 'ocean ecosystem' brings to mind waves, currents, fish, whales, etc. An *ecosystem* (or ecological system) is a region in which living organisms interact with their environment.

Is there a more formal definition of an ecosystem? Before we can define the term ecosystem in a formal way, we must get to know a few more basic terms. An *organism* is any living thing—an animal, a plant, or a microbe. A mosquito, a fish, a plant, a rabbit, a tree, an elephant, and a human being are all organisms. A *cell* is the basic unit of life in organisms. An organism like a bacterium consists of a single cell, while most organisms have many cells.

The organisms in this world can be classified into different species. You would know that human beings are one species, while roses are another. A *species* is a set of organisms that resemble one another in appearance and behaviour. The organisms in a species are potentially capable of reproducing naturally among themselves. Generally, organisms from different species do not interbreed and, even if they do, they do not produce fertile offspring.

The members of a species living and interacting within a specific geographical region are together called a *population*. Neem trees in a forest, people in a country, and goldfish in a pond are examples of populations. There is diversity in most natural populations. While there are broad similarities among the members, they do not all look exactly alike and they do not all behave in the same way.

The term population refers only to those members of a certain species that live within a given area. The term species includes all members of a certain kind, even if they exist in different populations in widely separated areas.

We need one more definition before we can come back to ecosystems. A *community* is the assemblage of all the interacting populations of different species existing in a geographical area. It is a complex interacting network of plants, animals, and micro-organisms. Each population plays a defined role in the community.

We are now ready for a more formal definition of an ecosystem.

An *ecosystem* is a community of living organisms (populations of species) interacting with one another and with the non-living physical and chemical environment. The interactions are such as to perpetuate the community and to retain a large degree of stability under varying conditions.

A puddle of water, a stream, a clump of bushes, a thick forest, and a large desert are all ecosystems. The ecosystems of this planet are interconnected and interdependent and together they make up the vast biosphere.

What exactly is the biosphere?

To understand what the biosphere is, we should first know what the Earth is like. The Earth has several spherical layers. The *atmosphere* is a thin envelope of air around the earth extending to about 50 km above its surface. That part of the atmosphere up to a distance of 17 km above sea level is called the *troposphere*; it contains the planet's air. Above the troposphere is the *stratosphere* that contains ozone. It is this ozone that supports life by filtering out the harmful ultraviolet radiation from the Sun.

The *hydrosphere* consists of the liquid water, ice, and water vapour, while the *lithosphere* is the earth's upper crust containing fossil fuels and minerals. Then we have the *biosphere*, in which all the living organisms interact with each other and with their environment. The biosphere includes most of the hydrosphere, and parts of the lower atmosphere and upper lithosphere. Thus, the biosphere is that portion of the planet and its environment which can support life.

How are ecosystems classified?

The terrestrial portion of the biosphere is divided into biomes. We group similar or related ecosystems to form biomes. A biome usually has a distinct climate and life forms adapted to that climate—deserts, grasslands, tropical rainforests, temperate forests, coniferous forests, and the tundra are examples. Simply put, a *biome* is more extensive and complex than an ecosystem. It is the next level of ecological organization above a community and an ecosystem.

The non-terrestrial part of the biosphere is classified into *aquatic life zones*: freshwater swamps, marshes, and bogs make up one type of aquatic zone, while lakes and rivers, estuaries, the inter-tidal zone, the coastal ocean, and the open ocean are the other aquatic zones.

There is no accepted system of classification of biomes and aquatic life zones. In fact, we often use the term ecosystem instead of biome and aquatic life zone.

What is the structure of an ecosystem?

Let us now study ecosystems in greater detail. An ecosystem has a living component, called the *biotic community* or *biota*, and non-living components, called *abiotic conditions*. The biotic community of species interact with each other and with the abiotic conditions in such a way as to perpetuate the species.

The biotic community includes plants, animals, and micro-organisms. Species in a community are mutually dependent in many ways. For example, in a forest ecosystem, the birds eat the fruits from the trees and build their nests in them. The birds in turn spread the plant seeds, helping the propagation of the tree species.

The abiotic conditions of terrestrial ecosystems include water, air, soil, nutrients, minerals, salinity and acidity levels, energy, sunlight, temperature, wind, rainfall, and climate. We could also include the latitude and altitude of the area and the frequency of fires there. In short, we define all the conditions under which the organisms live. The abiotic conditions both support and limit the biotic community.

In aquatic ecosystems, the abiotic conditions include the depth, turbidity, salinity and temperature of the water, chemical nutrients, suspended solids, currents, and the texture of the bottom (rocky or silty).

What is the boundary of an ecosystem? Where does the sea end and the land begin? What is the boundary line between a forest and the adjoining grassland? Natural ecosystems rarely have distinct boundaries. In fact, they are never truly self-contained and self-sustaining systems.

An ecosystem gradually merges with an adjoining one through a transitional zone called the *ecotone*. You will find in the ecotone a mixture of species found in both ecosystems. Often, the ecotone will have additional species not found in either ecosystem. Or, the ecotone may have greater population densities of certain species than in either of the adjoining communities. This phenomenon is called the *edge effect*. Ecotones could be studied as distinct ecosystems.

We can see the edge effect in coastal zones, where the land and the ocean meet. Because of the flow of tides, coastal zones do not have fixed boundaries in either the seaward or landward directions. These areas are rich in biological diversity and, in particular, they provide a habitat for organisms that can survive in water and on land. They are also home to unique plants like mangroves that thrive in salt water.

An estuary, where a river meets the sea, is even more clearly an ecotone. Here, there is a constant exchange of seawater and freshwater and the biodiversity is very high. An example is the Chennai estuary, where the river Adyar meets the sea. This area attracts a wide variety of organisms including many species of birds.

What sustains organisms?

Food gives organisms the energy to do biological work. It gives them the energy to grow, move, reproduce, and survive. Energy comes in many forms, but two fundamental laws of energy (or Laws of Thermodynamics) apply to all things in the universe.

What is the First Law of Thermodynamics? Energy can neither be created, nor destroyed. It can only be transformed from one form to another. For example, when we burn petrol in a car, the chemical energy of oil is transformed into mechanical energy.

What the First Law of Thermodynamics really says is that, with regard to energy, we cannot get something for nothing. An organism cannot create the energy it needs to live. It must capture this energy from the food it eats. It has to convert the chemical energy in its food into mechanical energy that it needs.

What is the Second Law of Thermodynamics? This law takes the use of energy one step further and gives us some bad news. It says that whenever we transform energy from one form into another, we will lose a part of it as heat. That is, all the energy is not available to do work. Some of it just disperses into the environment as heat. Never again can we use the energy lost, though it still exists in the physical environment.

The Second Law of Thermodynamics tells us that, with regard to energy conversion, we cannot even break even. During any conversion, some usable energy is degraded into a less usable form (heat) that disperses into the environment. Since energy is being converted all the time, it follows that, in the Universe, the total amount of energy available to do work decreases over time.

Does the Second Law of Thermodynamics contradict the First? The second law is consistent with the first. There is never any change in the total amount of energy in the Universe. What decreases is the amount available to perform useful work.

Many billions of years hence, all the energy may only be in unusable form. Everything will be at the same temperature, no work will be possible, and the universe will not be able to function. That is, however, too far away in time!

What are food chains and food webs?

The concept of a *food chain* should be intuitively clear to you. The insect is eaten by the frog, the frog becomes lunch for the snake, and the swooping hawk finishes off the snake and thus we have a food chain. It is just a sequence of organisms, in which each is food for the next. All organisms, dead or alive, are potential sources of food for other organisms.

Food chains overlap, since most organisms have more than one item on their menu. Again, an organism could be found on the menus of many other organisms. Thus, we have a complex network of interconnected food chains called a *food web* (see Figure 2.1).

If every organism must eat another organism for its survival, where does the food chain start? Obviously there must be at least one self-feeding organism that produces food and eats it too! Such an organism is called a *producer.* All green plants are producers.

Producers take simple inorganic substances from their abiotic environment and turn them into complex organic molecules using solar energy. They are the only organisms in an ecosystem that can trap energy from the Sun and make new organic material. All other organisms in this world are, directly or indirectly, dependent on the producers for their food.

The process that producers use to convert inorganic material into organic matter is called *photosynthesis* (see Figure 2.2). A green plant takes in carbon dioxide from the atmosphere, water from the soil, and energy from the Sun to make a kind of sugar molecule. In the process, it gives out oxygen.

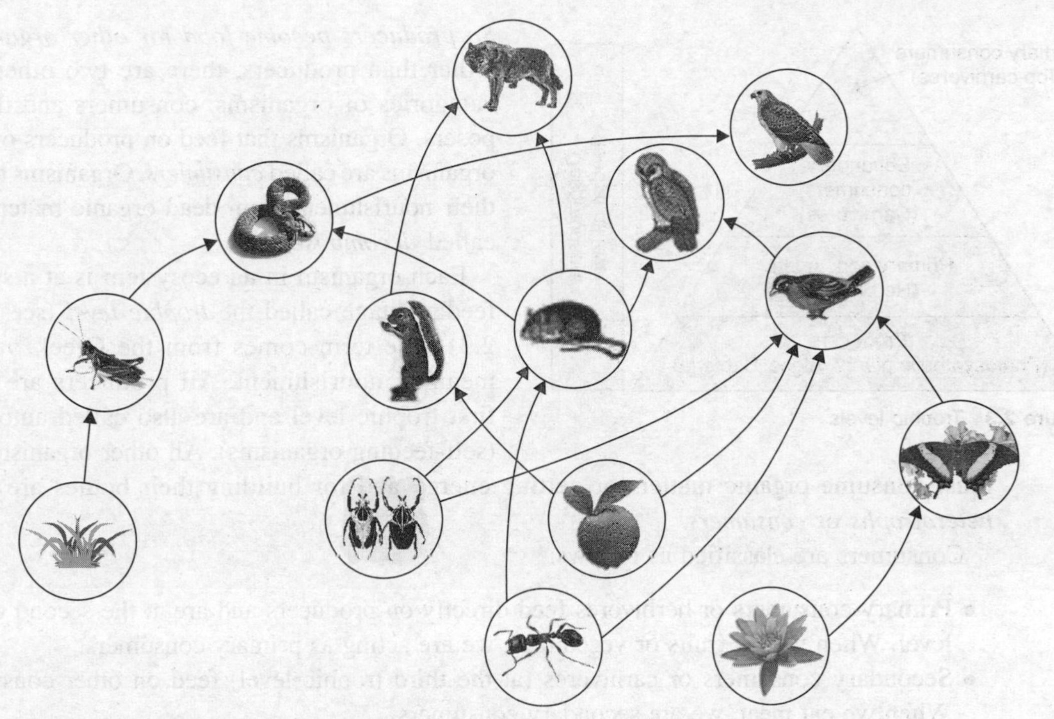

Figure 2.1 Example of a food web

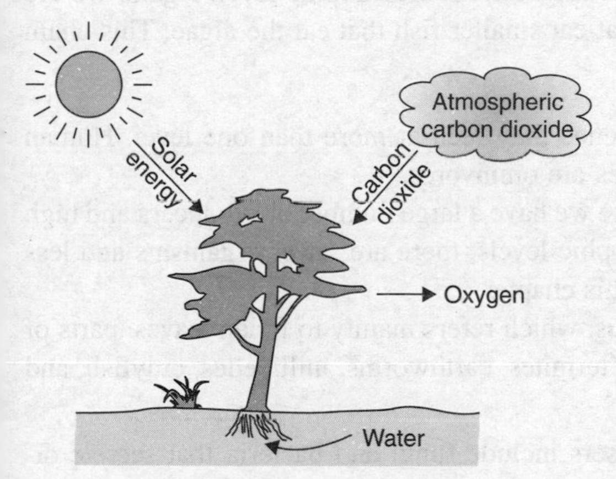

Figure 2.2 The process of photosynthesis

The reaction that takes place can be represented as follows:

$$\text{carbon dioxide + water + solar energy} \rightarrow \text{sugar + oxygen}$$

The green pigment chlorophyll in the plant traps the solar energy. Hence, we can say that photosynthesis takes place in the leaves of plants.

Thus, the world depends on the food made by the producers and, fortunately for us, producers are of all kinds and come in all sizes, from microscopic, single-cell algae to small plants and even gigantic trees.

Does photosynthesis take place in the ocean too?

In aquatic ecosystems, a different process called *chemosynthesis* takes place. Here specialized bacteria convert simple compounds from their environment into more complex nutrient compounds without sunlight. Instead, they use a different source of energy. Deep in the earth's core, heat is generated by the decay of radioactive elements. This heat is released at hot water vents in the ocean's depths. The bacteria use this geothermal energy for chemosynthesis. Later, these bacteria are consumed by many aquatic animals. Thus, in addition to green plants on land, certain bacteria in the ocean act as producers.

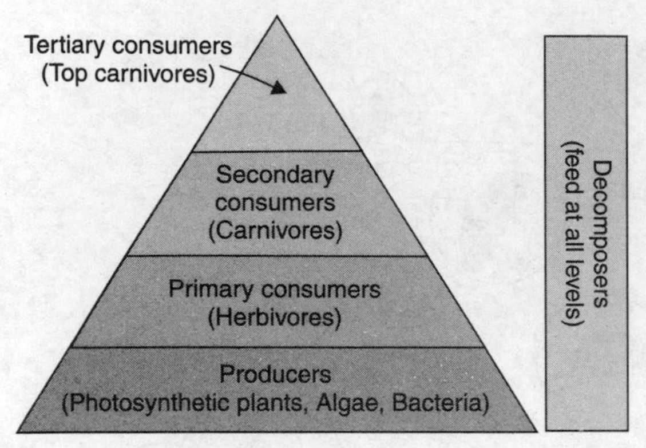

Figure 2.3 Trophic levels

So producers become food for other organisms? Other than producers, there are two other main categories of organisms: consumers and decomposers. Organisms that feed on producers or other organisms are called *consumers*. Organisms that get their nourishment from dead organic material are called *decomposers*.

Each organism in an ecosystem is at a specific feeding stage called the *trophic level* (see Figure 2.3). The term comes from the Greek *trophos,* meaning nourishment. All producers are at the first trophic level and are also called autotrophs (self-feeding organisms). All other organisms that must consume organic matter for getting energy and for building their bodies are called *heterotrophs* or *consumers*.

Consumers are classified as follows:

- Primary consumers or herbivores feed directly on producers and are at the second trophic level. When we eat fruits or vegetables, we are acting as primary consumers.
- Secondary consumers or carnivores (at the third trophic level) feed on other consumers. When we eat meat, we are secondary consumers.
- Tertiary consumers eat other carnivores and are at the fourth trophic level. Again, we are tertiary consumers when we eat the fish that eat smaller fish that eat the algae. This chain may go on to one or two more levels.

Omnivores eat both plants and animals and hence they feed at more than one level. Human beings, bears, foxes, pigs, rats, and cockroaches are omnivores.

The trophic levels form a pyramid. At the base we have a large number of producers and high energy. As we go up the pyramid to higher trophic levels, there are fewer organisms and less energy. This is explained in a later section in this chapter.

Detrivores are consumers that feed on detritus, which refers mainly to fallen leaves, parts of dead trees, and faecal wastes of animals. Ants, termites, earthworms, millipedes, crayfish, and crabs are examples of detrivores.

What is the role of decomposers? Decomposers include fungi and bacteria that secrete digestive enzymes, which decompose organic matter into simple sugars. The decomposers then absorb the sugars for their own nourishment. They release the resulting simpler inorganic compounds into the soil and water. The producers take in these compounds as nutrients.

Detrivores and decomposers are essential for the long-term survival of a community. They play the vital role of completing the matter cycle. Without them, enormous wastes of plant litter, dead animal bodies, animal excreta, and garbage would collect on Earth. Also, important nutrients like nitrogen, phosphorus, and potassium would remain indefinitely in dead matter. The producers would not get their nutrients, and life would be impossible.

How does food become energy in organisms? The organisms within an ecosystem are constantly growing, reproducing, dying, and decaying. They need energy for their activities.

When organisms eat food, they take in the carbohydrates that have chemical energy stored in them. This energy is released within the cells of the organisms through a process called *cell respiration*, which is the reverse of photosynthesis. The sugar molecules, in the presence of oxygen and water, are broken down into carbon dioxide and water and energy is released in the process:

sugar + oxygen + water → carbon dioxide + water + energy

Cell respiration makes the stored chemical energy available for biological work. All organisms, including green plants, respire to get energy.

What is an ecological pyramid?

For most ecosystems, the main source of energy is the Sun. Producers trap solar energy and store it as sugar, starch, fats, and proteins. When primary consumers eat the producers, the energy also moves up the trophic level. During this transfer, however, the Second Law of Thermodynamics comes into play and about 90 per cent of the energy is lost to the environment as unusable heat.

As we move up the trophic levels, the amount of usable energy available at each stage declines. Thus, we have a *pyramid of energy flow*. If the producer has 10,000 units of energy, the primary consumer receives only 1000 units and the secondary consumer gets only 100 units. The tertiary consumer is left with just 10 units. The energy flow is one-directional and the loss at each stage is simply released into the environment (see Figure 2.4).

What about matter? Each trophic level contains a certain amount of biomass, which is the dry weight of all the matter contained in the organisms at that level. As we move up trophic levels, biomass decreases drastically. At each level, we lose 90–99 per cent of biomass. We call this a *pyramid of biomass*.

There are two ways in which biomass decreases with each level. First, much of the food taken in by a consumer is not converted into body tissue. It is stored as energy to be used by the consumer when needed. Second, much of the biomass, especially at the producer level, is never eaten and goes directly to the decomposers (see Figure 2.5).

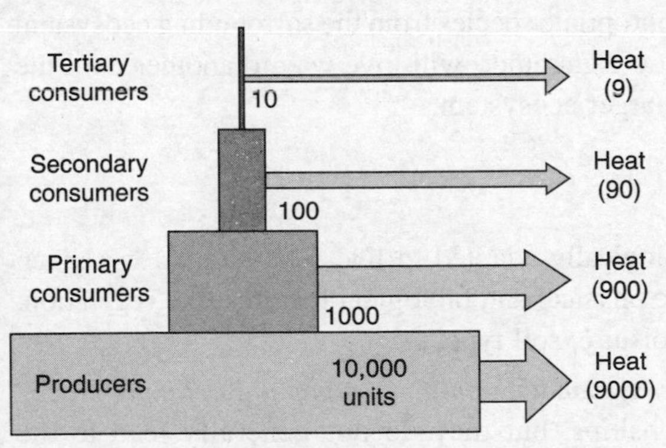

Figure 2.4 Pyramid of energy flow

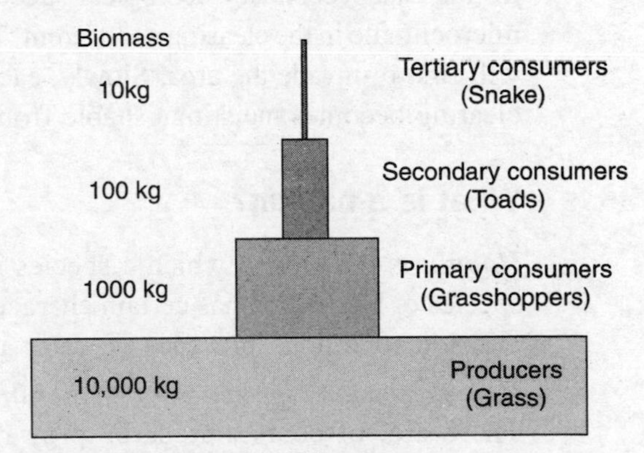

Figure 2.5 Pyramid of biomass

How does an ecosystem get established?

Consider the rocky surface of a barren hill that is left undisturbed. If there is some rain or moisture, moss starts growing in the minute cracks in the rocks. Slowly, the moss becomes a mat that holds any soil that comes its way. The seeds of small plants start growing on this soil and in turn the plants trap more soil. Shrubs and trees follow and the litter from the trees eliminates the original moss and even the small plants. Over time, more trees grow and ultimately there is a forest. As the big trees grow, many of the small plants under them disappear. This orderly process of transition from one biotic community to another is called *primary ecological succession*.

What is interesting about ecological succession? When a biotic community grows successfully its very success makes the area more favourable to another community. When the second community moves in and prospers, conditions may become unfavourable to the first and it may disappear.

Also note that for succession to take place, there must be a diversity of species present in the area. It is not a matter of a new species developing or even of an old species adapting to new conditions. The population of an existing species takes advantage of favourable conditions and prospers.

Ecological succession does not go on indefinitely. A stage is reached at which all the species are in dynamic equilibrium among themselves as also with the environment. This state is called *climax ecosystem*. During ecological succession, one biotic community gradually makes way for another, the second gives way to a third, and so on, until the climax is reached. Major biomes are examples of climax ecosytems.

Climax ecosystems could also change. If drastic climate changes occur or if alien species are introduced into the ecosystem, a process of adaptation and succession follows. Left undisturbed, an ecosystem always moves towards a state of dynamic equilibrium.

Why then do many hillsides remain barren? Ecological succession can happen only when an area is left undisturbed. Any interference by human beings disturbs the process. In the hill of our example, if shepherds take their goats every day to graze on the hillside, no growth would be possible.

What happens if a forest area is cleared by fire or by human beings and then left undisturbed? In this case, secondary ecological succession takes place. The old soil is still present, but the microclimate in the clearing is different. The appropriate species from the surrounding ecosystem will first re-invade the area. Slowly, each biotic community will give way to another until the clearing becomes indistinguishable from the larger ecosystem.

What is a habitat?

Habitat is the area to which a species is biologically adapted to live. The habitat of a given species or population has certain characteristic physical and biological features like vegetation, climatic conditions, presence of water and moisture, soil type, etc.

How do so many species live together in an ecosystem without fierce competition taking place? There are, of course, predator–prey relationships, but they do not generally lead to the extermination of species. The reason is that each species in an ecosystem has found its habitat and, what is more, its own *ecological niche* within the habitat.

This ecological niche is characterized by particular food habits, shelter-seeking methods, ways of nesting and reproduction, etc., of the species. It includes all aspects of the organism's existence—all the physical, chemical, and biological factors that it needs in order to live and reproduce. Where the organism lives, what it eats, which organisms eat it, how it competes with others, how it interacts with its abiotic environment—all these aspects together make up the niche. When different species live in the same habitat, competition may be slight or even non-existent, because each has its own niche.

The current ecological niche of an organism is called its *realized niche*. The organism may potentially be capable of living in a broader niche, using more of the environment's resources than in it does now. That would be its *fundamental niche*.

What are cycles in ecosystems?

We saw that energy flows through an ecosystem in one direction. Matter, on the other hand, moves through ecosystems in numerous cycles. The nutrients that organisms need to grow, live, and reproduce are continuously taken from the abiotic environment, consumed, and then recycled back to the environment.

There are several such *biogeochemical cycles* (with biological, geological, and chemical interactions), powered directly or indirectly by solar energy. They include the carbon, oxygen, nitrogen, phosphorus, and water cycles. With respect to matter, the Earth is essentially a closed system. Matter cannot escape from its boundaries.

How does water circulate on Earth? The *hydrologic* or *water cycle* (see Figure 2.6) is powered by the Sun and gravity. Solar energy causes water to evaporate from the ocean, rivers, lakes, and other water bodies. Plants extract water from the soil through their roots and transport it to their leaves from where it evaporates. This process is called transpiration.

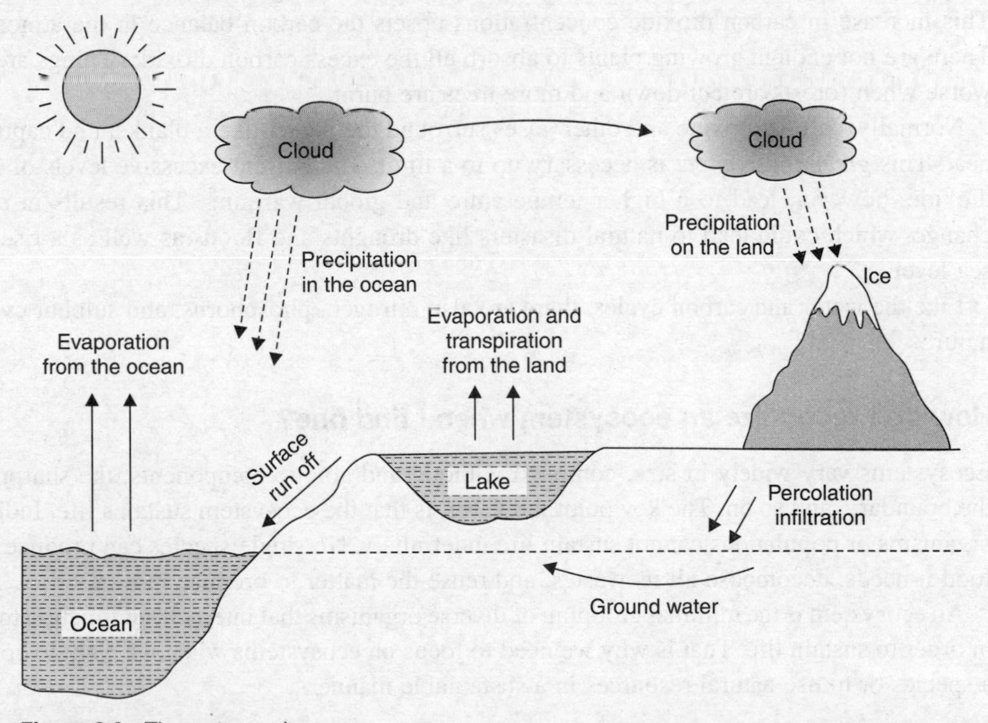

Figure 2.6 The water cycle

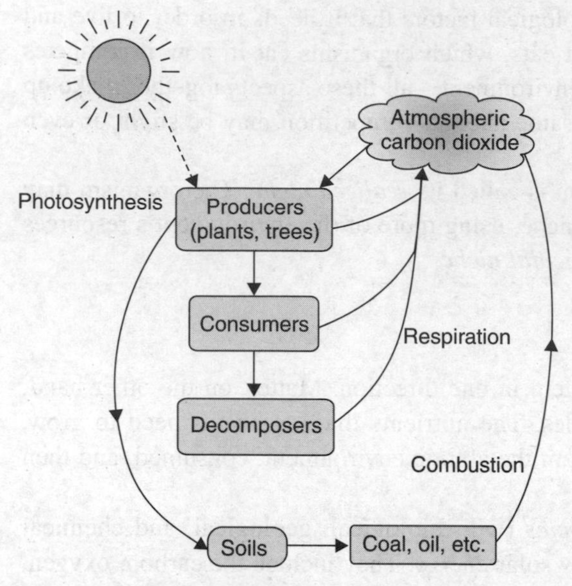

Figure 2.7 The carbon cycle

Wind and air move water vapour through the atmosphere. When the vapour cools, it condenses into tiny droplets that form clouds. The next stage is precipitation as rain or snow, mostly over the ocean and partly over land. The water that falls on the land may percolate through the soil and collect in aquifers, or it may flow back to the ocean as run-off though the rivers and streams.

We will return to the water cycle in Chapter 7.

What is the carbon cycle? Carbon dioxide constitutes just 0.035 per cent by volume of the atmosphere, and yet it is vital to life. Plants take carbon from carbon dioxide in the air and they use chlorophyll to gather energy from the Sun. From these inputs and water, plants make glucose as a basic building block. In this process of photosynthesis plants release oxygen. Animals breathe in this oxygen, digest their food (that comes from plants), and recombine the carbon and the oxygen to make carbon dioxide, which goes back to the plants. This is called the *carbon cycle* (see Figure 2.7).

When we burn fossil fuels like oil, the carbon in the fuel combines with atmospheric oxygen to form carbon dioxide. Since we burn a lot of oil, there are huge emissions of carbon dioxide. This increase in carbon dioxide concentrations upsets the carbon balance in the atmosphere. There are not enough growing plants to absorb all the excess carbon dioxide. Things are made worse when forests are cut down and more trees are burnt.

Normally, carbon dioxide and other gases surround the planet like a blanket and capture the heat. This *greenhouse effect* is necessary up to a limit. The current excessive levels of carbon dioxide, however, lead to a higher temperature and global warming. This results in climate change, which could lead to natural disasters like droughts and floods as well as a rise in the sea level.

Like the water and carbon cycles, there are also nitrogen, phosphorus, and sulphur cycles in nature.

How do I recognize an ecosystem when I find one?

Ecosystems vary widely in size, complexity, biotic and abiotic components, the sharpness of the boundary, and so on. The key point, however, is that the ecosystem sustains life. Individual organisms or populations cannot sustain life indefinitely. No single species can produce all the food it needs, decompose all its wastes, and reuse the matter to produce more food.

An ecosystem is the minimal grouping of diverse organisms that interact and function together in order to sustain life. That is why we need to focus on ecosystems when we want to conserve a species or to use natural resources in a sustainable manner.

What ultimately sustains life in an ecosystem? The Sun is the primary sustainer of life on Earth. It is a huge nuclear fusion reactor running on hydrogen. Energy from the Sun flows through materials and living organisms and eventually goes back into space as heat. Only a tiny part of the Sun's energy falls on Earth, but that is enough to:

- Warm the land and the atmosphere
- Enable plants to produce food through photosynthesis
- Evaporate water and cycle it through the biosphere
- Power the cycling of matter (carbon, nitrogen, etc.)
- Generate winds
- Drive the climate and weather systems

Ecosystems are surely wonders of nature, but what do they do for me? Why should I conserve them? We take nature and ecosystems for granted. We take from them whatever we want without worrying about the costs or repayment. Life on earth, including that of human beings, would be impossible without the help of ecosystems.

What services do ecosystems provide for us? We are just beginning to appreciate the extent of services provided by ecosystems. The main *ecosystem services* provided to us by the Earth and the Sun are:

- Maintenance of the biogeochemical cycles
  - hydrologic or water cycle including purification
  - carbon cycle
  - cycling of vital chemicals like carbon, nitrogen, phosphorus, and sulphur
- Modification of climate
- Waste removal and detoxification
- Natural pest and disease control
- Erosion control, soil building, and soil renewal

These services are in addition to the energy and materials that nature gives us:

- Energy:
  - solar energy (that accounts for 99 per cent of the total energy used on Earth)
  - non-renewable energy like fossil fuels
  - renewable energy like biofuels
- Materials like minerals
- Air and its purification
- Food (plants, animals, fishes, etc.)
- Biodiversity and the gene pool
- Renewable resources like forests, wildlife, grasslands, etc.

What is the value of these ecosystem services that nature provides for free? A team of ecologists, economists, and geographers have tried to answer this question. Led by Costanza, the team first divided the Earth's surface into 16 major biomes and aquatic life zones. They then listed 17 goods and services provided by nature in each of these areas and examined over 100 studies that have tried to put a money value on such services. Finally, they came up with an estimate of the monetary value of ecosystem services.

Their conclusion was that the ecosystem services were worth more than US$ 36 trillion per year (Costanza 1997). This is comparable to the annual Gross World Product that is estimated to be US$ 39 trillion. Some experts believe that this is a conservative figure and that the real worth of nature could be much more, even a million times more.

We can ask ourselves the question: What should be the value of the natural capital of the world for it to give us an annual return of US$ 36 trillion? The answer is a mind-boggling US$ 500 trillion! Costanza's estimate involves many assumptions and omissions. Yet, it draws our attention to the enormous value of ecosystem services. Any estimate is better than not attaching any value at all to what we take from nature. Sometimes we can get an idea of the value of ecosystem services by working out the cost of replacing nature by a man-made system (read Box 2.3 for an example).

Can we put a money value on life and nature at all? Some critics feel that placing a money value on ecosystem services is misleading and dangerous. They argue that they are infinitely valuable, since they are irreplaceable.

Beyond all these services, consider the rainbow in the sky, the Sun rising from the ocean, the dance of the whales, the grace of the deer, the shimmering oasis in the desert, or the snow on the mountain. What price can you put on these?

Box 2.3 Economics: Value of an ecosystem service

The citizens of New York used to be proud of their drinking water that was supposed to be tasty, pure, clean, and healthy. Some time ago, however, the quality of the water, obtained from the Catskill watershed, fell below the standards set by the Environment Protection Agency (EPA).

A study found the cause of the problem: the increasing levels of activity in agriculture, industry, and urban development were together:

- Diminishing the actual forested area of the watershed
- Altering the biogeochemical cycles and the species composition

Two options were considered to solve the problem:

1. Buy and restore the watershed so that the forest will continue to provide natural purification and filtration: The estimated cost was US$ 1 billion.
2. Build and maintain a new purification and filtration plant: The capital cost would be US$ 6–8 billion and the annual operating cost US$ 300 million.

The first option was the obvious choice. The case gives you some idea of the value of ecosystem services.

Source: Lubchenco (2001)

Box 2.4 Make a difference: Observe Earth Day

The first Earth Day was observed in the US on April 22, 1970. It was founded by Gaylord Nelson and organized by Denis Hayes. Twenty million Americans participated in the event. It marked the beginning of the environmental movement and a clear focus on reducing pollution. Since then Earth Day is observed every year in many parts of the world.

The Earth Day Network (EDN) promotes environmental awareness and year round progressive action, worldwide. It reaches over 12,000 organizations in 174 countries. Earth Day is the only event celebrated simultaneously around the globe by people of all backgrounds, faiths, and nationalities. More than half a billion people participate in the campaigns every year.

Earth Day has been an annual event for people around the world to celebrate the earth and our responsibility toward it. Volunteer. Go to a festival. Install solar panels on your roof. Organize an event where you live. Change a habit. Help launch a community garden. Communicate your priorities to your elected representatives. The possibilities are endless! Do something nice for the Earth, have fun, meet new people, and make a difference.

The official date of Earth Day is 22 April. Earth Day events are often scheduled for the weekends before or after 22 April. Many communities and groups choose to organize environmental activities over the longer period of Earth Week (the week before Earth Day) or even Earth Month (the whole month of April).

Source: The website *www.earthday.net*

A quick recap

- Ecosystems represent complex interactions among organisms and between them and the environment.
- Energy can neither be created, nor destroyed. It can only be transformed from one form to another; during any such transformation, a part of the energy is lost as less useful heat.
- Organisms have roles—producers, consumers, and decomposers—in the food chain; they maintain the flows of matter and energy.
- An ecosystem always moves towards a state of dynamic equilibrium.
- Every species has its ecological niche in an ecosystem.
- Water and carbon are continuously cycled in the biosphere.
- The Sun is the primary sustainer of all life on Earth.
- Ecosystems provide invaluable services to the biosphere, which includes human beings.

What important terms did I learn?

Abiotic conditions
Aquatic life zone
Biogeochemical cycle
Biome
Biosphere
Biota
Carbon cycle
Community
Consumers
Decomposers
Ecological niche
Ecological pyramid
Ecological succession
Ecosystem
Ecosystem service
Ecotone
Edge effect
Energy flow
First Law of Thermodynamics
Food chain
Food web
Habitat
Photosynthesis
Population
Producers
Pyramid of biomass
Pyramid of energy flow
Second Law of Thermodynamics
Species
Trophic level
Water cycle

What have I learned?

Essay questions

1. Write an essay describing the Himalayas as a mountain ecosystem and explain the ecological problems the region is facing today. Try to add to the information given in this chapter.
2. Examine the implications for human activities of the First and Second Laws of Thermodynamics.
3. Describe the processes of primary and secondary ecological succession.

Short-answer questions

1. Explain the difference between species, population, and community.
2. What is a biome? Give examples.
3. What are the main components of an ecosystem?
4. What are the abiotic factors in terrestrial biomes and aquatic life zones?
5. What is the edge effect and what is its importance?
6. Every organism is potential food for other organisms. Who then makes food first to get the chain going?
7. What is the main difference between a producer, a consumer, and a decomposer in an ecosystem? Explain through examples.
8. Explain the following statements in the context of the First and Second Laws of Thermodynamics:
   (i) 'You cannot get something for nothing'
   (ii) 'You cannot even break even'
9. Explain a food chain and a food web.
10. What happens to the energy flow as we move up the trophic levels?
11. How can we describe the ecological niche of a species?
12. What are biogeochemical cycles? Describe the water and carbon cycles.
13. List the important ecological and economic services provided by ecosystems. What is the estimated value of these services?

How can I learn by doing?

1. Choose a small ecosystem like a pond or a garden that is relatively undisturbed. Observe the ecosystem over three months or more and keep a log of your observations. Try to identify the producers, consumers, and decomposers.
2. Choose a small animal or bird that you can watch in your back garden or a nearby park. Observe the behaviour of the organism regularly and keep a journal of your observations. Watch what the organism eats, what it rejects, where it lives, the area over which it moves, how it reproduces, the organisms which are its predators, etc.

What can I do to make a difference?

Observe Earth Day on April 22 to create awareness in your neighbourhood or college about environmental issues. Box 2.4 gives more information.

Where are the related topics?

- Chapter 7 provides more information on the water cycle.
- Chapter 8 provides information on the carbon cycle.

What is the deeper issue that goes beyond the book?

As human beings, we are also part of ecosystems. We have, however, overcome the usual barriers and limiting factors faced by other species. We have interfered with the working of most ecosystems and are plundering the resources of ecosystems without any care for the consequences. What then would be the consequences of continuing this trend?

Where can I find more information?

- Costanza (1997) is the classic paper on ecosystem services and their value.
- Lubchenco (2001) provides a short discussion of ecosystem services including the case of the New York City water supply.
- Goldsmith (1997) and Tripathi (2004) describe the work of Sunderlal Bahuguna.

What does the Muse say?

Here is a hymn from the Atharva Veda:

Mother of all plants,
Firm Earth upheld by Eternal Law,
May she be ever beneficent and gracious to us
As we tread on her.

CHAPTER 3

Forest, Grassland, and Desert Ecosystems

Forests precede civilizations,
deserts follow them.
Chateaubriand

What will I learn?

After studying this chapter, the reader should be able to:

- recall the factors that give rise to different kinds of biomes in different regions;
- describe the main characteristics and importance of tropical rainforests;
- list the main features of temperate and coniferous forests as well as the Arctic tundra;
- give an account of the forest cover in India based on the Forest Survey;
- appreciate the tremendous impact of human activities and natural forces on the forests of the world;
- describe the main characteristics of grasslands and deserts;
- explain how desert organisms have adapted themselves to the harsh conditions of their environment; and
- recall the ecological services provided by mountain ecosystems.

Box 3.1 Crisis: Blue Mountains in peril

In 1818, the East India Company asked John Sullivan, the Collector of Coimbatore in South India, to investigate the origin of the fabulous tales that were being circulated concerning the 'Blue Mountains' to verify their authenticity and to send a report to the authorities. Sullivan set off towards the Nilgiris with a contingent of prisoners, elephants, dogs, ponies, and English huntsmen.

Above 1000 feet, the expedition had to abandon its animals and climb with cords and pulleys. After a week of hard ascent and several casualties on the way, the group reached a plateau of unparalleled beauty. This plateau became the hill resort of Ootacamund or Ooty.

The Nilgiris or the Blue Mountains are a mountain range at the junction of the Western and Eastern Ghats. With a maximum elevation of 2600 m above mean sea level, the climate ranges from humid to semi-arid and from warm to cold conditions. The most striking feature of the Nilgiris is their massive appearance.

The Nilgiri Biosphere Reserve (NBR), with an area of 5520 sq km, was established in 1986 and was included in the UNESCO Man and the

Biosphere Programme in 2000. (UNESCO stands for the United Nations Educational, Scientific, and Cultural Organization.) The NBR contains several dry scrub, dry and moist deciduous, semi-evergreen and wet evergreen forests, grasslands and swamps. It is the home of the largest known population of two endangered animal species, the Nilgiri Tahr and the Lion-tailed Macaque, and the largest South Indian population of elephant, tiger, gaur, sambar, and chital as well as a number of endemic (unique to this area) and endangered plants.

The biodiversity of the NBR includes 3300 flowering plants, 175 orchids, over 100 species of mammals, 350 species of birds, 80 species of reptiles and amphibians, 300 species of butterflies, and innumerable invertebrates. Many of these species are endemic to the region.

The British began replacing the pristine vegetation by tea and coffee plantations and this process has picked up greater momentum in recent decades. Half of all the cultivated area, about 30,000 hectares, is covered with tea, which now governs the economy of the district. Initially it brought an economic boom, but it has been an ecological disaster by promoting monoculture, excessive use of chemical pesticides and fertilizers, and an extremely dependent population. The eco-cultural degradation has been compounded further by the more recent crisis in the tea industry and stagnation in tourism.

The total population of NBR is more than one million, of which about one-third is tribal. The five Nilgiri peoples—Toda, Kota, Badaga, Kurumba, and Irula—were linked in a ritual, economic, and social symbiosis and they were known for their harmonious use of the environment. Over the past 40 years, however, environmental degradation in the area has had adverse consequences on the social, economic, and cultural life of these communities. They have been deprived of their traditional economy, culture, and lifestyle, which had a symbiotic relationship with the ecology of the Nilgiris. There were also conflicts with immigrant repatriates from Sri Lanka.

Landslides have become more frequent in recent years. Often poor villagers have seen their homes destroyed in these incidents. Deforestation along with the conversion of land for agriculture, tea plantations, and residential houses, even on steep slopes, are the reasons for the landslides.

More than ten years ago, concerned citizens of the region came together under the 'Save the Nilgiris' Campaign. They have been able to get the government to pass laws and take other measures to protect the ecosystem. Thanks to their efforts, the government set up the Hill Areas Conservation Authority for Tamil Nadu. They have also campaigned against the excessive promotion of tea and eucalyptus plantations. More than anything else, they have created awareness among the people on environmental issues.

The Nilgiris are still a vast reserve of biological and cultural heritage with many unique characteristics. The local people continue to depend on the forests for their livelihood and well-being. It is possible even now to prevent further destruction of this unique ecosystem.

Source: Various websites including *www.geocities.com/TheTropics/Cabana/3841* and *www.envis.tn.nic.in/NilgirisBio.htm*

What does the Nilgiris' story tell us?

In the previous chapter, we learned how ecosystems function and how they try to reach a state of dynamic equilibrium. However, human beings are now interfering with the ecosystems everywhere. The case of the Nilgiris is an example of the damage that human impact can cause.

Before we learn more about human impact on the environment, however, we should find out more about forests and other natural systems. That is the task of this chapter and the next. In this chapter, we will learn about forests, grasslands, deserts, and mountains.

Why do different regions have different biomes like forests, grasslands, and deserts?

Temperature and precipitation are the two major factors that account for the differences in biomes from region to region (note that the term *precipitation* covers all the forms in which water comes down on Earth, including rain, snow, hail, etc.). The mean values of temperature and precipitation, as well as the way they vary on an annual (and even daily) basis, will determine the kind of biomes we find in an area.

Different climates promote different communities of plants and animals. Again, the types and numbers of organisms may vary from location to location within a biome because of variations in the local climate and soil, and because of the natural and human-caused disturbances that occur in the area. Yet, the regions of a biome have broadly comparable biological communities.

Climate, plants, and animal species of regions vary with latitude and altitude. You can observe these changes as you move from the Equator to the Poles or from the base to the summit of a tall mountain. Again, in a thick forest, different communities are found at different levels.

What is wonderful is that every species has adapted to the climate and has found its ecological niche in the community. Left undisturbed, the biome reaches a stable state. It continues to be dynamic and undergoes changes that occur over long periods. Trouble occurs, however, when humans enter the ecosystem and cause rapid, and often devastating, changes.

How are different biomes distributed in the world?

It is difficult to accurately map land use in the world, but Table 3.1 gives some figures from the UN Food and Agricultural Organization (FAO).

Table 3.1 World land use pattern

Biome type	Percentage of land
Forest	32
Rangeland and pasture	26
Desert	20
Cropland	11
Tundra and wetlands	9
Urban areas	2
Total	**100**

Source: FAO in Miller (2004)

How far have human beings explored and disturbed the land? If we exclude uninhabitable areas of ice, rock, desert, and steep mountains, just 27 per cent of land has escaped human interference.

Let us learn more about the different biomes, beginning with forests. Forests can be classified in different ways and one way is to divide them into tropical rainforests, temperate forests, coniferous forests, and tundra.

What are tropical rainforests?

As the name implies, tropical rainforests are found in the hot and humid regions near the Equator. These regions have abundant rainfall (2000–4500 mm per year) that occurs almost daily, and the temperature varies little over the year. They are found in South and central America, western and central Africa, South East Asia, and in some islands of the Indian and Pacific Oceans.

Though they occupy just 2 per cent of the earth, tropical rainforests contain 50–80 per cent of all terrestrial plant species. The enormous biodiversity in plant life is supplemented by an abundance of amphibians, reptiles, birds, monkeys, and small mammals, as well as predators like tigers and jaguars.

These rainforests are marked by a variety of tall trees and a dense canopy. There are at least three distinct stories or layers of vegetation. The soils are thin, acidic, and nutrient-poor. Little organic matter accumulates since the litter that falls on the floor is very quickly decomposed and the minerals quickly absorbed by the roots.

Why are tropical forests considered so important? Apart from preserving immense biodiversity, tropical forests play a major role in recycling water. In a landmark research study conducted in the early 1980s in the Amazon, a team of Brazilian scientists led by Eneas Salati of the University of Sao Paulo found that a forest could return as much as 75 per cent of the moisture it received back into the atmosphere (*The Hindu* 1983).

We have always been told that more trees meant more rain, but this Brazilian study gave a scientific basis to this connection. The research showed that the forests could return sufficient amounts of water to cause the formation of rain clouds. Land that is covered by trees collects and returns to the atmosphere at least 10 times more moisture than deforested land does.

What are temperate forests?

Temperate forests have seasonal variations in climate; they are very cold in winter and warm and humid in summer. The annual rainfall is about 750–2000 mm and the soil is rich.

Temperate forests are found in western and central Europe, eastern Asia, and eastern North America. They have deciduous (leaf-shedding) trees like oaks, maples, ash, and beech, and some coniferous trees like pines. Shrubby undergrowth, ferns, lichens, and mosses are also found.

These forests contain abundant micro-organisms, and mammals like squirrels, porcupines, chipmunks, raccoons, hares, deer, foxes, coyotes, and black bear. Birds like warblers, woodpeckers, owls, and hawks, besides snakes, frogs, and salamanders, are common.

What are coniferous forests?

These forests derive their name from the abundance of coniferous trees like spruce, fir, pine, and hemlock. A coniferous tree produces hard dry fruit called cones. Coniferous forests also contain smaller amounts of deciduous trees like birch and maple. They occur in the northern parts of North America, Europe, and Asia.

In coniferous forests, winters are usually long and cold. The precipitation is often light in winter and heavier in summer. The soil in these forests is acidic and humus-rich and there is much leaf litter.

The main animals found in these forests are large herbivores like the mule deer, moose, elk, and caribou. There are also smaller herbivores like mice, hares, and red squirrels; and predators like lynx, foxes, and bears. They are often important nesting areas for many migratory birds like warblers and thrushes.

What kind of forests do we have in the Arctic?

The forests in the Arctic are called tundra. They occur in the extreme northern latitudes, where the snow melts seasonally. There is no equivalent in the southern hemisphere, since there is no land in the corresponding latitude.

The region has long and harsh winters and very short summers. Precipitation occurs mostly in summer and the annual average is just 10–25 cm.

The species diversity is low, but the few species that do exist here do so in large numbers. Mosses, lichens, grasses, and some dwarf trees are the main plants. The animal species include lemmings, weasels, arctic foxes, hares, snowy owls, and musk oxen. Many bird species migrate here in summer and feast on the abundance of insects—mosquitoes, black flies, deerflies, etc.

The tundra is a fragile ecosystem. Even a casual hike by humans across the land causes enormous damage, and regeneration is very slow. Large parts of the Arctic tundra have been affected by oil exploration and military use (read Box 3.2).

Box 3.2 Crisis: Choosing between pristine nature and petroleum

It is one of last areas where nature remains unspoiled and it is now in trouble. The story of the Arctic National Wildlife Refuge (ANWR) in Alaska is a classic case of the environment vs development dilemma.

In 1980, the US Congress created the ANWR—7.3 million hectares of undisturbed tundra, wetlands, and glaciers. It is home to an extremely diverse community of organisms, including polar bears, arctic foxes, peregrine falcons, musk oxen, snow geese, and many other animal species. It is also the calving area for large migrating herds of caribou. More than 150,000 caribou return here after wintering in Canada. All the tundra plant species, such as mosses, lichens, and dwarf trees are found here. Under a thin layer of soil there is permafrost, ice that does not melt even in summer.

The ecosystem is fragile and finely balanced. Any additional stress can do irreversible damage. Environmentalists have always argued for leaving the area undisturbed. Unfortunately for the region, however, it has deposits of oil!

The US Geological Survey estimates the oil reserves to be about 7.7 billion barrels, 2 billion of which could be extracted economically under current oil prices. There was always pressure to begin drilling in ANWR, but the US Congress had allowed only exploration, with the condition that its approval was needed for any drilling activity. Faced with insatiable demand, rising prices of crude oil, and uncertainties of imports, the US government and the oil companies are dead set on drilling in ANWR.

The pro-drilling groups point out that, in neighbouring Prudhoe Bay, oil is already being extracted and the infrastructure for drilling exists. No environmental impact has been reported. What is more, only 1.5 per cent of the total area will be encroached upon. There are sound economic reasons for extracting oil from ANWR and reducing the dependence on imports.

The environmentalists say that any drilling poses a permanent threat to the delicate balance of nature. Prudhoe Bay has in fact suffered damage—wolves and bears have been decreasing in number, resulting in an abnormal increase in the caribou population. What is more, the oil will not last long, but we would have already sacrificed nature. In fact, the production in Prudhoe Bay has peaked and is now declining.

As the arguments go back and forth, the tundra is waiting to know its fate.

What is the state of the forests in India?

The *State of Forest Report 2001*, prepared by the Forest Survey of India (FSI), Dehradun, estimates the country's forest cover at 676,000 sq km, constituting 20.55 per cent of the geographic area of the country. Of this cover, 417,000 sq km is dense forest, 259,000 sq km open forest, and 4490 sq km mangroves. The report claims that, between 1999 and 2001, the total forest cover increased by 6 per cent. Madhya Pradesh accounts for the largest forest cover of the country with 77,265 sq km followed by Arunachal Pradesh (68,045 sq km) and Chhattisgarh (56,448 sq km).

Environmentalists and voluntary organizations have raised questions about these statistics. Grassroots organizations are witness to the unending deforestation in many parts of the country and they say that an increase of 6 per cent in two years does not appear possible. The apparent increase could be due to the new assessment methodology used by the FSI.

What are the important types of forests in India?

About 80 per cent of India's forests are spread over four types. Table 3.2 gives the distribution and the locations. For a detailed account of the different types and their descriptions, see Sagreiya (1994).

Table 3.2 Distribution of Indian forests by type

Type	Percentage	Locations
Tropical moist deciduous	37	Andamans, Uttar Pradesh, Madhya Pradesh, Gujarat, Maharashtra, Karnataka, Kerala
Tropical dry deciduous	28	North–south strip from Himalayas to Kanyakumari
Tropical dry evergreen	8	Western Ghats, Assam, Andamans
Subtropical pine	7	Himalayas
Others	20	–
Total	**100**	

Source: Sagreiya (1994)

What is the impact of human activities and natural forces on the forests of the world?

In Chapter 9, we will examine in detail the impact of humankind as well as the forces of nature on forest resources. The major impact on the forest ecosystem is as follows:

- The clearing and burning of the forests for agriculture, cattle rearing, and timber extraction result in loss of biodiversity, extinction of species, and soil erosion, resulting in the loss of vital topsoil, and disturbance of the carbon cycle leading to global warming.
- The clear-cutting and conversion of forest land in hilly areas for agriculture, plantations, and housing lead to landslides and floods, which affect people in the forests and on the plains. It also increases the siltation of rivers.
- Many forests have been affected by acid deposition originating from industries.
- The harvesting of old growth forests destroys crucial habitat for endangered species.

Felling the last tree on Earth

- Pesticide spraying to control insects in forest plantations leads to poisoning all the way up the food chain and unintended loss of species like predatory hawks, owls, and eagles; this in turn leads to an increase in the pest population.
- Dams built in forest areas for hydropower and water drown huge areas, destroying species and depriving people of their lands; they could also induce earth tremors.
- In wilderness areas like the Arctic, oil exploration and military activities disrupt the ecosystem, contaminate areas, and lead to the decline of species.

What are grasslands?

Grasslands are regions where the average annual precipitation is high enough (250–1500 mm) for grass, and perhaps a few trees, to grow. The rainfall is, however, erratic and uncertain and hence large groups of trees do not grow. It is hot and dry in summer with a high incidence of fires.

Grasslands are the vast expanses of plains and rolling hills in the interiors of continents. They are found in central North America, central Russia and Siberia, sub-equatorial Africa and South America, southern India, and northern Australia.

The soil in the grassland areas is rich and deep, supporting many grass species, sparse bushes, and occasional woodlands. The animals found here include large grazing mammals like bison, antelopes, wild horses, kangaroos, giraffes, zebras, and rhinos, as well as predators like wolves, coyotes, leopards, cheetahs, hyenas, and lions. Birds, rabbits, and mongoose also flourish in the grasslands.

There has been large-scale conversion of grasslands into croplands since they are well suited to agriculture. They are also used as rangelands for grazing livestock.

The savannas are tropical grasslands with widely scattered clumps of low trees. Their climate is characterized by low rainfall and prolonged dry periods. The African savannas are known for their enormous herds of hoofed animals like gazelles, zebras, giraffes, and antelopes. There are predators like cheetahs, lions, hyenas, eagles, and hawks. During the hot periods, the larger animals migrate in search of water and grass. The smaller ones stay dormant or eat seeds.

How do we define a desert?

A desert is an area where evaporation often exceeds precipitation. The typical annual precipitation is between 25 mm and 50 mm, spread unevenly over the year.

Deserts are found mostly in a band 30 degrees north and south of the Equator and occupy about 30 per cent of the Earth's land surface. Their area is, however, increasing with continuous desertification taking place in many parts of the world. We will return to the topic of desertification in Chapter 10.

The largest deserts are in the interiors of continents, where moist sea air or moisture-bearing winds do not reach. There are also local deserts, which are rain-shadow areas on the downward side of mountain ranges.

How do plants and animals adapt to deserts? The desert soil has very little organic matter, but it is rich in minerals. The desert plants have adapted to the dry conditions and conserve water by having few or no leaves. The giant saguaro cactus, for example, has a stem that can expand to store water. Many have thorns or toxins to protect themselves from being grazed by animals.

Some plants have wax-coated leaves that minimize the loss of moisture. Others have deep roots that reach the groundwater. Alternatively, some have widely spread, shallow roots that collect water after any rain and store it in spongy tissues.

Desert animals are usually small in size and they remain under cover during the day and come out to feed at night. Many have a thick external shell that minimizes moisture loss due to evaporation, and they can lie dormant during the driest periods. Some of the animals found in the desert are frogs, reptiles, rodents, jack rabbits, foxes, and owls.

In deserts, during the day the Sun heats up the ground and the temperature is high. The night can be quite cold, since the lack of vegetation allows the heat from the ground to radiate away into the atmosphere very quickly.

Is there any human impact on deserts? You may think that a dry and hot desert cannot be damaged much by human activity. In fact, the desert ecosystem is quite sensitive and delicate. A heavy vehicle driven across the desert can do enormous damage to the thin topsoil and the sparse vegetation. The disturbed area will take a long time to recover because of the extremely slow rate of growth of vegetation and low species diversity.

Apart from soil destruction, there are other human disturbances to desert ecosystems. The creation of desert cities, depletion of ground water supplies, land disturbance, the pollution from mining, and the storage of toxic and nuclear wastes are some of the human activities that cause damage.

How can we describe the Thar Desert?

The Thar or the Great Indian Desert is spread over four states in India—Punjab, Haryana, Rajasthan, and Gujarat—and two states in Pakistan. It covers an area of about 446,000 sq km. Though smaller than the Sahara in Africa and the Gobi in Russia, it is the most populated desert in the world, with about 13 million people.

The average annual rainfall is between 100 mm and 500 mm and most of it occurs in a few showers during the monsoon. The only river in the region is the Ghaggar, which enters Rajasthan from Punjab and dries up in the desert.

The Thar has no oasis, native cactuses, or palm trees. Yet, a variety of flora and fauna thrive in this area. Flowering plants like shrubs and wild grasses, trees like the *khejra*, *babul*, and *rohida*, and fruit trees like the *ber* and the *pilu* are found here.

Sheep, goats, and camels are the common animals used by the people. The Asiatic wild ass, deer species like the black buck and the Indian gazelle, common Indian hare, red lynx, jackal, and wild dog are found here. About 23 species of lizard and 25 species of snakes are found here and several of them are endemic to the region. Local communities like the Bishnois have traditionally protected the trees and the animals (read Box 3.3).

Box 3.3 Inspiration: Giving one's life for a tree

The Bishnois of Rajasthan have been known for their concern for trees, birds, and animals. In 1731, Abhay Singh, the King of Jodhpur, wanted large quantities of wood to fire bricks to build his new palace. He sent his Minister, Giridhardas, with woodcutters to cut the trees in the forests near the Bishnoi villages.

Amritadevi, a mother of three, wanted to save the trees. She hugged a tree and begged Giridhardas and the cutters to stop. Her daughters too followed her example. Giridhardas ordered the men to proceed and they cut down the trees and the women. The news spread and more villagers came to the rescue of the trees. But the cutting continued and by nightfall, 363 people had given their lives to save the forest.

When the king heard the news, he was overcome with remorse. He then banned the felling of trees in the Bishnoi forests forever.

Recently, the Government of India has instituted the Amritadevi Wildlife Protection Award, to be given to village communities that show valour and courage in the protection of forests and wildlife.

The increase in the human and livestock population in the desert has led to the deterioration of the desert ecosystem resulting in degradation of soil fertility and vegetation. The desert is spreading due to overgrazing and deforestation of marginal land.

Are mountains distinct biomes?

Mountains have unique characteristics and can be studied as distinct biomes. They cover about 20 per cent of the Earth's land area. Of all the continents, Antarctica is the most mountainous. Elsewhere, thanks to its relative inaccessibility, a mountain often stands out as an island of rich biodiversity in the midst of a devastated landscape.

As you go up a mountain, you will see dramatic changes in climate, soil, and vegetation even over short distances. In a way, moving up a mountain is similar to moving from the Equator to the North Pole, since you can experience similar changes in biomes. Beginning with low deciduous forest, you could move through coniferous forests to alpine tundra and snow.

Mountains perform many ecological services for us. The majority of the world's forests are in the mountains. Apart from being home to endemic species of plants and animals not found elsewhere, they are also sanctuaries to animals driven away from the lowlands by human activities.

Mountains play a vital role in the water cycle. Their soil and vegetation absorb precipitation and slowly release the water through small streams. These streams often join together and become rivers. Thus the mountains act as a reservoir, storing water in the monsoon and releasing it during the dry season.

What is now happening to the mountain ecosystems? Man is out to plunder the mountains too: roads have been laid, forests have been cleared, rivers have been dammed, hill stations have been built, and minerals have been mined. The resulting degradation has many effects.

The barren slopes get further eroded when rains wash away the valuable topsoil that ends up as sediment in rivers, reservoirs, and the ocean. The mountains can no longer hold the water

A barren hillside

and hence there are floods downstream during the monsoon and no water in summer. The biodiversity, of course, disappears.

Alarmed by such degradation, the UN declared the year 2002 as the International Year of the Mountain to focus attention on their condition. There is also the International Centre for Integrated Mountain Development (ICIMOD) in Nepal that works for the sustainable development of mountains.

Can we end on a hopeful note?

In Box 3.1, we saw how the 'Save the Nilgiris' Campaign, a citizens' initiative, did produce positive results. Now read Box 3.4 for a famous success story in the annals of the Indian environmental movement.

Box 3.4 Resolution of a crisis: A pristine valley saved?

Situated in the Kundali Hills of the Western Ghats in Kerala, the Silent Valley forests are considered to be one of the last remaining virgin tropical evergreen forests in India, a climax ecosystem. Covering an area of 90 sq km, they have many rare plants and animals. The tree diversity here is very high and is comparable to some rainforests.

In the late 1970s, the Kerala government decided to build a hydel power project in the area to generate 240 MW of power and irrigate 100,000 hectares of land. The project would have, however, submerged 500 hectares of the forest.

Several organizations like the Bombay Natural History Society (BNHS) and the Kerala Sastra Sahitya Parishad (KSSP) formed a 'Save Silent Valley' Movement to urge the state government to abandon the project. International organizations like the World Wide Fund for Nature and the World Conservation Union supported the struggle.

The issue became highly politicised and the pro-dam groups came up with the slogan, 'Man or Monkey'. The reference was to the lion-tailed macaque that the conservationists were concerned about.

Finally, the continued agitation and the intervention of the then Prime Minister, Indira Gandhi, forced the government to shelve the project. In 1984, the area was declared a national park. It is now part of the core area of the Nilgiris Biosphere Reserve.

A story with a happy ending? Well, the struggle over Silent Valley is not over. In 2003, the Kerala government proposed another dam, 3.5 km downstream from the proposed site of the earlier one and 500 m below the border of the National Park. The debates and arguments have begun all over again!

A quick recap

- Terrestrial biomes are determined by temperature and precipitation.
- Forest ecosystems still cover the maximum area on land.
- Human beings have interfered with about 73 per cent of the land on Earth—even the Arctic has not escaped our attention.
- Tropical forests are the treasure chest of the world's biodiversity and they are being destroyed at a rapid pace.
- Forest ecosystems are experiencing serious ecological and social problems in India and the rest of the world.
- Deserts can be damaged easily and take very long to recover.
- Mountains are unique ecosystems providing valuable services.

What important terms did I learn?

Biosphere reserve
Coniferous forest
Grassland
Precipitation
Rainforest
Savanna
Temperate forest
Tundra

What have I learned?

Essay questions

1. How do biomes vary based on the temperature and precipitation of the region?
2. Describe the human impact on the world's forests.
3. What lessons can we learn from the two cases described in this chapter: The Nilgiris and the Silent Valley?

Short-answer questions

1. How do desert organisms survive the heat and lack of water?
2. Why are tropical rainforests important to us?
3. What are the ecosystem services provided by mountains?

How can I learn by doing?

If possible, spend your vacation in a forest area, observing the biotic species and abiotic conditions, the lives of the local people, and the efforts of environmental groups. Write a report on the environmental and social problems in the area and suggest solutions to the same.

What can I do to make a difference?

1. Join any group action (like the 'Save the Nilgiris' Campaign or the Silent Valley Movement) that is going on in your area.
2. Observe July 1–7 as Vanamahotsava Week. The idea of observing a Vanamahotsava Week was started in India several decades ago. As the name implies, it was to be a forest festival with the participation of the people. Over the years, however, the practice lost its original appeal and has now become a mere ritual. Take the lead in putting some fresh life into this festival and promote the planting and protection of trees in your college or in your neighbourhood.

3. Observe December 11 as International Mountain Day. The UN General Assembly has designated December 11 as International Mountain Day from 2003. The decision marked the end of 2002, observed as the UN Year of the Mountains. Each year the observance focuses on a theme. For example, the theme for 2003 was 'Mountains—The Source of Freshwater'. You can find out the latest theme from the website *www.mountainpartnership.org/imd/imd.html*.

Where are the related topics?

- Chapter 9 is devoted to forest resources.
- Chapter 10 gives more information on desertification.

What are the deeper issues that go beyond this book?

How does one use a forest without harming it? What can we learn from tribal communities like the Bishnois of India or the Native Americans in this regard?

Where can I find more information?

- Publications Division (2004) gives some statistics on India's forests.
- Sagreiya (1994) gives a more detailed account of forests in India, but the statistics are not current.
- Websites:
  - *http://envfor.nic.in/fsi/sfraa/sfr.htm* contains the *State of the Forest Report, 1999*
  - *http://www.wildlifeinindia.com/forests-of-india.html*
  - *http://edugreen.teri.res.in/explore/forestry/types.htm*

What does the Muse say?

Here is a hymn sung by the Badagas of the Nilgiris to Heththe, their ancestral goddess. (It was adapted and translated into English by the Rev. Philip K. Mulley.) It is taken from the website *http://www.geocities.com/TheTropics/Cabana/3841/badaga.html*.

Mother endearing, plighted fast unto us
Precious-ever, promises of Thee.
Come O Dodda Heththe! Heththe of Bereghanni
Maasi is thine name, Naalku betta is thine home.

When mountains so lofty, over us loom and lo!
Brilliant is the bloom of flowers so numerous
Wreathed in silver, the smile of thee
Vaulted in the sky, the radiance of moon so benign
Golden is Thine shade and sweet so it turns
Summers so many.

Nilgiri is thine abode, majestic its walls all around
Bestow on us, Mighty Mother, smother us with boons of life
Beseech we, of thee blessings of prosperity
Treasure ever, thine providence is
Measureless are offerings of thy bounty
Thy protection we behold, thy presence we adore.

CHAPTER 4

Aquatic Ecosystems

For all at last return to the Sea –
to Oceanus, the ocean river,
like the ever-flowing stream of time,
the beginning and the end.

Rachel Carson

What will I learn?

After studying this chapter, the reader should be able to:

- differentiate between terrestrial and aquatic ecosystems with respect to the factors that determine the types of organisms found;
- describe the main types of aquatic organisms;
- appreciate the importance of the ocean with respect to the ecosystem services it provides;
- explain the division of the ocean into zones;
- understand the importance of coral reefs and mangroves;
- appreciate the special features of Antarctica and the current state of its environment;
- recall the features of the UN Convention on the Law of the Sea (UNCLOS) and the importance of the Exclusive Economic Zone (EEZ);
- explain why the coastal zone is vital and what is happening to the coastal areas of the world;
- appreciate the importance of estuaries, and coastal and freshwater wetlands;
- describe the general characteristics of rivers and the spread of the river system in India; and
- differentiate between the zones of a large lake.

Box 4.1 Crisis: A unique aquatic ecosystem in peril

The island of Kurusadai, near Rameswaram, is a biologist's paradise. Even today you can see its dense foliage, shallow coral reefs, plants, and aquatic life with an extraordinary diversity of form and colour. This is just one example of the marine life in the Gulf of Mannar, one of the world's richest marine biodiversity regions.

The Gulf of Mannar Biosphere Reserve (GOMBR) is located in the southeastern tip of Tamil Nadu extending from Rameswaram to Kanyakumari. It covers an area of 10,500 sq km, includes 21 islands, and contains over 3600 species of fauna and flora. They include mangroves, seagrass beds, algae, coral reefs, many fish species, marine turtles, dolphins, dugongs, and many migratory birds.

This biological paradise is now in trouble. The habitat is getting degraded, the marine species are threatened, and the people's livelihood seriously affected. The three main threats to the ecosystem are the destruction of the marine resources by bottom trawling, the mining of corals, and the pollution from industries.

The trawlers scrape the seabed using huge nets weighed down by chains. In the process, they destroy the breeding grounds of a large number of marine species. Since the nets are of fine mesh, they also bring up large amounts of smaller fish (bycatch), thus decimating the marine life. The traditional fisherfolk, on the other hand, use small boats and baiting methods that do not harm the non-target species.

There is a perpetual conflict between the two groups. The trawlers are expected to operate beyond three nautical miles from the shoreline and the fisherfolk within this distance. Each group accuses the other of violating the distance restriction. Mutual arrangements regarding times and days of operation do not work.

Illegal coral mining goes on unchecked in the area. Several hundred boats mine huge quantities of coral every day using levers and dynamite to break up the reefs. The catch is sent to cement companies as raw material. About 65 per cent of the coral reefs in the area are already dead, mostly due to human interference. The dugongs have disappeared and many marine species are threatened.

Over 100,000 people in 90 villages and hamlets in the area depend on fishing for their livelihood. Over the years, the variety and catch of fish has steadily gone down. As a result, there is more indebtedness among the fishermen.

The Gulf of Mannar was declared a biosphere reserve in 1989 and was brought within the ambit of the Man and the Biosphere Programme of UNESCO in 2001. This implies that fishing or any other activity is prohibited in the area. Such a restriction is unfair and unenforceable, as it goes against the traditional rights of the local fisherfolk. The only way to save the biosphere is by involving the local people as partners.

The Global Environmental Facility (GEF) has now provided support to the biosphere reserve through the Gulf of Mannar Biosphere Reserve Trust. This body is responsible for the coordination of a management plan for the biosphere reserve in cooperation with government agencies, private entrepreneurs, and local people's representatives. Priority is being given to encouraging community-based management. Will such efforts solve the many ecological and social problems of this unique ecosystem?

Source: Menon (2003) and *website www2.unesco.org/mab/*

What do we learn from the Gulf of Mannar case?

Once again this is a case of a unique natural environment, population pressure, people's livelihood issues, and local and international conservation efforts. The story tells us how valuable the ocean is and how human impact is adversely affecting the ocean ecosystems. Our task in this chapter is to learn more about different types of aquatic ecosystems: the ocean, rivers, lakes, wetlands, and so on.

How do aquatic ecosystems differ from terrestrial ones?

Aquatic life zones, which include the ocean, rivers, lakes, and wetlands, differ in many respects from terrestrial biomes. Unlike the situation in terrestrial biomes, temperature and precipitation are not the factors that determine the types of organisms that thrive in an aquatic life zone. First, water itself tends to moderate temperature and second, any precipitation falling on water cannot obviously be a major influence. In aquatic life zones, the characteristics of water—salinity, light penetration, nutrient levels, waves, and currents—are the crucial factors. The two main categories of aquatic life zones are saltwater systems and freshwater systems.

What kinds of organisms do we find in aquatic life zones? There are three main types of organisms in aquatic life zones: plankton, nekton, and benthos. The free-floating plankton are micro-organisms that cannot swim easily and are buffeted about by the waves and currents. Plankton again are of two types: phytoplankton and zooplankton.

Phytoplankton are the photosynthetic producers of the ocean and they form the basis of the ocean's food web. The tiny zooplankton are the primary consumers that feed on phytoplankton. They in turn become food for newly-hatched fish and small organisms, and the food chain continues up the trophic levels.

The nekton are strong swimmers, and they include all the larger organisms like fish, turtles, and whales. The benthos are bottom-dwellers adapted to living on the floor of the water body. Some fix themselves to one spot like sponges, oysters, and barnacles. Others burrow into the sand like worms and clams. Some others move about on the floor, like crawfish and brittle stars.

How large is the ocean and how important is it?

This planet should really be called the Ocean and not the Earth, because the ocean covers 71 per cent of the Earth's surface. The ocean has an area of about 361 million sq km, an average depth of 3.7 km, and a total volume of 1,347,000 million cu. km. Seawater contains about 35 grams of salt per kg.

Although we talk of different oceans in the world, there is just a single global ocean. For convenience, we have divided the ocean into the Pacific, the Atlantic, the Indian, and the Arctic. The largest geographical division is the Pacific Ocean that covers one-third of the Earth's surface and contains more than 50 per cent of its water. The Pacific also has the deepest part of the ocean, the Mariana Trench, with a maximum depth of 11,000 m, which is 2200 m more than the height of Mt Everest!

The global ocean is important to us in many ways. The biodiversity and resource-base of the ocean far exceeds that of the land. The ocean contains more than 250,000 species of plants and animals. This could well be an underestimate, since hardly 5 per cent of the ocean has been explored and mapped in detail. Every day, we are discovering new wonders about the ocean, its behaviour, the services it renders, and the products we get from it.

The ocean provides many ecosystem services like regulation of climate and rainfall, cycling of nutrients, absorption of carbon dioxide, waste treatment, a nursery for many species, storage of biodiversity and genetic resources, and protection from storms. The products obtained from it include food items (fish, seaweed, and other items), fishmeal, oil and gas, minerals, medicines, and building materials. In addition, it provides services like transportation, recreation, and employment.

How do we divide the ocean into meaningful zones?

The large marine ecosystem is divided into two major zones: the coastal zone and the open ocean (see Figure 4.1). The coastal zone extends from the high tide mark on land to the edge of the continental shelf, which is the submerged part of the continent. At this point, there is a sharp increase in the depth of water and from here onwards it is open ocean. The coastal zone is discussed in a later section of this chapter.

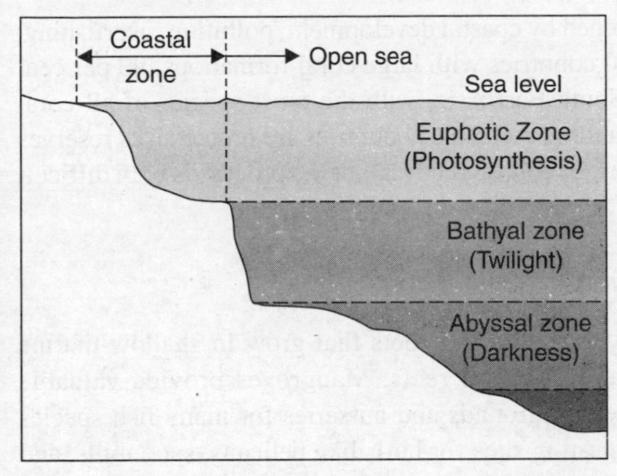

Figure 4.1 The zones of the ocean

The open ocean is divided into three zones by depth. The euphotic zone is the upper part where there is enough light for the phytoplankton to carry out photosynthesis. Large fish like sharks and swordfish are found here. The bathyal zone lacks sunlight and hence contains no producers. Zooplankton and smaller fish that populate this zone have to move to the upper levels at night to feed. The abyssal zone is the cold and dark zone at the bottom, reaching in places to a depth of 4000–6000 m. It is too dark for photosynthesis and yet there are producer organisms at this depth near hydrothermal vents. They are bacteria that can withstand temperatures of 200ºC, and they produce food through a chemical reaction.

What are coral reefs?

Coral reefs are one of the natural wonders of the ocean. On many beaches these beautiful and colourful creations are on sale. Where do they come from and who makes them? Coral reefs are found in the shallow coastal zones of tropical and sub-tropical oceans, where light can penetrate. Apart from being aesthetically appealing, they are also among the world's oldest, most diverse, and most productive ecosystems.

Corals are formed by huge colonies of tiny organisms called polyps. They secrete calcium carbonate (limestone) to form a protective crust around their soft bodies. When they die, their outer skeletons remain as a platform for others to continue building the coral. The intricate crevices and holes in the coral catacombs become the home for 25 per cent of all marine species.

The colour of the coral comes from zooxanthellae, the tiny single-celled algae that live inside the tissues of the polyps. In return for the home provided by the polyps, zooxanthellae produce food and oxygen through photosynthesis.

Why are coral reefs important? Coral reefs are complex ecosystems that perform many ecological services. When polyps form their shells, they absorb some carbon dioxide as part of the carbon cycle. These reefs help protect the coastal zone from the impact of the waves and from storms. They act as nurseries for hundreds of marine organisms. In their biodiversity and intricacy of relationships, they can be called the rainforests of the ocean.

On the economic side, they provide fish, shellfish, building materials, medicines, and employment to people. Through tourism, they bring in valuable foreign exchange and give us the pleasure of enjoying the world under the sea.

Coral reefs are of course in trouble. They are very vulnerable to damage because they grow very slowly, get disrupted easily, and are very sensitive to variations in temperature and salinity.

Coral bleaching is the biggest threat faced by all colonies. When a reef becomes stressed, it expels the zooxanthellae, loses its colour and food, and ultimately dies. An increase of even one degree in the water temperature can trigger bleaching. Sediment run-off from land that smothers the reef and prevents photosynthesis is further stress. Any pollution and disease could also lead to bleaching. Direct physical damage can also occur.

Of the world's reefs, 60 per cent are threatened by coastal development, pollution, overfishing, and warmer temperatures. Of more than 100 countries with large coral formations, 90 per cent are experiencing the decline of the reefs. South East Asia, with the most species of all coral reefs, is the most threatened region. Three hundred reefs in 65 countries are protected as reserves and 600 more are under consideration. Protected reefs do recover, but the process is both difficult and expensive.

What is the importance of mangroves?

Mangroves are unique salt-tolerant trees with interlacing roots that grow in shallow marine sediments. Often they are found just inland off coral reefs. Mangroves provide valuable ecosystem services. Their roots are the breeding grounds and nurseries for many fish species like shrimp and sea trout. The branches are nesting sites for birds like pelicans, spoonbills, and egrets. They stabilize the soil, preventing erosion, and provide protection to the coast during cyclones. They are more effective than concrete barriers in absorbing wave action.

Mangroves occur all along the Indian coast covering 6700 sq km, which is about 7 per cent of the world's total mangrove area. Some of the important mangrove forests are found in the Andaman and Nicobar Islands, the Sunderbans (West Bengal), Bhitarkanika (Orissa), Pitchavaram (Tamil Nadu), and Goa.

In India and elsewhere, mangroves are fast disappearing due to coastal development, logging, and shrimp aquaculture. Between 1985 and 2000, the world lost half its mangroves.

Before we move to the coastal zone and the fresh water lifezones, let us look at a unique part of the ocean, that is, the Southern Ocean.

What is special about Antarctica and the Southern Ocean?

Antarctica is an icy continent, surrounded by the Southern Ocean. It is a fragile ecosystem and is relatively unexploited. The Antarctic ice cap contains 70 per cent of the world's freshwater. A thin layer on the ocean supports krill (very small shellfish), which is consumed by other fish, seals, penguins, and whales.

Antarctica is a sensitive indicator of environmental changes. Between 1947 and 2000, the mean temperature of the Antarctic Peninsula went up by 2.5 ºC in summer and 5.6ºC in winter. The reason for this may be global warming from the burning of fossil fuels. With increasing temperature, huge pieces of the ice shelf have been breaking off. Over a period of time, this melting of ice could raise global sea levels.

Seals and whales were first hunted in Antarctica, leading to the near-extinction of the southern fur seal, the elephant seal, and the blue whale. The pollution from the rest of the world has now reached Antarctica. Radioactive particles from atomic testing as well as pesticides like DDT and other chemical residues have been found in these waters. Tourism that is being promoted now will also have ecological consequences.

No country owns Antarctica, though several countries have made claims on parts of it. In 1991, 41 countries (including India) signed the Antarctic Treaty. They agreed to ensure that Antarctica is used for peaceful purposes, for international cooperation in scientific research, and does not become the scene or object of international discord.

The treaty, which seeks to establish Antarctica as a zone free of nuclear testing and radioactive waste, includes agreed measures for the conservation of Antarctic fauna and flora.

Several countries including India have been conducting scientific research and exploration for oil and minerals in Antarctica. The first Indian expedition to Antarctica was sent in 1981 and since then Indian scientists have established two research stations, Dakshin Gangotri and Maitri, on the continent. Joint teams representing a number of scientific organizations in the country have participated in more than 20 expeditions to the continent.

Are inland seas also special?

These are large saline lakes that have salinities in excess of 3 grams per litre and occur on all continents. They include the Caspian Sea, Dead Sea, Black Sea, and Aral Sea. They are sensitive ecosystems and their environmental importance is only now becoming clear. Most of the large inland seas of the world are facing environmental degradation. They were once healthy waters used for fishing, transport, etc. Now, however, human activities like overfishing, pollution, and diversion of inflow are decimating many of these water bodies. (Read about the Aral Sea in Box 4.2.)

Box 4.2 Crisis: Death of a sea

The Aral Sea, on the border between Uzbekistan and Kazakhstan, was once the world's fourth largest freshwater lake. In 1960 it was spread over 68,000 sq km. Over the past 40 years, however, it has lost over 75 per cent of its volume and over 50 per cent of its area. How did such degradation happen?

The lake has no outflow other than evaporation and the inflow is from the rivers, Amu Darya and Syr Darya. In the 1950s, the USSR, in its insatiable drive for development, decided to divert the water from these two rivers to irrigate the desert area around the lake for planting cotton. The Soviets also constructed the world's longest irrigation canal (over 1300 km long) to take the water into Turkmenistan.

Millions of hectares of cotton were planted and soon USSR became a net exporter of cotton. This success came at a cost to the Aral Sea. By the early 1980s, 95 per cent of the inflow had been diverted and by then the sea had shrunk to a considerable extent.

The fish catch used to be 44,000 tonnes per year, providing a livelihood for 600,000 people. Today, most of the 24 fish species are gone and the fishing industry has collapsed. These are hard times for the 35 million people living in the watershed.

The shoreline has receded by an incredible 100 km! Maynak, which was a seaport 25 years ago, is today 30 km from the shoreline. Abandoned boats dot the landscape. A toxic dust-salt mixture rises like a storm from the exposed seabed and gets deposited on the farmland, damaging crops. The dust also causes many illnesses including throat cancer and tuberculosis. The weather has changed for the worse, with summers getting warmer and the winters colder.

There is worse to come. The Soviet Union had buried hundreds of tons of deadly bio-warfare organisms on an island in the Aral Sea. Soon the island will join the mainland and expensive decontamination will have to commence.

In 1994, the five republics of the region established a joint fund to prevent the complete disappearance of the Aral Sea. The World Bank and the United Nations Environment Programme (UNEP) have given funds to improve the environment. Reviving the lake is a hard task, however. Even if all the diversion of water for irrigation is stopped now, it will take 50 years for the lake to fill up! The dilemma is that, with the collapse of the fisheries, irrigation for agriculture has become even more important. Do we want water on the farms or water in the lake?

What will be the fate of the Aral Sea? At best, the further recession of the sea can be stopped. At worst, it will become a part of history.

We cannot possibly spoil the vast ocean!

The traditional view of the ocean is that of an unbounded domain with inexhaustible wealth. The apparent vastness of the ocean also gives the impression that it can accommodate all users and all kinds of uses and that any amount of waste will be absorbed by the marine system. We now know, however, that the ocean is a finite resource and that human activities are already having an extremely adverse impact on the ocean. The world over, fish-catches have been steadily declining because of pollution and overfishing, with large trawlers and factory ships. Many fish species have disappeared forever.

The deep seabed has mineral as well as biological resources and many countries are now developing techniques to tap this wealth. The insatiable demand for petroleum has led to large-scale extraction of oil from the ocean with many negative consequences. The main shipping lanes of the ocean now carry very heavy traffic with consequent problems like large oil spills from giant tankers.

The dumping of hazardous waste, including nuclear material, into the ocean is raising pollution levels to great heights. Near the shore, heavy discharges from industry and sewage systems are also adding to the pollution. In fact, it is estimated that 77 per cent of marine pollution originates from the land.

Who controls the ocean activities? For a long time the ocean area beyond the territorial limits of countries was governed by the concept of the 'Freedom of the High Seas'. This meant that any country or anyone could exploit the wealth of the ocean without any controls. Now, however, the concept has been replaced by that of the 'Common Heritage of Mankind'.

In a landmark agreement, many countries came together to draft the UN Convention on the Law of the Sea (UNCLOS), which came into force in November 1994. Apart from setting the territorial limit as 12 nautical miles (from the shoreline), this Convention establishes for every country an Exclusive Economic Zone (EEZ) up to a distance of 200 nautical miles from the shoreline (a nautical mile is equal to 1.85 km). Every country has the exclusive right to exploit the resources of its EEZ. What is beyond the EEZ is the International Area of the Seabed.

According to UNCLOS the ocean has to be used only for peaceful purposes and should be managed in the interests of all, including future generations. To control the ocean resources and to resolve disputes, two bodies have been established: the International Tribunal for the Law of the Sea, Hamburg, and the International Seabed Authority, Jamaica. The Seabed Authority controls the exploitation of resources in the International Area.

UNCLOS gives a fair deal to small islands and developing countries. This was due to the dedicated efforts of people like Elisabeth Mann Borgese (read Box 4.3).

India has a coastline of over 7000 km and an EEZ of two million sq km, which is two-thirds of our land area. The Seabed Authority has granted exclusive rights to India to explore 150,000 sq km in the central Indian Ocean.

Box 4.3 Inspiration: Mother of the ocean

She trained as a concert pianist, became a professor of political science, but spent the greater part of her life working for the peaceful and sustainable use of the ocean.

Elisabeth Mann Borgese was born in Munich in 1918 as the second youngest of six children. Her father was Thomas Mann, the great German writer and Nobel laureate. When she was fifteen, the family emigrated from Nazi Germany to Switzerland and later to the US.

In 1972, Elisabeth founded the International Ocean Institute (IOI) as a non-governmental organization to work for the ocean. The IOI is now a global network of 25 operational centres around the world. One of them is located at the Indian Institute of Technology Madras at Chennai. The IOI centres have trained hundreds of people on ocean-related issues and themes. Elisabeth organized an Annual Conference on the theme 'Peace on the Oceans' for over 25 years. She wrote many books on the ocean.

She took an active part in the UN deliberations for drafting the Law of the Sea. In these negotiations, she fought hard to protect the interests of the developing countries and the small islands. She continued this campaign later at the UN and many other fora. Her tireless work encouraged world leaders to rethink our relationship to the sea and our management of marine resources. She was often called 'The Mother of the Ocean'.

Elisabeth showed that, to help conserve the environment, one need not only hug the trees or join the protest marchers. One can work behind the scenes, patiently influencing the policy makers and working out just clauses in international agreements. She inspired many with her vision of the oceans as the 'common heritage of humankind' and by her ceaseless efforts to make this vision a reality.

Elisabeth continued her mission even in her eighties, travelling all over the world in the cause of the ocean. Like Elzeard Bouffier (see Module 2: Prologue), she went on working until she died suddenly and quietly in February 2002.

What is the coastal zone?

The coastal zone is the area of interaction between land and ocean. There is no clear definition of this zone. On the seaward side, it is commonly taken to extend up to the gently sloping, shallow edge of the continental shelf. On the landward side, we could define the limit of the zone as the high tide line or even up to the watershed (see Figure 4.2). Thanks to the edge effect, this zone shows very high levels of biodiversity. Ninety per cent of all marine species are found in this zone.

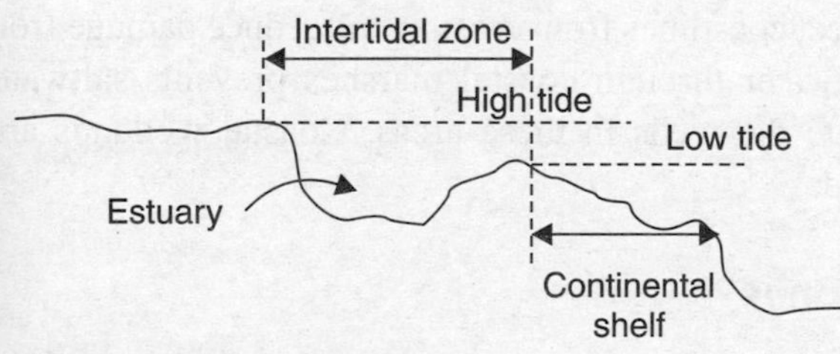

Figure 4.2 The Coastal Zone

The coastal zone is easily affected by human activities. A general trend in the world is for the population to move towards the coastal zone. One estimate is that 60 per cent of the world's population now lives within 60 km of the coast. This proportion is expected to increase to 75 per cent by 2025. Even conservative estimates suggest that at least 40 per cent of the population lives within 100 km of the coast. One incontrovertible fact is

Industrial effluents being pumped into the sea

that the rate of increase of coastal population is higher than that of the general population.

Most of the large cities of the world are to be found on the coast. Twelve of the 20 highly populated urban areas in the world are within 150 km of the coast and these include Mumbai and Kolkata. Coastal megacities have very large ecological footprints. In most cases, they dump enormous amounts of untreated sewage and industrial pollutants into the ocean.

What is the intertidal zone?

Within the coastal zone, the intertidal zone is the area of shoreline between the low and high tide marks. It is the transition between the land and the ocean and is marked by high levels of sunlight, nutrients, and oxygen. As a result, it is biologically very productive with a variety of ecological niches, but it is also a zone of high stress for the organisms that live there.

If the zone is sandy, the organisms must adapt themselves to a constantly shifting environment. Since there is no protection against wave action, the organisms are usually continuous and active burrowers. If the zone is rocky, the organisms are exposed to wave action during high tide and the process of drying and warming up during the low tide. They usually have a means of attaching themselves to rocks, such as thread-like anchors or glue-like secretions.

Why are estuaries and coastal wetlands important?

Estuaries and coastal wetlands include river mouths, bays, mangrove forests, and salt marshes. Here the seawater mixes with freshwater and nutrients from rivers, streams, and the run-off. These ecosystems experience wide daily and seasonal fluctuations in temperature and salinity levels because of tidal rhythms, variations in freshwater inflow from the land and from rivers, and the impact of storms.

The organisms in these ecosystems have to adapt to the changing conditions. Yet, these ecotones are among the most fertile ecosystems, often more productive than the adjacent ocean or river. The main reasons for this are the flow of nutrients from the land, the removal of waste products by tidal action, and the availability of sunlight.

Coastal saltwater wetlands are the ocean's nurseries, where many fish species spend the first part of their lives. These wetlands also protect coastlines from erosion and reduce damage from storms and cyclones. The flow of groundwater through coastal marshes prevents saltwater intrusion that would otherwise contaminate the wells in these areas. Coastal wetlands are, however, in steep decline all over the world.

How do we define freshwater life zones?

Aquatic zones with less than one per cent dissolved salts are freshwater life zones. They include flowing bodies like streams and rivers and standing bodies like lakes, ponds, and inland wetlands.

Freshwater ecosystems perform many services like climate moderation, nutrient recycling, flood control, waste treatment, groundwater recharge, providing a habitat for species, and preserving a pool of genetic resources and biodiversity. On the economic side, they provide food, drinking water, irrigation, hydropower, transportation, recreation, and employment.

Freshwater ecosystems are influenced by the local climate, soil, and resident communities as well as by the surrounding biomes. Anything that happens upstream will have an impact on them.

What are the general characteristics of rivers and streams?

The nature of any flowing-water ecosystem like a stream or a river varies greatly between its source and its mouth (i.e., where it meets the ocean or flows into another water body like a lake). A river often begins as a stream high in the mountains in a forest, which absorbs precipitation and releases it slowly. In other cases, the river begins with the melting of snow or a glacier.

The source or headwater stream flows downstream rapidly over rocks and at times it may even become a waterfall. Once it reaches the plains, it slows down. Other tributaries may flow into the river. Finally, it reaches the delta, the low-lying plains at the mouth, before flowing into the ocean. The delta region is fertile thanks to the sediments carried along by the river.

Human activities are taking a heavy toll on the world's rivers. About 60 per cent of the 240 large rivers are strongly or moderately fragmented by dams, diversions, and canals.

What is the status of India's rivers? More than 70 per cent of India's territory drains into the Bay of Bengal via the Ganga-Brahmaputra river system and a number of large and small peninsular rivers.

The Ganga and the Brahmaputra, together with their tributaries, drain about one-third of India. The Ganga, considered a sacred river, is 2500 km long from its origin to its mouth. Although its deltaic portion lies mostly in Bangladesh, the course of the Ganga within India is longer than that of any of the country's other rivers.

Since the Ganga is fed by the melting of snow in the Himalayas, the variability of its flow is considerably less than that of the exclusively rain-fed peninsular rivers. This consistency of flow enhances its suitability for irrigation and navigation.

Although the total length of the Brahmaputra (about 3000 km) exceeds that of the Ganga, only 750 km of its course lies within India. The Brahmaputra, like the Indus, has its source in a trans-Himalayan area about 60 miles southeast of Lake Manasarovar in the Tibet Autonomous Region of China. The narrow Brahmaputra basin in Assam is prone to flooding because of its large catchment area, parts of which experience exceedingly heavy rainfall.

Bathing in the polluted Ganga

The Indus basin (in northwestern India) and a completely separate set of basins to the south (in Gujarat, Madhya Pradesh, and Maharashtra) drain into the Arabian Sea. India shares the Indus drainage basin with China, Afghanistan, and Pakistan. The Indus and its tributaries the Sutlej, Jhelum, Chenab, Ravi, and Beas are the major rivers in the northwestern basin. Despite low rainfall in the Punjab plains, the moderately high run-off from the Himalayas ensures a year-round flow in the Indus and its tributaries, which are extensively utilized for canal irrigation.

The two most important west-flowing rivers of peninsular India are the Narmada and Tapti. The Narmada and its basin are undergoing large-scale, multipurpose development.

The major rivers that drain into the Bay of Bengal are the Mahanadi, Godavari, Krishna, and Kaveri. Except for the Mahanadi, the headwaters of these rivers are in the high-rainfall zones of the Western Ghats, and they flow along the entire width of the plateau before reaching the Bay of Bengal. The Mahanadi's source is at the southern edge of the Chhattisgarh Plain.

The peninsular rivers experience considerable variations in flow between the dry and wet seasons. However, thanks to relatively steep gradients, they rarely give rise to floods of the type that occur in the plains of northern India.

The question of interlinking Indian rivers is discussed in Chapter 7 of this book.

How can we study large lakes?

Lakes are natural bodies of standing freshwater. They are formed when precipitation, run-off, or groundwater seepage fills up depressions in the land formed by geological changes. A large lake has four zones: the littoral, limnetic, profundal, and benthic.

The littoral zone is the area of sunlit shallow water near the shore of the lake. It extends to the depth at which rooted plants stop growing. Due to the edge effect, this is a zone of high biodiversity and photosynthesis, with the nutrient supply coming from the surrounding land. Some of the organisms found in this zone are frogs, turtles, worms, fish, and insect larvae.

The limnetic zone is the open water away from the shore of the lake. It extends down till the depth to which sunlight penetrates and permits photosynthesis. The vegetation in this zone is less than that in the littoral zone, but there are plenty of producers making food for most of the consumers. Phytoplankton, zooplankton, and larger fish are all found here.

The profundal zone, found in large lakes, is the deep water, which is too dark to permit photosynthesis. There are no plants or algae found here, but food drifts down from the other zones. Bacteria decompose the dead organisms that reach this zone, using up oxygen and releasing minerals in the process. The benthic zone is the bottom of the lake. The organisms that live here can withstand the low temperatures and low oxygen levels found here.

What about Indian lakes? There are very few large natural lakes in India. Many of our large ones are manmade reservoirs formed by dams. Wular Lake in Jammu and Kashmir is the largest natural freshwater lake in India. The Sambhar is a salt lake in the Great Indian Desert of Rajasthan. The inflow into such salt lakes comes from intermittent streams, which experience flash floods during occasional intense rain and become dry later. The lake water also evaporates leaving a layer of salt, which can be extracted.

The large lakes of the world are also facing environmental degradation (to read about Lake Victoria see Box 4.4).

Box 4.4 Crisis: You reap what you sow

Lake Victoria is in East Africa and is shared by Kenya, Tanzania, and Uganda. The second largest freshwater lake in the world and the source of the great Nile river, Lake Victoria is in serious ecological trouble leaving the local people in utter distress.

Until the 1980s, the lake had nearly 400 species of small colourful fishes called cichlids, not found anywhere else on Earth. Each species had a slightly different ecological niche in the lake: some grazed on algae; others fed on dead organic matter at the bottom; some ate insects, shrimps or even other cichlid species.

Cichlids provided a protein diet as well as a livelihood to more than 30 million people around the lake. It was a happy story until an unsuspected villain came along in 1960. A new fish species, the Nile perch, was introduced in the lake to stimulate the local economy and the fishing industry.

The Nile perch was a large and voracious predator that fed on cichlids. Initially, there were enough cichlids for man and predator. Over the next 20 years, however, the perch population steadily increased. As the harvest of Nile perch increased, the catch of other fishes started plummeting.

A strange combination of factors led to the decimation of the fish. The perch of course was thriving at the cost of the cichlid. Further, the local people started cutting the forests nearby for fuelwood, which they needed for drying the large Nile perch. This in turn led to soil erosion, which caused turbidity in the lake. Chemical fertilizers and untreated sewage from the land also washed into the lake.

Thanks to the nutrients in the fertilizer, there were frequent algal blooms and cultural eutrophication, leading to low oxygen levels at the lower depths of the lake. The cichlids were driven to the shallower waters, where the predators ate them up.

The local people could not afford the expensive perch and the small fishers were driven out by mechanized boats. Poverty and malnutrition increased. Matters became worse when water hyacinth invaded large areas of the lake blotting out sunlight, reducing oxygen levels further, and hindering the movement of small boats. The stagnant water promoted diseases spread by snails and mosquitoes.

How is the Nile perch doing? With the cichlids gone, it now feeds on tiny shrimp and its own young. It may also disappear after its 40-year domination. In effect, the fisheries have collapsed. The men from the area have been migrating in search of livelihood, leaving behind families in poverty, disease, and destitution. The World Bank is funding a Lake Victoria Environment Management Programme, but progress is slow.

Source: Miller (2004) and Raven and Berg (2001)

What are freshwater wetlands and why are they important?

Wetlands are land surfaces covered or saturated with water for the whole or a part of the year. They are shallow ecosystems with black and rich sediments and abundant nutrients. They have vegetation adapted to thrive in saturated conditions.

There are three types of wetlands: swamps, marshes, and bogs. Marshes have no trees, but they contain grasslike plants (cattails, sedges, and reeds). Swamps have water-tolerant trees like the red maple and cedar. Bogs are waterlogged areas saturated with groundwater or rainwater and may contain sphagnum moss and low shrubs.

Marshes and swamps are marked by the full penetration of light and consequently high photosynthetic activity. Biomass production and species diversity are much higher than in the surrounding uplands. They provide a habitat for migrating waterfowl and other birds, beavers, otters, muskrats, and game fish.

Traditionally, wetlands were considered wastelands, which bred mosquitoes. They were drained and developed for agriculture, fish farming, or human habitation. Their value as providers of ecosystem services is only now being realized.

Wetlands control flooding by holding excess water and releasing it slowly. They help in recharging groundwater and in purifying water by trapping and holding pollutants in the soil. They are the major breeding, nesting, and migration staging areas for the waterfowl and shorebirds.

They add moisture to the atmosphere, which could fall back as rain. They can even be used as sewage treatment plants that require little technology or maintenance. If they are managed well, they do not breed mosquitoes because of the presence of fish, insects, and birds.

Even as we realize their importance, however, they are shrinking rapidly all over the world. Almost half of them have disappeared over the past century and those that remain have been fragmented by 'development'. What is worse, some wetlands have been used as toxic waste dumps and landfills.

Is there any good news for a change?

Box 4.5 gives an account of some success achieved in restoring the Chilika Lake in Orissa.

Box 4.5 Cure: New life for an old lake?

With an area of about 1000 sq km, the Chilika Lake in coastal Orissa is the largest brackish water lagoon in Asia. Rich in fauna, it has been home to a variety of residential and migratory birds, many fish species, and the Irrawady dolphin.

It is one of the largest wintering grounds for migratory waterfowl from all over the world. From October to December every year, birds migrate here from places as far away as Siberia and Australia.

Chilika was recognized as a biodiversity hotspot and was designated a Ramsar site in 1982. This refers to the Convention on Wetlands, an intergovernmental treaty adopted in 1971 in the Iranian city of Ramsar. There are more than 1200 Ramsar sites in the world.

During the later part of the twentieth century, Chilika had been facing a number of ecological and social problems. Heavy siltation due to massive deforestation in the catchment area was reducing the lake area by 2 sq km every year. The blocking of its opening to the sea reduced the salinity level and many fish species were affected. The number of fish species dropped from 126 in 1930 to 69 in 1995. The poor drainage of water led to waterlogging in the agricultural lands around the lake.

200,000 fisherfolk and 800,000 people in the catchment area depend on the lake for their livelihood, and the increasing population strained the resources further. The annual fish catch dropped from 8800 tons to 1600 tons during the 1990s.

Illegal prawn culture has been a problem, since this activity near the shore affects the fish. In 1999, the fishermen took direct action to destroy some shrimp farms and, in the police firing that followed, some of them were killed.

In 1993, Chilika had been placed on the Montreux Record, a list of Ramsar sites requiring priority conservation attention. By the end of 1990s, the lake appeared to be dying, with no hope of restoration. Positive developments were, however, in the offing.

The Government of Orissa created the Chilika Development Authority (CDA) for the restoration of the lagoon. The CDA commissioned the National Institute of Oceanography (NIO), Goa and the Central Water and Power Research Station (CWPRS), Pune to study the problems and suggest solutions.

The consultants pinpointed the decrease in the salinity level as the major cause of the problems. Acting on their recommendation, the CDA opened an artificial mouth connecting the lake and the sea in 2000. This intervention seemed to have a positive effect, since the fish catch started increasing—4900 tons in 2000–2001 and 11,800 tons in 2001–2002. The Irrawaddy dolphin was seen again and the sea-grass meadows reappeared. The increase in the salinity level also led to the shrinking of the undesirable invasive weed population.

The better exchange of water between the lagoon and the sea brought in more nutrients and flushed away the waste products. As a result, the productivity of the ecosystem increased. Thanks to quicker drainage, the surrounding villages were not affected even in the unprecedented floods in 2001.

The local people had been involved in the planning and implementation of the restoration work. Training programmes on appropriate land use practices and soil moisture conservation were conducted and women's self-help groups were formed. The Orissa Government also banned aquaculture in the lake area.

Impressed by the achievements, the Ramsar Bureau removed Chilika from the Montreux Record in 2002 and also conferred the Ramsar Wetland Conservation Award for 2002 on CDA for the successful restoration work involving the local community.

A success story? Not quite, say some environmentalists. The increased salinity appears to have affected the migratory birds and their numbers are declining. In addition, large-scale poaching of the birds has also been reported. Illegal shrimp culture, overfishing, and the use of fine mesh nets are some of the other continuing problems. Chilika is perhaps waiting for a second rebirth.

Source: Pattnaik (2003) and the website *www.angelfire.com/in/aquiline/chilika.htm*

Box 4.6 Make a difference: Observe World Wetlands Day

The World Wetlands Day is observed every year on February 2, which marks the date of the signing of the Ramsar Convention on Wetlands in 1971, in the Iranian city of Ramsar. On this day, government agencies, non-governmental organizations, and citizens undertake action aimed at raising public awareness of the values and benefits of wetlands in general and of the Ramsar Convention in particular.

Each year the Ramsar Bureau announces a theme for the year's celebration. The theme for 2004 was, 'From the Mountains to the Sea: Wetlands at Work for Us'. The Bureau will send on request posters and other material to help you observe World Wetlands Day. The Ramsar website contains reports from various countries on how the day was observed. You can get ideas from these reports. If you send them a report on your activities, that will also be posted on the website.

The activities can include lectures and seminars, nature walks, children's art contests, community clean-up, radio and television interviews and letters to newspapers, the launch of new wetland policies, visits to Ramsar sites, etc.

For more information, access the Ramsar website *www.ramsar.org*.

Box 4.7 Make a difference: Observe National Maritime Day

April 5 marks the National Maritime Day of India. On this day in 1919, the *SS Loyalty*, the first ship of the Scindia Steam Navigation Company, sailed for the United Kingdom. National Maritime Day is being observed since 1964.

Though the day is celebrated more by seafarers and port authorities than by others, you could use it as an opportunity to create awareness about the ocean. In fact, in 2004, a Sagar Mela was organized in Mumbai on this day, which was attended by the Prime Minister.

Use your creativity to bring life to an observance that has become a mere annual ritual! Perhaps you can get ideas from Canada, which observes June 8 as Oceans Day.

A quick recap

- Many large inland seas and freshwater lakes are facing severe environmental problems.
- The ocean is a treasure house of biodiversity and the provider of a whole range of ecological services.
- Coral reefs and mangroves are vital organisms that are under threat.
- The relatively unexploited Antarctica is now under pressure.
- The use of the ocean is now regulated by the UNCLOS.
- The coastal zone with its immense biodiversity is being degraded by development and population pressure.
- The importance of estuaries and wetlands is now being appreciated.

What important terms did I learn?

Abyssal zone
Bathyal zone
Benthos
Bog
Coastal zone
Continental shelf
Coral reef
Euphotic zone
Exclusive Economic Zone
Headwaters
Intertidal zone
Mangrove
Marsh
Nautical mile
Nekton
Phytoplankton
Plankton
Polyp
Swamp
Wetlands
Zooplankton
Zooxanthellae

What have I learned?

Essay questions

1. Write essays on the importance of
   (a) the ocean
   (b) the coastal zone with regard to economic and ecological services rendered
2. What are the general conclusions you have drawn about the environmental problems of aquatic ecosystems from the cases described in this chapter?

Short-answer questions

1. What are the factors that determine the types of organisms found in aquatic ecosystems?
2. List the main types of marine organisms. Who are the producers and consumers in the ocean?
3. Describe the main features of UNCLOS.
4. Explain why the following are important:
   (a) Coral reefs
   (b) Mangroves
   (c) Estuaries and coastal wetlands
   (d) Freshwater wetlands

How can I learn by doing?

Locate a wetland and study it as an ecosystem over a period of (at least) three months. Trace the inflow and outflow of water. Find answers to questions such as: What are the plant and animal species found in the ecosystem? What kind of services does the wetland provide? What are the threats faced by the system and how can they be handled? Write a report on your study.

What can I do to make a difference?

1. If you are in a coastal town, join voluntary groups engaged in activities like beach clean up.
2. If there are ponds, lakes, or wetlands in or near your area, there could be groups trying to save them. If there is no group and if a water body is in peril, organize a group to make representations to the government and to create awareness in the community.
3. Observe February 2 as World Wetlands Day (Box 4.6).
4. Observe April 5 as National Maritime Day (Box 4.7).

Where are the related topics?

- Chapters 5 and 6 cover biodiversity (including aquatic biodiversity) in detail.
- Chapter 7 gives an account of the use of freshwater sources.
- Chapter 12 discusses freshwater and marine pollution.

What is the deeper issue that goes beyond the book?

There are still vast areas of ocean to be explored. How should we regulate exploration and exploitation of the ocean's resources?

Where can I find more information?

- Menon (2003) gives an account of the Gulf of Mannar.
- For the Chilika Lake story see:
  - Pattnaik (2003) and the website *www.chilika.com*, both of which give an official account of the restoration project.
  - The site *http://www.angelfire.com/in/aquiline/chilika.htm* provides an account of the agitation against shrimp farms.
- The UNESCO website offers more information on the Man and the Biosphere Programme.
- The Ramsar website, *www.ramsar.org*, for more information on the wetlands and the Ramsar Convention.

What does the Muse say?

Here is a translation of an Eskimo song:

The great sea has set me in motion
Set me adrift
And I move as a weed in river
The arch of sky
And mightiness of storms
Encompasses me
And I am left
Trembling with joy.

MODULE 3

Biodiversity

Prologue

A fable for tomorrow

RACHEL CARSON

A strange silence had descended on the small town. The birds—crows, myenas, doves, warblers, parrots, and others—had stopped singing or making any sound. In fact, fewer and fewer birds were seen sitting on the trees or flying in the sky. A few birds perching here and there looked too sick to fly.

The town was not always like this. Not long before, it was a green and lively little town. Flourishing farms surrounded the town and trees completely covered the hillsides. Every season brought its unique range of colours, and the sky, clouds, leaves, and flowers exhibited a variety of shades. Even the roads were beautiful with ferns and wild flowers on both sides.

Thousands of colourful birds came to feed on the seeds and their diverse songs and sounds brought joy to all. One could hear foxes howling in the hills or watch the deer silently crossing the fields on misty mornings. There was an abundance of fish in the clear brooks and streams. Chicken, cattle, and sheep all seemed to thrive on this land. The town and all the life in it seemed to be in perfect harmony with nature.

Drawn by the beauty and harmony of this place, many people came to settle in the town. They built houses, set up farms, and established industries. It seemed to be a paradise on earth.

Then came a change, slowly but surely. A strange sickness began to attack the animals. The chicken went down first, followed by the sheep and the cattle. Hordes of them died even as the people watched in disbelief.

The chicken that survived laid eggs but could not hatch them. Most of the animals also turned infertile. Some flowers bloomed in the trees, but there were no bees or birds to pollinate them, and hence no fruits appeared.

The fields began to wither and no grain could be harvested. The trees on the hillsides began to turn brown and die. The roadsides too became dry and brown as if a fire had struck them. There was no more fish to be caught in the brooks.

Unknown diseases now spread among the people. Many adults and children died inexplicably. It was then that the people also noticed that the songbirds had disappeared and many bird species could no longer be seen. It was in fact a silent spring.

Patches of a white granular powder could be seen in the gutters and on rooftops, which, some weeks back, had fallen on the roofs, lawns, fields, and rivers like snow.

Was it a curse on the town? Was it biological warfare unleashed by an enemy? Not really. The people themselves had caused the blight!

This fable is based on the haunting first chapter of the book *Silent Spring* written by Rachel Carson in 1962. The book described how the use of pesticides, DDT in particular, could affect people's health and also destroy wildlife to such an extent that the spring season could arrive without the song of the birds. Perhaps no other book on the environment in the twentieth century had the impact that this book had on people's consciousness. It appeared right at the beginning of the environmental movement and became a cult book.

Rachel Carson was a marine biologist, who had already written several popular books on the sea, before she turned her attention to the harmful effects of chemical pesticides. *Silent Spring* created a big debate and the chemical industry tried to malign her and discredit her data. Carson, however, was thorough in providing a solid scientific basis for her claims and ultimately her book led to the banning of DDT in the US.

Carson did not live long after the publication of the book. Shortly before her death from cancer in 1964, she remarked, 'Man's attitude toward nature is today critically important simply because we have now acquired a fateful power to alter and destroy nature. But man is a part of nature, and his war against nature is inevitably a war against himself.'

What can we expect in Module 3?

In this module we will examine the tremendous diversity of life and nature, appreciate its importance to the world and to us, and trace the steep decline in the world's biodiversity. We will also find out how the world is making feeble attempts to conserve biodiversity.

CHAPTER 5

Introduction to Biodiversity

For each of the Big Five [extinctions],
there are theories of what caused them,
some of them compelling, but none proven.
For the sixth extinction, however,
we do know the culprit.
We are.
Richard Leakey

What will I learn?

After studying this chapter, the reader should be able to:

- explain species, genetic, and ecosystem diversities;
- list the factors that determine the level of biodiversity in a community;
- appreciate the value of biodiversity and distinguish between instrumental, aesthetic, and intrinsic values;
- give an estimate of the number and distribution of species in the world and describe the state of our knowledge of biodiversity;
- give an account of India's biodiversity and the threats it faces;
- explain local, ecological, and biological extinction;
- distinguish between background and mass extinction;
- appreciate the current rapid loss of biodiversity in the world and a possible sixth mass extinction;
- explain the reasons for, and the impact of, the loss of biodiversity;
- recall the impact of the Green Revolution on biodiversity; and
- appreciate why we have to conserve biodiversity.

Box 5.1 Crisis: Seeds of plenty or seeds of suicide?

Prathap, Mahendra Reddy, Adinarayana—the names of just a few of the many small farmers of Andhra Pradesh who have committed suicide in recent times. In most cases, an intolerable debt burden drove them to take their lives. They had all borrowed money for the inputs into their farming—seeds, fertilizers, and pesticides.

Until about the middle of the twentieth century, such small farmers in India needed very little input from outside. They grew a variety of crops and kept aside some part of the harvest as seeds for the next year. They used organic manure and natural pesticides. Even if one crop failed, there were others to save the farmer from ruin.

Life was not easy, but the farmers were not in deep crisis either. However, India's population was growing and food was being imported. This was also the problem in many poor countries around the world.

The solution came in the form of the Green Revolution with the promise of plenty. The key element was new seeds called High Yielding Varieties (HYVs). Developed first in Mexico and then taken to many countries, the new seeds greatly increased the yield per hectare.

It was certainly a revolution in agriculture, but it came as a whole package. The HYVs needed large inputs of fertilizers, pesticides, and water. Further, the small number of laboratory-developed HYVs replaced the very large number of traditional varieties cultivated by farmers.

With the active encouragement of the government and international agencies, the Green Revolution was enthusiastically adopted in India. Food production increased and the country became self-sufficient.

The Green Revolution, however, changed traditional agricultural practices. The farmers had to buy new seeds, fertilizers, and pesticides every year. Soon, the traditional varieties disappeared, leaving just a few HYVs. In other words, diversity gave place to uniformity. The seeds came from the National Seeds Corporation or other certified government agencies.

From about 1998, another major change occurred with the entry into India of the international seed companies. These companies need mass markets and they aggressively promote the use of a few varieties of seeds.

The small farmer is now in the clutches of a new breed of entrepreneurs. They provide all the inputs at high prices, give credit at very high interest rates, and buy the crops at low prices. A crop failure due to spurious or low quality seeds, a pest attack, or drought is often the last straw for the farmer. The farmer is perpetually in debt and often the only escape is suicide.

Source: Sainath (2004), Joshi (2004), and Rosset et al. (2000)

What is the basic cause of the farmers' problems?

The Green Revolution increased yields, but created problems like soil degradation, falling productivity over time, increased use of fertilizers, a greater propensity for disease, and excessive withdrawal of groundwater. At the basis of all the problems, however, was the loss of biodiversity. That is the topic of this chapter and Chapter 6 that follows.

What is biodiversity?

Biodiversity or biological diversity refers to the numbers, variety, and variability of living organisms and ecosystems. The term includes all the terrestrial, marine, and other aquatic organisms. It also covers diversity within species, between species, as well as the variations among ecosystems. In addition the field of biodiversity is concerned with the complex ecological interrelationships of species. Biodiversity is the Earth's primary life support system and is a precondition for human survival.

What is species diversity? Species diversity refers to the number of plant and animal species present in a community or an ecosystem. It varies a great deal between ecosystems. For example, species diversity is very high in tropical rainforests and coral reefs and low in isolated islands. You will find a large number of different plants and animals in an ecosystem with high diversity.

Genetic diversity is the variety in the genetic make-up of individuals within a species. Ecosystem diversity is the variety of habitats found in an area. It refers to the variety of forests, deserts, grasslands, aquatic ecosystems, etc., that occur in an area.

The following factors determine the degree of diversity in an ecosystem or community:

- Habitat stress: Diversity is low in habitats under any stresses like harsh climate or pollution.
- Geographical isolation: Diversity is less in an isolated region like an island. If a species on an island disappears due to random events, it cannot be easily replaced. Organisms from the mainland would find it difficult to reach the island and colonize it.
- Dominance by one species: The dominant species consumes a disproportionate share of the resources. This does not allow many species to evolve and flourish.
- Availability of ecological niches: A complex community offers a greater variety of niches than a simple community and this promotes greater diversity.
- Edge effect: We saw in Chapter 2 how there is always greater diversity in ecotones or transition areas between ecosystems.
- Geological history: Old and stable ecosystems like rainforests that have not experienced many changes have high diversity. An ecosystem like the Arctic has undergone many changes and this does not allow many species to establish themselves.

What is the value of biodiversity?

First, let us look at biodiversity purely from the human angle. Clearly, many organisms have instrumental value for us. They provide us with the following:

- Food that is directly eaten by humans including grains, vegetables, fruits, meat, and fish. This is the consumptive value of biodiversity.
- Goods of various kinds like fuel, timber, paper, and medicines—most of our current food crops came from wild tropical plants. The majority of the world's poor depends even now on traditional medicines derived from plants. Many of the new drugs developed by the pharmaceutical industry are also derived from plants and animals. Many useful microbes have been identified: while some are capable of cleaning up oil spills and toxic wastes others can extract metals from ores. These are some examples of the productive uses of biodiversity.
- Ecological services: We have already discussed in Chapter 2 the diverse ecological services provided by nature.
- Recreation: The biodiversity of the planet enables activities like wildlife tourism, nature photography, trekking, and birdwatching.
- Genetic resources: Biotechnology and genetic engineering use the genes of organisms to make new types of crops, medicines, etc.
- Option value: We have the option of paying now for the future use of nature. For example, we might work now for the establishment of a wildlife park so that we (or succeeding generations) can use and enjoy the facility later.

What is the medicinal value of biodiversity? Hundreds of plants are still used in the traditional medicine of developing countries. The local people have a vast and unique knowledge of plants and their medicinal values.

Modern pharmaceutical companies in industrialized countries depend heavily on plants in developing countries in their search for new drugs. Some examples of plants from which very effective drugs have been developed are:

- Rosy periwinkle from Madagascar for leukaemia
- Cinchona for malaria
- Rauwolfia serpentina for hypertension
- Coca for anaesthesia

The commercial value of plant-derived medicines is estimated to be US$ 50 billion a year. Yet, all the plants are taken from the poor countries without any compensation being paid to them.

What value can we attach to the sheer beauty of nature and its biodiversity? That is the non-utilitarian or aesthetic value of biodiversity. Just being close to nature gives many of us enjoyment and even spiritual solace. Writers, poets, artists, and composers derive inspiration from nature for their creative work.

Finally, we can look at biodiversity from the ecosystem point of view. It has then an intrinsic or ethical value. In this view, every species has a value and role in nature. It has a right to exist, whether or not it is known to be useful to humans. In fact, since humans have so much power over nature, they should conserve all species. The final ethical argument is that all life is sacred and must be protected.

How many species are there in this world?

We do not know exactly how many species inhabit the Earth. Estimates range from about four to 100 million. The best guess is about 10–14 million. Most of the species in the world are of insects and micro-organisms not visible to the naked eye.

How many species have been identified and named? So far about 1.8 million species (not including bacteria) have been identified, named, and catalogued. These include 270,000 plant species, 45,000 vertebrates, and 950,000 insects. Roughly 10,000 new species are identified every year.

Overall, our knowledge of species, biomes, and ecosystems is poor. Even out of the identified (1.8 million) species, only a third have been studied to any significant level. Among these, we understand the exact roles and interactions of just a small number of species. The new field of bioinformatics attempts to collate all the data on biodiversity using computers and information technology (read Box 5.2).

Box 5.2 Career: Bioinformatics

Bioinformatics is an emerging field with new opportunities. It combines biology with computer science and thus uses the power of information technology to study and conserve biodiversity.

Our knowledge of the world's biodiversity is still very poor and what little we know is also scattered in museums and research institutions all over the world. There is an urgent need to collate all the available information on species names, descriptions, distributions, status of populations, habitat requirements, interactions with other species, etc. This is one of the tasks of bioinformatics.

The main objectives of bioinformatics are to:

- Build computer databases to organize and store biological information
- Develop computer tools to find, analyze, and visualize the information
- Communicate the information using the Internet

Two major bioinformatics programmes are the Species 200 Project, which collects baseline data set for studies on global biodiversity and the Global Biological Information Facility, which puts together 300 years of information from museums.

Source: Miller (2004).

Indigenous knowledge of biodiversity is increasingly recognized as being very important. What was dismissed as primitive knowledge and quackery is now seen as a valuable storehouse of knowledge of biodiversity. In agriculture, farmers have experimented over hundreds of years to develop local crop varieties and animal breeds best suited to the local conditions.

In medicine, local communities and medicine men in the villages possess a tremendous amount of knowledge about the uses of herbs and other plants in curing illnesses. In many cases, indigenous knowledge is far ahead of what scientists know. There are initiatives now to use such knowledge and compensate the concerned community (read Box 5.3).

Box 5.3 Solution: Compensating indigenous knowledge

In December 1987, a team of scientists was on a botanical expedition in the Western Ghats in Kerala. They had taken with them a few members of the Kani tribe as their guides. The scientists noticed that the guides were eating a fruit that seemed to keep them energetic even during tough treks. When the scientists tried it, they too felt a 'sudden flush of energy and strength'.

Initially, the Kanis were reluctant to reveal any information about the plant, saying that it was a sacred tribal secret not to be revealed to outsiders. After considerable persuasion, the tribals showed the team the plant *Aarogyappacha* as the source of the fruit.

The scientists, who were from the Tropical Botanic Garden and Research Institute (TBGRI), secured specimens of the plant, and conducted investigations. They found anti-stress and other beneficial properties in the plant's active ingredients. Using *Aarogyappacha* and three other medicinal plants, they formulated a drug and gave it the name *Jeevani*.

TBGRI gave the right to manufacture the drug to a private company, Arya Vaidya Pharmacy (AVP) for a licence fee of Rs 1,000,000 and a royalty of 2 per cent. The Institute, however, wanted the Kanis to get a part of the benefits as compensation for sharing their knowledge of the plant and its properties. The Kanis were to receive half the fee and half the royalty. This was the first case of an indigenous community receiving compensation in exchange for their traditional knowledge of plants and their uses.

The Kanis formed a trust, which would use the money for promoting the welfare of the community, preparing a biodiversity register to document their knowledge, and promoting conservation of biodiversity.

A model for all to follow? Well, there have been problems with this arrangement. The deal had been made with the Kanis of one panchayat area and the Kanis in other areas felt that they should have been consulted too. Another criticism was that the medical knowledge of the tribe had been taken away without involving the Kani medicine men. The amount of compensation was also considered to be too low.

It was also not clear whether the Kanis had the right to grow and harvest the plant in the forest area that they did not own. The plant was not on the list of minor produce that the Forest Department had allowed the locals to exploit. There was also no protection against outsiders coming into the game.

The Kani-TBGRI case shows that, even with good intentions, it is not easy to ensure that indigenous communities get benefits from sharing their knowledge of biodiversity. The case, however, is a landmark in this field.

Source: Anuradha (2000).

Where is all this biodiversity?

The vast majority of all species are found in the developing countries. About 50–75 per cent of all species are to be found in the tropical moist forests that account for just 6 per cent of the land area. A handful of soil in a tropical forest contains hundreds of species and more than a million individual organisms.

Large herds of zebra and wildebeest in Africa

In the tropics and the sub-tropics, where most of the developing countries are situated, there has always been evolutionary activity, giving rise to rich biodiversity. Biodiversity is less in the colder northern regions because the recurrent ice ages there slowed down the proliferation of lifeforms.

The 19 most biodiverse nations of the world are listed in Table 5.1.

Almost all the plants eaten today in Europe originated in the developing countries (see Table 5.2). Thus, genetic diversity needed to maintain the world agricultural system is mainly found in the developing world. Most of the medicinal plants are also found only in the these countries.

Table 5.1 The mega-diversity countries of the world

Australia	Madagascar
Brazil	Malaysia
Cameroon	Mexico
China	Myanmar
Colombia	Peru
Costa Rica	Philippines
Ecuador	South Africa
Ethiopia	Venezuela
India	Zaire
Indonesia	

Table 5.2 Origins of Food Plants

Plant	Place of origin
Potato	The Andes, South America
Wheat	Turkey and Afghanistan
Bean	Central America
Coffee	Ethiopia
Soya, Cucumber, Orange	China
Rice	India

What about aquatic biodiversity?

We know very little about aquatic biodiversity, since we have explored and mapped less than 5 per cent of the ocean. Corals reefs, estuaries, and the deep-ocean floor contain enormous amount of biodiversity. The deep ocean species are mostly microbes living under the sea floor in a dark world without oxygen. Biodiversity is high near the coasts too, thanks to a variety of producers, habitats, and nursery areas.

What is the level of biodiversity in India?

India is one of the 19 mega-biodiversity countries of the world and, so far, about 70 per cent of the total area has been surveyed for biodiversity assessment. Till date 45,000 wild species of plants and 81,000 wild species of animals have been identified here. Together they represent 6.5 per cent of the world's biodiversity.

The actual numbers must be much higher since many biologically-rich areas like the North East have not been fully explored. The rich biodiversity is attributed to the presence of a variety of ecosystems and climates.

Of the plants found in India about 18 per cent (including many flowering plants) are endemic. At least 166 crop species and 320 species of their wild relatives have originated in India. We also have a large variety of domesticated species. For example, the number of rice varieties alone ranges between 50,000 and 60,000.

What is the biogeographic classification of India?

India has been divided into 10 biogeographic zones (Rodgers and Panwar 1988):

1. Trans-Himalayan: Extension of the Tibetan Plateau including the high-altitude cold desert in Ladakh and Lahaul-Spiti.
2. Himalayas: The entire mountain chain, diverse biotic provinces, and biomes.
3. Gangetic Plain: The Ganga river system.
4. North Eastern zone: The plains and the non-Himalayan hill ranges
5. Desert: The arid area west of the Aravalli hill range, the salt desert of Gujarat, and the sand desert of Rajasthan.
6. Semi-arid zone: The area between the desert and the Deccan Plateau including the Aravalli range.
7. Western Ghats: The hill ranges and plains running along the western coastline.
8. Deccan Peninsula: The south and south-central plateau, south of the river Tapti.
9. Islands: The Andaman and Nicobar Islands.
10. The coasts and the Lakshadweep Islands.

Each zone contains distinct species of flora and fauna.

What is meant by the extinction of species?

By extinction, we mean the complete disappearance of a species, that is, not a single member of the extinct species is then found on Earth. It is an irreversible loss and is called biological extinction.

Before a species becomes biologically extinct, it goes through stages of local and ecological extinction. Local extinction means that the species is no longer to be found in the area that it once inhabited. It is, however, present elsewhere in the world. Ecological extinction of a species means that so few members of a species are left that the species can no longer play its normal ecological role in the community.

Extinction is the ultimate fate of all species. Since multi-cellular organisms evolved on Earth 570 million years ago, about 30 billion species have lived on this planet. Today, there are only about 14 million of them. This means that 99.9 per cent of all species that have ever lived are now extinct!

Over the life of the Earth, physical and environmental conditions have been changing, sometimes gradually and at other times rapidly. When such changes occur, the affected species must adapt itself, or move to a more favourable area, or become extinct.

There are mainly two ways in which species become biologically extinct. Background extinction refers to the gradual disappearance of species due to changes in local environmental conditions. Thus, a process of natural and low-level extinction goes on continuously.

The rate of background extinction has been generally uniform over long geological periods. On occasions, however, a second type, called a mass extinction, has occurred on Earth.

What is a mass extinction? A mass extinction is characterized by a rate of disappearance significantly higher than background extinction. It is often a global, catastrophic event, with more than 65 per cent of all species becoming extinct over some millions of years. This is a brief period in geological terms, compared to the 4.6 billion years of the Earth's existence.

Scientists believe that there have been five mass extinctions over the past 500 million years. In each case, the character of ecological communities changed dramatically. After each such extinction, it took 20 to 100 million years for global biodiversity to recover.

The most severe extinction occurred about 225 million years ago, when 95 per cent of all marine species vanished. The most famous (and the last) mass extinction, however, took place about 65 million years ago, when the giant dinosaurs, which had ruled the world for 140 million years, were wiped out.

Why did mass extinctions occur? Both biological and environmental factors would have led to mass extinctions. The suggested causes include global cooling, falling sea levels, predation, and competition. A more recent theory, with considerable supportive evidence, is that a giant comet or asteroid hit the Earth. This event created huge clouds of dust that blocked out sunlight for so long that the majority of species died.

What is happening now to the world's species?

Before humankind became very active, the world was annually losing one out of every million species. In the early twentieth century, we were perhaps losing one species a year. Now, we are losing one to 100 species a day and soon the rate of extinction is likely to be 1000 each day. These are estimates, but it is definite that we are losing species rapidly.

Over the next 50–100 years, as the population and resource use grow exponentially, the rate of biodiversity loss will also increase sharply. Biologists estimate that 20 per cent of the current species will be gone by 2030 and 50 per cent by the end of the century. Distinguished scientists like Edward O. Wilson and Richard Leakey are convinced that we are facing the sixth mass extinction.

According to Norman Myers, a biodiversity expert,

> Within just a few human generations, we shall—in the absence of greatly expanded conservation efforts—impoverish the biosphere to an extent that will persist for at least 200,000 human generations or twenty times longer than the period since humans emerged as a species. (Miller 2004)

Many species that we never knew about may have become extinct. Again, we can never be sure that a given species has finally become extinct. Members of a species may not be sighted for years and all of a sudden, one member may be found somewhere. As Middleton (1999) puts it, 'Absence of evidence is not evidence of absence'.

What are keystone species? Keystone species play roles affecting many other organisms in an ecosystem. They determine the ability of a large number of other species to survive. When a keystone species becomes extinct, it could result in a cascade of extinctions of other species.

An example of a keystone species is a top predator like the grey wolf. If wolves go locally extinct (say, due to hunting) in an ecosystem, the populations of deer and other herbivores would increase exponentially. The grazing pressure then drives many plants to extinction. Next, the small animals and insects that depend on the plants disappear.

What are indicator or sentinel species? They are very sensitive indicators of environmental problems. They give us an early warning of problems that could potentially affect other species. Frogs and other amphibians are examples of sentinel species (read Box 5.4).

Box 5.4 Danger ahead: Where have all the frogs gone?

Scientists are puzzled and concerned at two events concerning frogs occurring in recent times: first, the sharp worldwide decline in frog populations and, second, the large-scale occurrences of deformities among frogs in some places.

Why should we worry about mere frogs? Because frogs and other amphibians are sentinel species, that is, they are very sensitive indicators of environmental problems. They give us early warning of problems that could potentially affect other species.

Amphibians are organisms that can live both on land and in water, and they include frogs, toads, and salamanders. These cold-blooded organisms first appeared 350 million years ago. There are 5100 known species of amphibians from small frogs to salamanders 1.5 m long. Since frog species live in small, specialized habitats, they are indicators of the environmental conditions in a variety of places.

Frogs breathe through moist, absorptive, and permeable skins that are sensitive to pollutants in the environment. They lay gelatinous and unprotected eggs in ponds and these are very vulnerable to radiation and pollution.

Hundreds of amphibian species are vanishing or declining everywhere. According to the World Conservation Union (IUCN), 25 per cent of all known species are extinct, endangered, or vulnerable. In the US, 92 of the 242 native amphibian species are in decline. Worldwide, 32 species have become extinct in the last few decades and more than 200 species are in decline.

What worries scientists is the fact that the decline is not restricted to areas where there is a clear degradation of the environment or high levels of pollution. It is happening even in remote areas in pristine environments and also in protected parks and reserves. For example, all the seven native species in the Yosemite National Park are in trouble.

In 1995, a group of school children in Minnesota, in the US, discovered a large number of deformed frogs in a pond. In fact, 50 per cent of the leopard frogs in the pond were deformed, while the normal deformity rate is just 1 per cent. Soon, there were similar reports from most other parts of the US. Deformed frogs die early, before they can reproduce, thus leading to a decline in numbers.

Scientists believe that there are a number of causes of the problems with frog species. Agricultural pesticides in the water, increase in ultraviolet radiation in the atmosphere, infectious diseases, parasites, habitat loss, alien predators, and other such factors could be acting together on the amphibian populations.

The rapid disappearance of the tough, but sensitive, amphibians indicates a general deterioration of the global environmental health conditions. Further, the adult amphibians play an important role in ecosystems. For example, they eat more insects than birds do. If they die of pesticide poisoning, many organisms like reptiles and birds that eat them will also be affected. Amphibians are also providers of pharmaceutical products.

It is the frog today. Whose turn is it next?

Source: Raven and Berg (2004) and Miller (2004).

What are the causes of biodiversity loss?

The major causes for the decline in biodiversity are the following:

Habitat loss and degradation

- Destruction of biodiversity-rich areas like tropical forests
- Destruction of coral reefs and wetlands
- Ploughing of grasslands
- Freshwater fish species threatened because dams and water withdrawals have radically altered river systems
- Pollution of freshwater streams, lakes, and marine habitats

Habitat fragmentation

For a species to survive it requires (among other things) a minimum area in the ecosystem. For example, large animals like elephants and lions require large areas to move about. Due to human impact, many large, continuous areas of habitat are being reduced in extent or divided into a patchwork of isolated fragments. This has many effects:

- Species become divided into smaller populations that cannot sustain themselves.
- Smaller fragments means more edge area and this makes the species more vulnerable to predators and competitors as well as to wind and fire.
- Fragmentation creates barriers that limit the ability of the species to disperse and colonize new areas.
- Migratory birds face the loss of their seasonal habitats.

Commercial hunting and poaching

The illegal world trade in rare and endangered species of plants, birds, and animals is estimated at US$ 8 billion per year, second only to arms smuggling. In this market, a live mountain gorilla fetches US$ 150,000 and a panda pelt US$ 100,000.

More than 37,000 plant and animal species are affected and these include the rhinoceros, tiger, leopard, gorilla, butterfly, frog, tortoise, orchid, cactus, mahogany, etc. In addition, exotic pets and decorative plants are sold to collectors.

The poachers, mostly poor people in developing countries, depend on this trade for their livelihood. They collect specimens indiscriminately, killing young and old, male and female, often using very cruel methods. On an average, for each animal captured alive, 50 others are killed. What is worse, most of the live animals that are captured die in transit.

The tragedy is that the poacher finally gets very little, with most of the money going to middlemen. The country of origin also does not benefit in any way, since no taxes or duties are paid. The country only loses its biodiversity.

The Convention on International Trade in Endangered Animal and Plant Species (CITES) is an attempt to prevent this trade (see Chapter 6 for more information).

The mass killing of birds and animals for sport or for commercial purposes has driven some species like the passenger pigeon to extinction (read Box 5.5), and endangered others like the whale (read Box 5.6).

Box 5.5 Catastrophe: Billions today, extinct tomorrow

On March 24, 1900, a young boy in Ohio, in the US, shot a bird. What was special about it? Well, it was the last wild passenger pigeon on Earth. The bird was never seen again in the wild, though there were still some pigeons in zoos. In 1914, Martha, the last passenger pigeon in the world, died in Cincinnati Zoo and you can see the stuffed body in the National Museum of Natural History.

You might still wonder why we should worry about the extinction of a single bird. The passenger pigeon was not just another bird. It was the most abundant bird ever to have inhabited the planet. In the first half of the nineteenth century, there were 10 billion of them, in huge flocks, some containing more than two billion birds! In those days one could see a migrating flock of passenger pigeons that darkened the sky for three full days as they flew past!

How did they disappear? Their meat was good to eat, their feathers made good pillows, and their bones became fertilizer. What is more, it was easy to shoot them, since they moved in such large flocks and nested in long and narrow colonies. Their habitat and breeding grounds were also affected as forests were cleared.

From the middle of the nineteenth century, the hunting of passenger pigeons became big business. Traps, shot guns, dynamite, and even artillery were used to kill them in huge numbers. There was the case of one trapper alone killing three million birds.

In a few decades, billions of birds perished and the species became extinct. There cannot be a more tragic story of humankind's impact on the living organisms of this world.

Source: Miller (2004) and Middleton (1999).

Box 5.6 Crisis: The Big Blue in danger

It is the world's largest animal. When fully grown, it could be 30 m long and weigh the equivalent of 25 elephants. Its heart is the size of a small car and a child can swim through its artery. Today, it is an endangered species.

The blue whale is a remarkable animal. For eight months it is in the Antarctic eating millions of krill (very small fish) by filtering them from seawater. Then it moves to warmer waters, where its young are born.

There were 200,000 blue whales before commercial whaling began. Today, the numbers are down to 3000–4000. It has been hunted to near extinction for its oil, meat, and bone. A blue whale matures sexually only when it is 25, and even then it produces offspring only every two to five years. Conservationists think that the number left is too few for the population to recover and avoid extinction.

Blue whales were caught in large numbers when they were in groups feeding in the Antarctic. Even otherwise, their size makes them easy to spot and kill. An additional threat is the declining krill population in the Antarctic due to the melting of polar ice.

Whales in general have been under threat. Between 1925 and 1975, 1.5 million whales were killed, placing many species on the road to extinction. The International Whaling Commission (IWC) was set up in 1946 to fix annual hunting quotas for countries. The IWC, however, had no powers to enforce the quotas.

In 1986, the IWC finally declared a moratorium on whale hunting. There was a sharp decline in the numbers killed. Japan and Norway have continued to hunt and Iceland resumed its practice in 2002. These countries want the IWC ban on hunting and the CITES ban on trading to be lifted citing large figures for current populations.

Conservationists question the IWC figures and argue that any relaxation will lead to overharvesting of all the species. They feel that these peaceful, intelligent, and social animals should be protected for ecological and ethical reasons.

The future of whales remains a question mark.

Source: Miller (2004)

Introduction of non-native species

When a non-native species is introduced into an ecosystem and it has no predators, competitors, parasites, or pathogens to control its numbers, it can reduce or wipe out many local species. Hundreds of non-native species have been accidentally or intentionally introduced in coastal waters, lakes, and wetlands. Unintentional introduction occurs as stowaways in aircraft, through the ballast water of oil tankers and cargo ships, or as 'hitchhikers' on imported products like wooden packing crates.

In some cases, non-native species are deliberately introduced to get certain benefits. The introduction may initially be useful, but could cause problems later if the introduced species proliferates at the expense of local ones.

In 1859, rabbits were introduced in Australia for sport shooting. Since the environment was favourable to them and there were no predators, their population exploded. They destroyed vast areas of rangeland, native wildlife, and land used for sheep ranching. Finally, a disease virus was deliberately introduced to check their growth. We do not know what else the virus did!

Other causes of a decline in biodiversity are

- Growing population and migration of farmers from overpopulated areas to fragile ecosystems

- Pollution or conversion of wetlands
- Over-exploitation of resources, like overfishing in the ocean and excessive harvesting of medicinal plants
- Construction of large dams that flood large biodiversity-rich areas

Many species are affected by a combination of these reasons. An example is the decline of birds (read Box 5.7).

Box 5.7 Crisis: Where have the birds gone?

If you like to watch the beauty of an eagle's flight, to listen to the song of a nightingale, or to enjoy the chatter of a parrot, savour the feeling. Soon you may not have this pleasure.

About 1200 bird species, out of 9800 known ones, are likely to become extinct within this century. Over 900 species are already either endangered or critically endangered. About 128 species have vanished over the last 500 years and 103 of them since 1800.

Birds perform valuable ecosystem services like seed dispersal, insect and rodent control, scavenging, and pollination. Many birds are sentinel species that warn us of current or impending problems. They could indicate high acidity levels in water, chemical contamination, arrival of new diseases, and the effects of global warming.

Colombia, which has more than 1800 species, is the most bird-diverse country. India, with about 1250 species, ranks among the top 10 bird-diverse countries of the world.

The major threats to birds are:

- Habitat loss and fragmentation
- Introduction of non-native species: bird-eating snakes, rats, cats, mongooses, insects, and pathogens
- Hunting and capture
- Collection for the pet trade
- Longline fishing: Thousands of sea birds are caught in hooks and drown
- Oil spills
- Pesticides and herbicides
- Skyscrapers, towers, and power lines that kill millions of migratory birds

Bird lovers and birdwatchers all over the world are trying to save these winged wonders. The future, however, looks bleak for birds everywhere.

Source: Youth (2003) and Miller (2004)

How do we declare species as being threatened or endangered?

For a species to survive and flourish, enough numbers must be present in the habitat to make reproduction possible. If density and population size fall below threshold values, the numbers start going down.

A threatened or vulnerable species is one that is still found in a reasonable size in its natural habitat, but whose numbers are declining. However, unless conservation measures are taken, it is likely that it will move into the next category, that is, the endangered list.

A species is declared as endangered when the number of survivors is so small that it could soon become extinct over all or most of its habitat. That is, unless it is protected, it will move into the critically endangered category, before it goes extinct.

What is a biodiversity hotspot?

If an area is unusually rich in biodiversity (mostly endemic), and this flora and fauna are under a constant threat of overexploitation, it is called a biodiversity hotspot. There are about 25 such hotspots in the world, mostly in the tropical forests.

The biodiversity hotspots together cover just 1.4 per cent of the global land area, and yet they hold about 60 per cent of the world's biodiversity. One-third of all known species of the world are endemic to these hotspots.

A major reason why these hotspots are under great threat is that more than a billion people live near these areas in poverty. What is more, the rate of population growth in many of these areas is greater than the world average.

Is the biodiversity of India under threat?

At least 10 per cent of India's plant species and a larger percentage of its animal species are threatened. The cheetah and the pink-headed duck are amongst the well known species that have become extinct. More than 150 medicinal plants have disappeared in recent decades. About 10 per cent of flowering plants, 20 per cent of mammals, and 5 per cent of birds are threatened. Hundreds of crop varieties have disappeared and even their genes have not been preserved.

The biodiversity hotspots in India are the Eastern Himalayas and the Western Ghats. Both contain a large number of endemic species, which are now under threat.

What is the status of plant biodiversity in India? India had tremendous diversity in wild plants and to this farmers had added a large variety of crop species. There were thousands of rice and wheat varieties. As we saw in the beginning of the chapter, the Green Revolution encouraged farmers to plant the new HYV seeds, replacing the indigenous ones. Over the years, the traditional species have disappeared. In place of 30,000 varieties of rice, Indian farmers now plant just 12 HYVs.

What will be the impact of biodiversity loss?

The poor people in the developing countries, who are dependent on biodiversity for their daily survival, will feel the impact first. Soon, however, the industrialized countries will also start experiencing the effects of this loss. Most of their food crops, medicines, textiles, spices, dyes, and paper originate from plants in the developing countries. The destruction of the rainforests means that less carbon dioxide is absorbed and natural climate-control mechanisms are lost. This has a major impact on the world's climate.

Why should we care? Many scientists and environmentalists believe that the mass extinction that is now underway could be the biggest of our environmental problems. Almost all the other problems are potentially reversible. The loss of biodiversity is not.

To quote Norman Myers again,

> We could push back the deserts, restore the topsoil, and allow the ozone layer to be repaired in a century or so. We could restore climate stability in the wake of global warming within a thousand years. But once a species is gone, it is gone for good (Miller 2004).

We will see, in the next chapter, what the world is doing to conserve biodiversity.

A quick recap

- The Earth's biodiversity is its primary life support system and vital to our survival.
- Biodiversity has instrumental, aesthetic, and ethical value.
- Our knowledge of the world's biodiversity is limited.
- Most of the biodiversity exists in the tropical developing countries and India is one of the mega-diversity countries of the world.
- Biological extinction goes on all the time and most of the species that ever lived are now extinct.
- Five mass extinctions have occurred so far and the sixth one is underway due to several human-induced reasons.
- A large number of species are in the threatened or endangered categories and need urgent conservation.

What important terms did I learn?

Background extinction
Bioinformatics
Biological extinction
Ecological extinction
Ecosystem diversity
Endangered species
Genetic diversity
Habitat fragmentation
Indicator species
Keystone species
Local extinction
Mass extinction
Sentinel species
Species diversity
Threatened species

What have I learned?

Essay questions

1. Extinction is the ultimate fate of all species. Why then are we concerned about the decline of biodiversity?
2. Describe the different types of extinctions and explain how the coming sixth extinction is different.

Short-answer questions

1. A witty person said that, 'To a first approximation, all species are extinct today'. Obviously, this statement is incorrect. What is the correct situation about the extinction of species?
2. What are the suggested reasons for the five mass extinctions that have occurred so far?
3. How did the Green Revolution affect the diversity of food plants?
4. What are the factors that determine the level of biodiversity in an ecosystem?
5. What are the uses of biodiversity?
6. Distinguish between background extinction and mass extinction.
7. What are the human-induced causes of biodiversity loss?
8. What is the difference between threatened species and endangered ones?

How can I learn by doing?

Select a small park or garden in your area. Prepare a biodiversity register: list all the species found in the place; find their scientific names with the help of a botanist and a zoologist. Interview long-term users of the place and find out if any loss of biodiversity has been observed or whether they have noticed new species. Write a report describing your observations and your recommendations for conserving the biodiversity of the place.

What can I do to make a difference?

1. When you go to the market, buy local varieties of food items, fruits, and vegetables in preference to those that have come from abroad or from a place very far from your town. Buying local species ensures that the diversity is conserved.

2. Change your food habits and start eating indigenous varieties of rice and wheat. You will be conserving biodiversity and at the same time eating more healthy food.

Where are the related topics?

- Chapter 6 deals with biodiversity conservation.
- Forest conservation is covered in Chapter 9.
- The Green Revolution is discussed again in Chapter 10.

What are the deeper issues that go beyond this book?

Can you think of a situation in which human-induced extinction of a species would be morally justified? Should we allow the patenting of life forms?

Where can I find more information?

- See the BBC documentary, *The State of the Planet*, made by David Attenborough for an excellent account of biodiversity issues.
- All the books written by the Harvard University biologist Edward O. Wilson are very readable (for example, Wilson 2002).
- The book by Leakey and Lewin (1996) describes the five mass extinctions.

What does the Muse say?

Here are the words of Lao Tzu, the Chinese philosopher of the sixth century B.C., from his book, *Tao Te Ching*:

In harmony with Tao
The sky is clear and spacious
The earth is solid and full
All creatures flourish together
Content with the way they are
Endlessly repeating themselves
Endlessly renewed

When humanity interferes with Tao
The sky becomes filthy
The earth becomes depleted
The equilibrium crumbles
Creatures become extinct.

CHAPTER 6

Biodiversity Conservation

If all mankind were to disappear,
the world would regenerate back to the rich state
of equilibrium
that existed ten thousand years ago.
If insects were to vanish,
the environment would collapse into chaos.

Edward O. Wilson

What will I learn?

After studying this chapter, the reader should be able to:

- explain the meaning of in-situ and ex-situ conservation of biodiversity;
- list the advantages and disadvantages of creating reserves for in-situ conservation;
- describe the working of seed banks and point out their drawbacks;
- discuss the role of zoos and botanical gardens in biodiversity conservation;
- realize the importance of involving local communities in conservation;
- recall the main international conventions and agreements that seek to promote biodiversity conservation;
- explain the main features of the Convention on Biological Diversity (CBD) and its positive and negative implications;
- discuss the measures taken in India to conserve biodiversity and the country's response to the CBD; and
- explain what biotechnology does and describe the expected benefits and associated problems.

Box 6.1 Crisis: An annual dance of death?

Every year, they visit us in thousands. After travelling thousands of kilometres in the sea to reach the Indian coast, they lay millions of eggs. Thousands of them, however, die gory deaths and are washed ashore. This is the tragic story of the Olive Ridley turtles and their annual migration to the Gahirmatha coast of Orissa.

The Olive Ridley is a small, hard-shelled marine turtle. The most dramatic aspect of its life history is its habit of forming large nesting aggregations, called 'arribadas'. These nesting concentrations occur in Mexico, Costa Rica, Nicaragua, Panama, and India.

It is a threatened or endangered species throughout its range and there is a ban on all Olive Ridley products under the Convention on International Trade in Endangered Animal and Plant Species (CITES: discussed later in this chapter). In India, the killing or capture of the turtle is punishable under the Wildlife Protection Act, 1972.

The turtles face many threats as they move long distances to their nesting sites:

- In areas where recreational boating and ship traffic is intense, propeller and collision injuries are common.
- Turtles develop problems with their respiration, skin, blood chemistry, salt gland, and other functions due to a variety of reasons:
  - ♦ Consuming marine debris like plastic bags, plastic and Styrofoam pieces, tar balls, balloons and raw plastic pellets (such items often contain toxic chemicals)
  - ♦ Moving through oil spills
  - ♦ Ingestion of pesticides, heavy metals, and other toxic chemicals from various sources on land and sea

Orissa is one of the Olive Ridley mass nesting sites hosting as much as 50 per cent of the world's population. Gahirmatha supports perhaps the largest nesting population with an average of 400,000 females nesting in a given year.

The turtles start congregating off the coast by October–November every year. The couples mate in the sea and the females climb on to the sandy beaches to lay their eggs between January and March. A turtle nests up to three times per season, typically producing 100–110 eggs on each occasion.

After the females depart, it takes about 40 days for the eggs to hatch. During this period, the eggs are in danger. Some are washed away, others dry up in the sun, and yet others are eaten by birds and animals. The hatchlings again have a high mortality rate and it is a miracle that many of them make their way to the sea without their mothers to guide them.

Olive Ridley hatchlings making their way to the sea

The biggest threat to the turtles in Orissa is the mechanized fishing trawler. The turtles are either cut open by the rotary blades of the trawlers or suffocate in the fishing nets. More than 100,000 Olive Ridleys have been killed in this way, along the Orissa coast, in the past 10 years. Most of the trawlers operate illegally and they do not care to use the turtle exclusion device (TED) in their nets, which would allow the animals to escape.

There are attempts to save the turtles by the Forest Department as well as by environmental organizations. However, they lack the manpower and infrastructure facilities to monitor 280 km of the coastline. Gahirmatha itself is a marine sanctuary, but this has not helped matters much.

The latest threat to the turtles is the planned exploration for oil 90 km from their nesting site. It is tough to be an Olive Ridley turtle in the Bay of Bengal these days. How long will they continue to come here?

Source: Das (2001) and the websites *www.indianjungles.com/020704.htm* and *www.orf.org/turtles_oliveridley.html*

What is the status of world biodiversity?

Let us review what we have learnt about biodiversity in the previous chapter:

> The biosphere is an extremely complex system with millions of known and unknown species—plants, animals, birds, bacteria, and viruses. Thousands of interactions take place every moment among these organisms and every link in this web is important for keeping the whole system stable.

Humankind has been interfering with the system far too much. They have cut many of the links in the web by destroying, with unknown consequences, many species. Even if we are overestimating the loss of biodiversity, it is clear that the rate of extinction has gone up sharply due to human activities. We are losing more species than ever before. Instead of arguing over the numbers or waiting to get reliable data, we must take precautionary steps to prevent further losses.

The case of the Olive Ridley turtle shows how difficult it is to conserve biodiversity.

What is meant by biodiversity conservation?

In biodiversity conservation, we study how human activities affect the diversity of plants and animals, and develop ways of protecting that diversity. Conservation ranges from protecting the populations of a specific species to preserving entire ecosystems.

There are two main types of conservation. In-situ (on-site) conservation, which tries to protect species where they are, that is, in their natural habitat. Ex-situ (off-site) conservation attempts to preserve and protect the species in a place away from their natural habitat. In general, in-situ conservation is more cost-effective. In many cases, however, the ex-situ approach may be the only feasible one.

How is conservation done in-situ?

In-situ conservation requires the identification and protection of natural areas that have high biodiversity. This includes the establishment of natural parks and reserves. Over the past few decades, there has been an increase in the number of reserves and the area covered.

The main objective is to preserve large areas of undeveloped land so that the ecosystems and biodiversity can continue to flourish and evolve. Large animals like elephants need large reserves. Large reserves are also less vulnerable to disturbances, since their edge length is short relative to the area covered. If a reserve is split into patches, then corridors are provided to enable movement of the animals.

In-situ conservation through reserves has its limitations. Many reserves do not receive the level of protection and management they need. Widespread encroachment by poachers, settlers, and others continues in many parks. Flora, fauna, as well as resources like wood and minerals continue to be exploited.

The local people living near the reserve are often seen as adversaries and are not involved in the conservation process. The chances of success are higher if the local people are consulted in the planning and design of the reserve. Ideally, the locals must be made partners in conservation. They should be allowed to use part of the reserve, or a buffer zone, for grazing, cutting timber, agriculture, fishing, etc. They could also be trained as guides and wildlife experts to restore degraded areas.

What is on-farm conservation? A special case of in-situ approach is on-farm conservation, which is a method farmers have been practising all along. Faced with the problems of gene banks as well as in-situ methods, we have rediscovered the important role of the farmer in biodiversity conservation. In this approach, there is a farmer–scientist partnership to collect, maintain, and improve the traditional plant varieties.

It is impossible to preserve all biodiversity in-situ. Given the population and other pressures, we cannot set aside the required huge land area. Ex-situ conservation is equally important. In some cases, where the population of a species has dwindled to extremely low levels, ex-situ conservation may be the only way out.

How is ex-situ conservation practised?

In this approach, we conserve biodiversity in an artificial setting. This includes the storage of seeds in banks, breeding of captive animal species in zoos, and setting up of botanical gardens, aquariums, and research institutes.

How do seed banks work? There are more than 100 seed banks in the world and they hold more than four million seeds maintained at low temperatures and low humidity levels. The majority of the banks are in the industrialized countries or indirectly controlled by them.

The banks store a very large amount of plant genetic material in a small space. The seeds are supposed to be safe from habitat destruction, climate changes, and general destruction. They can even help reintroduce extinct species.

There are, however, many problems with seed banks. Seeds are dried out before storage and some seeds cannot tolerate this process. No seed remains alive indefinitely and every seed must be periodically germinated and new seeds collected for storage. This is an expensive and difficult process. It is said that a very large number of seeds in banks have not been germinated or tested for long and that many may be dead.

Accidential fires or power failures can permanently damage the seeds. One safeguard against such events is to subdivide the sample and store them in different banks. A more intractable problem is that the seeds in a bank have not been evolving in relationship to outer circumstances. When they are later reintroduced into the field, they may be less fit for survival.

Most of the seed banks concentrate on about 100 plants that give 90 per cent of our food. There is a need for many more banks to store many more species, particularly in the developing countries. In spite of the problems and limitations, however, seed banks remain an important method of preserving biodiversity for the future. If they are not managed properly, however, they may be giving us a false reassurance that we are conserving biodiversity.

What is the role of zoos in biodiversity conservation? Along with many other animals, zoos often preserve a few individuals of a critically endangered species. If an animal breeds in captivity, the zoo may ultimately reintroduce the species into protected reserves.

Zoos need large spaces and huge funds. Only a small percentage of endangered species can be protected in zoos. The public tends to support the saving of large or popular species like the tiger, elephant, and panda. There is not much interest in protecting smaller or less attractive species, even if they are known to be very important for the ecosystem.

What about botanical gardens? There are over 1,600 botanical gardens in the world, holding about four million plants. They cover about 80,000 species or 30 per cent of all known species. The largest one is Kew Gardens in England, which has 25,000 species. When Britain was an imperial power, large numbers of plant species were taken from the colonies to the Kew Gardens.

Botanical gardens are of significant educational value to both scientists and students. Like zoos, botanical gardens also face problems of funding and space. They increasingly focus on rare and endangered species.

How can we use indigenous knowledge to conserve biodiversity? As discussed in Chapter 5, the best approach is to:

- Involve local communities in the conservation process as partners.

- Make use of their traditional knowledge (especially in medicine) for documenting the value of diverse organisms and for developing new products.
- Compensate communities for the knowledge they have shared (as in the Kani—TBGRI case described in Chapter 5).

There is an international movement to conserve biodiversity through measures such as:

- Documenting indigenous knowledge
- Assisting local communities to grow medicinal plants, make medicines and market them
- Saving and propagating traditional seeds
- Reintroducing traditional food items through restaurants

Examples of organizations participating in this movement are Seeds of Change in the US and Association Kokopelli in France.

What is being done to protect wildlife?

Many countries have declared certain areas as reserves, and have restricted human activities in those areas. There are several thousands of national parks, sanctuaries, wildlife refuges, and wilderness areas in the world. They are partially or fully protected and they cover about one billion hectares.

Many of the reserves are too small to sustain large species or they are too poorly protected. The developing countries have the maximum biodiversity, but they do not have enough funds and, in many cases, the expertise needed to manage protected areas well. Most reserves in the poorer countries suffer encroachment by land grabbers, loggers, miners, and poachers.

Why is it difficult to protect marine biodiversity? We continue to view the sea as a huge natural resource that can absorb unlimited amounts of waste and pollution. Unfortunately, the damage to marine biodiversity is not visible to most people. At the same time, there is a tremendous pressure to expand human activities on the coast (see Chapter 4 of this book).

What are the international efforts in biodiversity conservation?

In 1980, three organizations—United Nations Environment Programme (UNEP), the World Conservation Union (IUCN), and the World Wide Fund for Nature (WWF)—prepared a World Conservation Strategy. It was a plan to:

- conserve biodiversity,
- preserve vital ecosystem processes on which all life depends for survival, and
- develop sustainable uses of organisms and ecosystems.

In 1991, they published a new version of this document entitled, 'Caring for the Earth'.

Following the publication of the World Conservation Strategy and the holding of conferences like the 1992 Earth Summit, several international agreements and global initiatives have focussed on biodiversity conservation. The major ones are discussed below.

The Convention on International Trade in Endangered Animal and Plant Species (CITES): This convention has been in force since 1975 and more than 150 countries have signed it. Even though enforcement is difficult, CITES has helped reduce trade in many threatened species including elephants, crocodiles, and chimpanzees (read Box 6.2).

Box 6.2 Solution: Stop the trade, save the species

In 1989, the African elephant was declared an endangered species. The elephants were disappearing primarily due to the ivory trade. By 2000, however, the elephant population had recovered in Namibia, Botswana, and Zimbabwe. This is one of the success stories of the Convention on International Trade in Endangered Animal and Plant Species (CITES).

Beginning with 21 countries in 1975, CITES currently has 152 signatories. The Convention bans the hunting, capturing, and selling of endangered or threatened species. It lists over 900 species that cannot be traded as live specimens or wildlife products. It restricts the international trade of 29,000 other species that are potentially threatened.

Under the Convention, each country is bound to pass laws according to the CITES guidelines and many countries have done so. Enforcement of the laws is, however, lax in many countries, and even when a violator is caught, the penalties are mild. Further, a member country can exempt itself from protecting any of the listed species. Trade in endangered species still goes on, especially in countries that have not joined CITES.

Returning to the African elephant, the recovery of the population is also causing a problem. In some places, the area is not enough for the large herds. They uproot many small trees, which in turn affect many other species. The local people want to hunt some of the elephants to keep the numbers down and to get economic benefits. In 1997, CITES moved the African elephant to a less restrictive, but a potentially threatened list, allowing the sale of stockpiled ivory to Japan.

Any ban on hunting and trading can only be a short-term measure. The better way is to educate consumers to stop buying the illegal products. If the prices and the market decline, the illegal trade will also stop.

Source: Raven and Berg (2004) and Miller (2004)

The UN Convention on the Law of the Sea (UNCLOS):
Thanks to UNCLOS (discussed in Chapter 4 of this book), countries now have jurisdiction over their territorial seas as well as their Exclusive Economic Zones (EEZs). In effect, 36 per cent of the ocean surface area and 90 per cent of the fish stocks come under the control of individual countries. This provides a great opportunity for each country to protect its marine biodiversity (or conversely to overexploit it!).

The International Convention for the Control and Management of Ships' Ballast Water and Sediments:
When a ship unloads at a port, its hull must be filled with water to maintain the ship's stability. As the ship then moves to another port, thousands of organisms travel with the water. When the ship takes in a new load at its destination, the ballast water and the organisms are released into the local waters. We saw in Chapter 5 how non-native species can destroy many local species. The Convention on ballast water, adopted by 74 countries in February 2004, introduces strict control over ballast water. All ships made after 2009 should have facilities to treat the ballast water before releasing it in foreign waters.

The Convention concerning the Protection of the World Cultural and Natural Heritage: This convention, adopted by UNESCO in 1972, seeks to encourage the identification, protection, and preservation of cultural and natural heritage around the world considered to be of outstanding value to humanity.

UNESCO's World Heritage mission is to:

- help the countries safeguard World Heritage properties by providing technical assistance and professional training;
- provide emergency assistance for World Heritage properties in immediate danger;
- support States Parties' public awareness-building activities for World Heritage conservation;
- encourage participation of the local population in the preservation of their cultural and natural heritage; and
- encourage international co-operation in the conservation of cultural and natural heritage.

A number of sites in the World Heritage list are nature parks and reserves that conserve biodiversity. UNESCO's Man and the Biosphere Programme is a major effort in biodiversity conservation (read Box 6.3).

Box 6.3 Solution: Saving biospheres

Alarmed at the rapid degradation of many ecosystems of the world, the United Nations Educational Scientific and Cultural Organization (UNESCO) began the Man and the Biosphere Programme in 1971. The plan was to establish at least one (ideally five or more) biosphere reserve in each of the Earth's 193 bio-geographical zones. Each protected biosphere should be large enough to prevent gradual species loss and should combine conservation and sustainable use.

Each reserve must be nominated by the national government and should meet the basic requirements. It will contain three zones:

- Core area: The most important and fragile part of the ecosystem, which should be legally protected from all human activities, except research and monitoring.
- Buffer Zone 1: The area in which non-destructive research, education, recreation, sustainable logging, agriculture, livestock grazing, hunting, and fishing could be permitted as long as the activity does not harm the core area.
- Buffer Zone 2: The transition zone, in which conservation could be combined with more intensive and yet sustainable forestry, agriculture, grazing, hunting, fishing, and recreation.

As of October 2004, there were 440 biosphere reserves in 97 countries. There are 47 reserves in the US, 31 in Russia, and 24 in China. The three Indian biospheres are in the Nilgiris, the Gulf of Mannar, and the Sunderbans. The major problem has been raising funds from local and outside sources to keep each biosphere going.

Other treaties that promote biodiversity conservation are the Convention on Wetlands (see Chapter 4 of this book) and the Convention on the Conservation of Migratory Species of Wild Animals (1983).

UNEP has initiated many regional agreements to protect large marine areas shared by countries. Under the UNESCO Man and the Biosphere Programme, many of the designated reserves are coastal or marine habitats. In addition, the IUCN supports 1,300 Marine Protected Areas covering about 0.2 per cent of the ocean area.

Bhutan is one country that has taken very positive steps to conserve its rich biodiversity (read Box 6.4).

Box 6.4 Solution: Thunder Dragon Conserves its Biodiversity

Bhutan or the Thunder Dragon, with its thick forest cover and immense biodiversity, is called the 'oxygen tank' or the carbon sink (see carbon cycle, Chapter 2) of the world. Out of an area of 40,000 sq km, 72 per cent is under forest cover. This eco-system supports 7,000 species of plants, 165 species of mammals, and 700 species of birds. No wonder it is one of the 18 biodiversity hotspots of the world.

The diversity can be explained by Bhutan's wide range in altitude, topography, and climate. There are three climatic zones: sub-tropical, mid-mountain, and alpine. While the southern foothills get as much as 5000 mm of annual rainfall, the northern regions receive as little as 500 mm. Numerous streams originate in the mountains, becoming six rivers that flow into India.

The Royal Government of Bhutan is fully committed to the conservation of the country's rich biodiversity. As early as 1964, Bhutan established two protected areas. In a major revamp In 1993, the country set up four national parks, four wildlife sanctuaries, and one protected area covering 10,513 sq km. Again, in November 2000, the government extended protection to a network of biological corridors, connecting key tiger habitats (with a population of over 100 tigers). Now 35 per cent of the country's area is protected.

The government is determined to maintain a forest cover of at least 60 per cent with the cooperation of local communities. However, they face many challenges, like the fuelwood and timber needs of a growing population, the smuggling of timber into India, and the entry of Tibetans who collect valuable medicinal plants. Yet, Bhutan is a rare biologically diverse country that has an active programme of conservation. It may show the way to the rest of the world.

Source: Johnsingh and Yonten (2004)

What is the Convention on Biological Diversity?

After five years of international negotiations, the Convention on Biological Diversity (CBD) was approved in 1992 at the Earth Summit in Rio de Janeiro, and it came into force in 1993. As of October 2004, 188 countries are parties to the convention. India ratified it in 1994. The US signed the convention in 1993, but has not ratified it.

The convention has three main goals:

- The conservation of biodiversity
- The sustainable use of the components of biodiversity
- Sharing the benefits arising from the commercial and other utilization of genetic resources in a fair and equitable way

Under the Convention, governments undertake to conserve and sustainably use biodiversity. They are required to develop national biodiversity strategies and action plans, and to integrate these into broader national plans for environment and development.

Other treaty commitments include:

- Identifying and monitoring the important components of biological diversity that need to be conserved and used sustainably.
- Establishing protected areas to conserve biological diversity while promoting environmentally sound development around these areas.
- Rehabilitating and restoring degraded ecosystems and promoting the recovery of threatened species in collaboration with local residents.
- Respecting, preserving, and maintaining traditional knowledge of the sustainable use of biological diversity with the involvement of indigenous peoples and local communities.
- Preventing the introduction of alien species that could threaten ecosystems, habitats, or indigenous species; controlling and eradicating alien species already introduced.
- Controlling the risks posed by organisms modified by biotechnology.
- Promoting public participation, particularly when it comes to assessing the environmental impact of development projects that threaten biological diversity.
- Educating people and raising awareness about the importance of biological diversity and the need to conserve it.
- Reporting on how each country is meeting its biodiversity goals.

While CBD is a step forward, it has some drawbacks and implementation problems:

- It excludes the existing gene bank collections. Thus, four million seeds, which are also of high commercial value, are outside the control of CBD.
- It encourages bilateral agreements, even though many biodiversity issues are regional or global. This has led to some poor countries signing agreements with rich countries or big corporations giving away their rights over indigenous biodiversity.
- As with many international agreements, the implementation of CBD has also been slow and poor. The Convention does not provide for severe penalties for violations, nor does it have an enforcement mechanism.

What actions have we taken to conserve India's biodiversity?

The first protected area in the country was the Corbett National Park, established in 1936. Currently, there are more than 500 national parks, sanctuaries, and biospheres covering about 4.5 per cent of the land.

In the national parks, habitations and private ownership of land are not permitted. Traditional activities like grazing and fuelwood collection are also prohibited. On the other hand, in sanctuaries some activities are permitted.

A biosphere belongs to the third category of protected areas. Here the wild flora and fauna are protected, but the people are allowed to live in the area and carry on their traditional activities. Table 6.1 lists the various biosphere reserves of India.

Table 6.1 Biosphere Reserves of India

S. No.	Date notified	Name of the site	Area in sq. km	Location (State)
1	01.08.86	Nilgiri	5,520	Parts of Wynad , Nagarhole, Bandipur and Mudumalai, Nilambur, Silent Valley, and the Siruvani Hills (Tamil Nadu, Kerala and Karnataka)
2	18.01.88	Nanda Devi	5,860.69	Parts of the Chamoli, Pithoragarh, and Almora districts (Uttaranchal)
3	01.09.88	Nokrerk	820	Part of Gora Hills (Meghalaya)
4	14.03.89	Manas	2,837	Parts of the Kokrajhar, Bongaigaon, Barpeta, Nalbari, Kamprup, and Darang districts (Assam)
5	29.03.89	Sunderbans	9,630	Parts of the Brahamaputra and Ganga deltas (West Bengal)
6	18.02.89	Gulf of Mannar	10,500	Indian part of Gulf of Mannar between India and Sri Lanka (Tamil Nadu)
7	06.01.89	Great Nicobar	885	Southernmost islands of the Andaman and Nicobar Islands
8	21.06.94	Similpal	4,374	Part of Mayurbhanj district (Orissa)
9	28.07.97	Dibru-Saikhowa	765	Parts of the Dibrugarh and Tinsukia districts (Assam)
10	02.09.98	Dehang Debang	5,112	Parts of Siang and Debang Valley (Arunachal Pradesh)
11	03.03.99	Pachmarhi	4,926.28	Parts of the Betul, Hoshangabad, and Chindwara districts (Madhya Pradesh)
12	07.02.00	Kanchanjanga	2,619.92	Part of Kanchanjanga Hills (Sikkim)

Source: www.teriin.org/biodiv/status.htm

Of the 26 Indian sites in the World Heritage list, five come under the natural heritage category. These are the Kaziranga National Park and the Manas Wildlife Sanctuary, both in Assam, the Keoladeo Ghana National Park in Rajasthan, the Sunderbans National Park in West Bengal, and the Nanda Devi National Park in Uttaranchal.

In addition, there are special projects to protect certain animal species like the tiger and the elephant. For example, Project Tiger attempts to save the Indian tiger from extinction. From 40,000 in 1900, the number had declined to 1,800 in 1972 and the tiger was declared an endangered species. Since then, 25 tiger reserves have been set up in 14 states covering more than 33,000 sq km. As a result, the tiger population has more than doubled.

Snakes and crocodiles are being protected from extinction by the efforts of a dedicated individual, Romulus Whitaker (read Box 6.5).

Box 6.5 Inspiration: Snake and croc man

The Irulas in Tamil Nadu are expert snake catchers. For long they had been catching snakes and selling them to the flourishing snakeskin trade. When the trade in snakeskin was banned, they lost their livelihood.

At that point, Romulus Whitaker entered their lives. He helped them set up the Irula Cooperative Society for extracting snake venom and selling it to the institutes that make lifesaving anti-venom. It was a win-win formula, since the killing of snakes stopped and the Irulas had a profitable occupation.

Romulus Whitaker came to India when he was seven. From his childhood, he had a natural affinity for snakes and in fact for all wildlife. 'It was like as if I was born with it, it was like some kind of reincarnation stuff, though I don't really believe in it,' he says.

When he was a school student in Kodaikanal, he used to wander in the Palani Hills and picked up the observation skills needed for dealing with wildlife. Later, he learnt snake-catching from the Irulas and crocodile-catching from the natives of Papua New Guinea.

In 1972, Whitaker and some friends set up the Madras Snake Park, which is today on the must-see list of every tourist to Chennai. Most of the snake-keepers in the Park are Irula tribals. The Park has 31 species of Indian snakes, all the three species of Indian crocodiles, four species of exotic crocodiles, one Aldarba tortoise, three species of Indian turtles, and five species of lizards.

Many species of reptiles including endangered species like the Indian python have been bred in captivity in the Park. The offspring have been either released into the wild or made available for exchange with other zoos.

The Park is of great educational value, giving people information about the habits of reptiles. It tries to dispel erroneous notions about snakes and also gives scientific advice on the treatment of snakebite.

Whitaker next established the Crocodile Bank, also in Chennai. It was originally conceived in the early 1970s because the crocodile was facing extinction in the country at the time. It was to be a 'gene bank' for crocodiles and was thus named 'Crocodile Bank' and not 'Crocodile Farm'.

The idea of a farm gives the idea that it is a place where the crocodiles would be killed or skinned. So he had to call it something very different because it was purely to conserve crocodiles first in captivity and subsequently release them into the wild.

The Crocodile Bank served the vital purpose of supplying a breeding populace for restarting the species in Tamil Nadu as well as in the other parts of the country. Later, it became an international crocodile bank, gathering species, which were endangered in other countries as well, like Siamese crocodiles from Thailand and crocodiles from Mexico.

The Bank now has crocodiles from all over the world and has become a gene pool for all species. At the same time it is also a vibrant educational institution, receiving over 500,000 visitors every year. For the first time they encounter crocodiles as non-menacing, interesting, and potentially viable animals, in terms of their ecological diversity.

Whitaker has made many documentary films, including the award-winning, King Cobra. Whitaker has this to say about conservation:

> I would say that the custodians of wildlife and forests are not the forest department and they are not us or anybody in Delhi or Chennai or Mumbai or Bangalore. It is the people who live there, whether you like it or not. They'll ruin it or make it okay, depending on how well we interact with them. If we act as policemen and keep telling them they are doing a hell of a job and ruining the countryside, nothing is going to happen basically. I mean they will keep destroying the trees to make money out of it. Give them a proper framework with a plan that is sustainable and make it profitable for them.

Whitaker's life and work show that a single individual can make an enormous contribution to biodiversity conservation through passion and dedication.

What about ex-situ conservation in India? There are 35 botanical gardens and 275 zoos, deer parks, safari parks, and aquaria. Two institutions engaged in conservation are the National Bureau of Plant Genetic Resources, New Delhi and the National Bureau of Animal Genetic Resources, Karnal.

How is indigenous knowledge being used to conserve biodiversity? Many local communities and voluntary organizations are now promoting conservation and the sustainable use of biodiversity.

In 1996, the village of Pattuvam in Kerala created history by declaring its absolute ownership over all the genetic materials currently growing within its jurisdiction. The villagers had earlier prepared a detailed register of all the species found within the village. The village also set up a

Forum for the Protection of People's Biodiversity, which would henceforth have to be consulted by any person or company seeking access to the register and the genetic material it lists.

There are efforts all over India to document local biodiversity and indigenous knowledge. Some of the organizations involved in such work are: the, Foundation for the Revitalisation of Local Health Traditions (read Box 6.6); the Green Foundation, Bangalore; the Centre for Indian Knowledge Systems, Chennai; Beej Bachao Andolan, Tehri Garhwal, and Navadanya, Dehradun. Annadana (in Auroville near Pondicherry), is the base of the South Asian Network of Soil and Seed Savers established with the help of Association Kokopelli of France.

Box 6.6 Solution: Conserve biodiversity, save health

If you wish to find traditional home remedies for the health problems of your family, subscribe to the magazine *Amruth,* published by the Foundation for the Revitalisation of Local Health Traditions (FRLHT) in Bangalore. This is one of the many activities of FRLHT, founded by Darshan Shankar.

The vision of FRLHT is to revitalize the Indian medical heritage. FRLHT has identified three thrust areas to fulfil this vision:

- Conserving natural resources used by Indian Systems of Medicine.
- Demonstrating the contemporary relevance of the theory and practice of Indian Systems of Medicine.
- Revitalizing social processes (institutional, oral, and commercial) for the transmission of traditional knowledge of health care, for its wider use and application.

FRLHT holds the view that in an era of globalization, India should make fuller use of her rich and diverse medicinal plant knowledge for her own needs and share the knowledge on fair terms with the rest of the world.

Some of the programmes are:

In-situ conservation: The setting up of a network of 54 Medicinal Plant Conservation Areas (MPCA) across different forest types and altitude zones in Tamil Nadu, Kerala, Karnataka, Andhra Pradesh, and Maharashtra, with community participation.

Ex-situ conservation: The establishment of 19 Medicinal Plant Conservation Parks (MPCP) and nurseries by NGOs and research institutes in Tamil Nadu, Kerala, and Karnataka. These parks and nurseries serve as community conservation education centres and as repositories of the region's medicinal plant resources and local health knowledge.

Conservation research: Development of a Bio-cultural Herbarium and Raw-Drug Library of Indian Medicinal Plants, databases of Indian medicinal plants, seed storages, and herbaria.

In 2003, FRLHT won the prestigious alternative medicine award of Columbia University, US, for its 'outstanding' role in promoting traditional medicine systems and conservation of Indian medicinal plants.

Source: www.frlht-india.org

What has India done under the Convention on Biological Diversity (CBD?

In May 1994, India became a party to the CBD. In January 2000, the government released the National Policy and Action Strategy on Biodiversity. This document seeks to consolidate the ongoing efforts at conservation and sustainable use of biodiversity and to establish a policy

Foreigners running away with medicinal plants

Exploitation by pharma companies

and programme regime for the purpose. The Indian Parliament passed the Biodiversity Bill in December 2002. The main intent of this legislation is to protect India's rich biodiversity and associated knowledge. It seeks to check biopiracy and to prevent the use of our biodiversity by foreign individuals and organizations without sharing the benefits with us. One thrust area is the conservation of medicinal plants.

India has prepared a National Biodiversity Strategy and Action Plan (read Box 6.7).

Box 6.7 Plan: Biodiversity conservation

As required by the Convention on Biological Diversity (CBD), a National Biodiversity Strategy and Action Plan (NBSAP) for India was prepared during 2000–2003. The NBSAP process was carried out by the Ministry of Environment and Forests (MoEF), Government of India, under sponsorship of the Global Environment Facility (GEF) through the United Nations Development Programme (UNDP). In a unique arrangement, Kalpavriksh (KV), a non-governmental organization, undertook its technical coordination.

Based in Delhi and Pune, KV is an action group working on environmental education, research, campaigns, and direct action since 1979. It has been working on a number of local, national, and global issues.

The NBSAP process involved consultations and planning with thousands of people across the country, including tribal (*adivasi*) and other local communities, NGOs, government agencies, academics and scientists, corporate houses, students, the armed forces, and other sections of society.

In order to reverse the erosion of biodiversity in India, the NBSAP focuses on three basic goals: conservation of biodiversity, sustainable use of biological resources, and equity in conservation and use.

Some important measures suggested in the NBSAP are:

- Restoration and regeneration of degraded ecosystems
- Recognition of community rights
- Development of alternative intellectual rights systems appropriate for indigenous knowledge
- Balancing of local, national, and international interests related to biodiversity
- Respect for cultural diversity
- Preventing deprivation of indigenous people from natural resources

Source: The website of Kalpavriksh, *www.kalpavriksh.org*

What is biotechnology?

Genes are the basic units through which an organism passes on its characteristics to its offspring. The pattern of genes in a plant or animal determines its potential physical shape, growth, and behaviour.

Biotechnology or genetic engineering manipulates the genes in an organism to change its characteristics. It can move a favourable gene from one organism to another. For example, biotechnology can make a plant resistant to specific pests or diseases. It can also produce new varieties of plants with some desired characteristics.

What are the benefits expected from biotechnology? Biotechnology claims to bring benefits in food, agriculture, and health. In agriculture, some possible products are the following:

- New crop varieties to double or triple yields per hectare with less inputs.
- Super plants that would produce their own fertilizer and pesticide.
- Plants adapted to grow in poor soils.

What is the role of biotechnology with reference to biodiversity conservation?

Traditionally, farmers were always improving plant characteristics through methods like cross-breeding. This process, however, took many generations. Biotechnology enables a quicker and more focused way of modifying the genetic structure of plants and animals. It also provides techniques for moving genes between organisms that do not exchange genes naturally.

Biotechnology could lead to new and improved methods for preserving plant and animal diversity. By increasing the value of biodiversity, it could lead to better conservation.

What are the problems with biotechnology? Biotechnology is largely under the control of private industry in the richer countries. Their objective is to make products for the global market. Hence, their research favours uniformity over diversity.

Biotechnology companies prefer mass production of genetically identical seeds. Such seeds may not be appropriate to the small farmer in a developing country, whose needs are location-specific.

A more serious problem, however, is the patenting of seeds. Traditionally, the farmer had full control over the seeds. He retained a part of the harvest as seeds for the next year. Now, seeds are increasingly becoming commodities that the farmer has to buy every year.

Can seeds be patented? Many countries like the US allow the patenting of seeds. There have been major controversies when American firms tried to patent Indian varieties like Basmati rice and turmeric. The patenting of plants or any life form is a complex issue.

Biotechnology can have a serious impact on developing countries in other ways too. For example, sweeteners developed from corn and other crops grown in the US have replaced sugar imports from the Caribbean and the Philippines. This has been a disaster for those economies. New substances, thousands of times sweeter than sugar, are now being developed using genes taken from tropical plants. They may mean the end to the sugar industry itself.

What are the basic issues in protecting wild flora and fauna?

There is often a conflict between conservation efforts and the livelihood issues of local communities. People often capture or kill endangered animals for two related reasons: the poverty of the people; combined with the willingness of the rich to pay huge sums of money for the products. What choice does the poor person have?

Should we try to protect all the endangered and threatened species?

From an ethical point of view, all life is sacred and must be protected. From a practical point of view, we are forced to decide which species we want to save. Money, scientific information, and trained personnel to do the job—all these are in short supply. Clearly, we will be able to save only a limited number of species.

How do we choose the species we would try to save? We need some criteria on which to base our choice. From a human point of view, we could choose those species that give us benefits now or that are likely to be useful in the future. The problem is that we do not have enough knowledge about the possible uses of thousands of species.

From the point of view of nature, we could select those that are of the most value to the ecosystem, that is, the keystone species (see Chapter 5 of this book). This may turn out to be good for the humans too. Another way is to simply choose species that have the best chance of survival. Given our constraints, it may be better to concentrate on a few strong species instead of many weak ones.

Box 6.8 Make a difference: Observe the International Day of Biological Diversity

The International Day of Biological Diversity is observed every year on May 22, supported by the Secretariat for the Convention on Biological Diversity. It is an opportunity to strengthen people's commitment and action for the conservation of the world's biological diversity.

You could celebrate the day by organizing:

- Meetings and lectures on the theme of biodiversity
- An excursion to a biodiversity-rich area with experts as guides
- Activities for children
- Environmental clean-up activities

You could also plan the day as a people's event with colourful activities such as street rallies, bicycle parades, green concerts, essay and poster competitions in schools, media events, tree planting, and public events in parks, nature conservancies, and botanical gardens.

Each year the Secretariat chooses a theme and provides posters and promotional material. The theme for 2004 was 'Biodiversity: Food, Water, and Health for All'.

Source: The website of the Convention on Biological Diversity (CBD), *www.biodiv.org*

A quick recap

- We can protect species in their habitat or away from their habitat and each method has its place.
- While there has been an increase in the number and area covered by the world's reserves, the management of many of the reserves is not satisfactory.
- Seed banks, zoos, and botanical gardens are important for biodiversity conservation, but they are beset with many difficulties.
- There are a range of international conventions and agreements to promote biodiversity conservation, the most important being CBD.

- Under CBD, India has prepared the National Biodiversity Strategy and Action Plan.
- India has taken measures like the creation of biosphere reserves, protected areas, zoos, and botanical gardens for conserving biodiversity.
- Biotechnology promises immense benefits for agriculture, food production, and health, but brings with it many problems too.

What important terms did I learn?

In-situ conservation
Ex-situ conservation
Biotechnology

What have I learned?

Essay questions

1. Describe the major international efforts to save biodiversity. What has been India's role in these efforts?
2. Describe the measures taken by India to save her biodiversity.

Short-answer questions

1. Distinguish between in-situ and ex-situ conservation. Explain the advantages and disadvantages of each approach.
2. What are the threats faced by the Olive Ridley turtles on the Orissa coast? What are the difficulties faced in the effort to save them from being killed?
3. What are the main provisions of the Convention on Biological Diversity (CBD)?
4. What are the benefits and problems of biotechnology?

How can I learn by doing?

Choose a small village and prepare a detailed biodiversity register listing the species found within the boundaries of the village. Take the help of a biologist to identify the plants. Before starting the project, discuss the issues of biodiversity and biopiracy with the villagers and explain the advantages of preparing a register. Involve the villagers, especially the youth, in the project.

What can I do to make a difference?

1. Do not buy any item produced by killing endangered animal species. This includes products made from ivory.
2. Find out more about citizens' movements aimed at conserving biodiversity and join one of them as a volunteer. For example:
   (a) If you are in south India, join the movement to save the Silent Valley.
   (b) If you are in Orissa, join the groups trying to save the Olive Ridley turtles.
   (c) If you are in Chennai, find out if you can help the Snake Park or the Crocodile Bank. You can also join Turtle Watch on the Chennai beach.
3. Observe May 22 as the International Day of Biological Diversity (refer to Box 6.8).

Where are the related topics?

- Chapter 5 covered the basic issues of biodiversity and threats to the world's biodiversity.
- Chapter 9 discusses forest resources.
- Chapter 10 covers food resources.

What are the deeper issues that go beyond the book?

1. Is it at all possible to conserve all the species? If that is not possible, how do we choose the species we should protect?
2. Diversity is not just about plants and animals. There is also human cultural diversity. The communities of this world differ in their agricultural practices, food habits, religion, spirituality, beliefs, medical systems, languages, literature, art, music, etc. This cultural diversity is also being threatened. Languages are becoming extinct, traditional knowledge of plants and their medicinal uses is being forgotten, and indigenous seeds are disappearing. Should this cultural diversity be conserved and if so, how can we do it?

Where can I find more information?

Access the following websites:

- *www.biodiv.org*, the official site of the CBD .
- *www.teriin.org/biodiv/status.htm*, the site of The Energy and Resources Institute (TERI) gives an account of India's biodiversity and a list of biosphere reserves.

What does the Muse say?

Read, reflect, and puzzle over the following poem by the British poet Gerard Manley Hopkins (1844–89):

Inversnaid

This darksome burn, horseback brown,
His rollrock highroad roaring down,
In coop and in comb the fleece of his foam
Flutes and low to the lake falls home.

A windpuff-bonnet of fawn-froth
Turns and twindles over the broth
Of a pool so pitchblack, fell-frowning
It rounds and rounds Despair to drowning.

Degged with dew, dappled with dew
Are the groins of the braes that the brook treads through,
Wiry heathpacks, flitches of fern,
And the beadbonny ash that sits over the burn.

What would the world be, once bereft
Of wet and of wildness? Let them be left,
O let them be left, wildness and wet;
Long live the weeds and the wilderness yet.

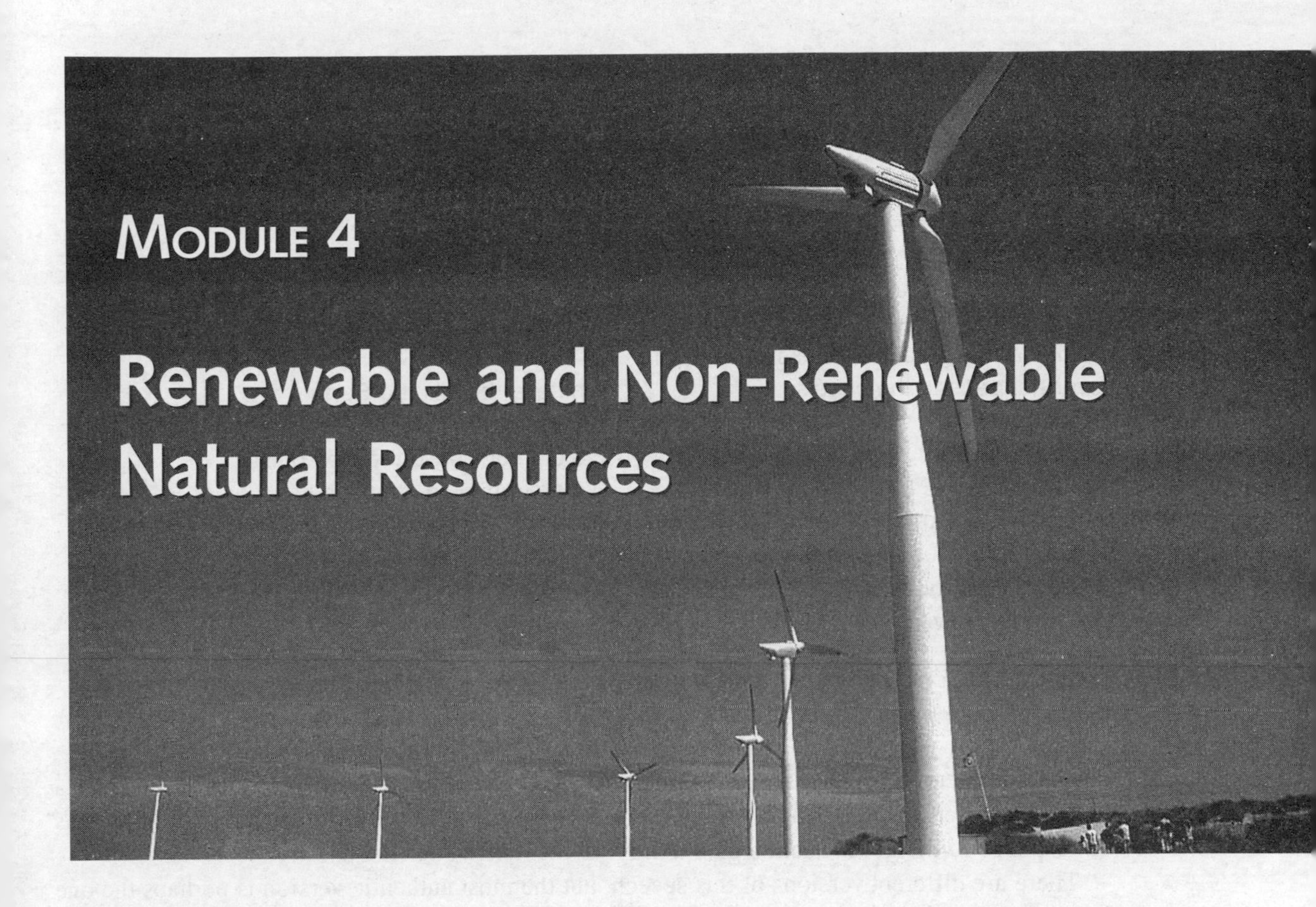

MODULE 4

Renewable and Non-Renewable Natural Resources

Prologue

How can one sell the air?

TED PERRY
(INSPIRED BY CHIEF SEATTLE)

How can you buy or sell the sky, the warmth of the land?
The idea is strange to us.
If we do not own the freshness of the air and the sparkle of the water,
how can you buy them?
Every part of the Earth is sacred to my people.
Every shining pine needle,
every sandy shore,
every mist in the dark woods,
every clear and humming insect
is holy in the memory and experience of my people.
Our dead never forget this beautiful Earth,
for it is the mother of the red man.
We are part of the Earth and it is part of us.
The perfumed flowers are our sisters,
the deer, the horse, the great eagle,
these are our brothers.
The rocky crests,
the juices in the meadows,
the body heat of the pony, and the man,
all belong to the same family.

The air is precious to the red man,
for all things share the same breath—
the beast, the tree, the man,
they all share the same breath.

What is man without the beasts?
If all the beasts were gone,
man would die from a great loneliness of the spirit.
For whatever happens to the beasts,
soon happens to man.
All things are connected.

Whatever befalls the Earth
befalls the sons and daughters of the Earth.
If men spit upon the ground,
they spit upon themselves.

This we know—
the Earth does not belong to us
we belong to the Earth.
This we know.
All things are connected
like the blood which unites one family.
All things are connected.

Whatever befalls the Earth
befalls the sons and daughters of the Earth.
We did not weave the web of life
We are merely a strand in it.
Whatever we do to the web,
We do to ourselves.

Source: Read Gifford and Cook (1992) for the full text of the different versions and an account of how they came about.

These are excerpts from a speech supposed to have been made by Chief Seattle, head of a Native American tribe, in 1854 when a representative of the US government visited him. It was in response to the government's plan to buy up the native's territory.

There are different versions of this speech, but the most authentic version is perhaps the one published by Henry Smith in 1887. The most popular version, however, is the one written by Ted Perry, a professor at the University of Texas. He wrote it in 1970 as a narration for a film on pollution and ecology and never claimed it as being authentic.

The Perry version caught the imagination of environmentalists and activists all over the world and became very popular during the environmental movement of the 1970s. It captures the spirit of modern environmentalism so well that it does not matter that Chief Seattle did not exactly utter those words. The excerpts given above are from the Perry version.

What awaits you in Module 4?

This module describes how humans are using and abusing the natural resources: water, energy sources, forests, land, and minerals. We discuss the degradation caused to these natural resources though human activities and the possible measures for reversing the decline. As in the other modules, there are many stories of crises, solutions, and hopes.

CHAPTER 7

Water Resources

If there is magic in this planet,
it is in water.

Loren Eiseley

What will I learn?

After studying this chapter, the reader should be able to:

- appreciate the unique nature of water as a natural resource;
- describe how the water cycle gives us fresh water every year;
- explain how water is a case of 'scarcity amidst plenty';
- identify the special problems with regard to water supply for cities;
- understand the pattern of water use for different activities;
- appreciate the potential problems in providing irrigation water;
- enumerate and describe the worldwide conflicts over water;
- give a picture of the water problems in India and the attempted solutions;
- list measures for water conservation;
- discuss the pros and cons of the proposal to interlink Indian rivers; and
- implement water-saving measures in her or his own life.

Box 7.1 Crisis: The wettest place on Earth, yet no water to drink

When the monsoon comes to this place, it does not rain, it pours. It could pour continuously for over two months, and you may not see the sun for 20–25 days at a stretch. With an average annual rainfall of 11.5 m, it is listed in every geography book as the wettest place on earth.

Some years it rains much more than the average. In 1974 the place received 24.5 m of rain, and the record for one day was 1563 mm. Yet, Cherrapunji, in the state of Meghalaya in North East India, faces severe drought conditions for the rest of the year!

Pouring rain in Cherrapunji

How is it that not a drop of the 11.5 m of rain remains to quench the thirst of the people? The answer lies in the destruction of forests. Once upon a time, the hills around Cherrapunji were covered with dense forests. These forests soaked up the heavy rainfall and released it slowly the rest of the year. Over the years, however, the forests were cut down. The heavy rains washed away the topsoil, turning the slopes into deserts. It is now a mining town, with whole families and immigrants making a living out of the extraction of coal and chalk. The dusty air chokes the life out of people.

There is no reservoir to store the water. For over 20 years, the residents have depended on a piped water supply that comes from afar. This supply is always erratic and undependable.

The people of Cherrapunji have traditionally celebrated the arrival of the rains every year with an elaborate festival. The rain gods do respond generously for four months. Yet the people do not have enough water for the next eight months!

Source: Joost de Haas, *Drinking the Sky*, a documentary film made for *BBC Earth Report*

Why is water a unique resource?

Water is a prerequisite for the existence of life. Human beings, animals, and plants cannot survive without water; in fact, the human body is mostly water. We can go without food for up to 30 days or more, but we cannot go without water even for a few days. When the water content in the body drops by just 1 per cent, we experience thirst. If it drops by 10 per cent, there is danger of death.

Water is the most critical limiting factor for many aspects of life, such as economic growth, environmental stability, biodiversity conservation, food security, and health care. In most cases there is no substitute for water. An energy source can be replaced by another, but water as a resource is largely irreplaceable.

The good news about water is that we can reuse it. It may change its form, but we can always retrieve it. In fact, the molecules in the water we use have been around for millions of years. The Earth holds the same quantity of water as it did when it was formed.

The paradox of the water situation is that there is scarcity amidst plenty. As we shall see, there is a lot of water on Earth and yet there are millions of people facing acute water scarcity.

How much water is there in this world?

The total water in the world is estimated to be 1400 million cu. km. If all this water were spread over the surface of the earth, it would be 3 km deep. This is a lot of water. Unfortunately, 97 per cent of this water is found in the ocean, and is too salty to drink.

Of the remaining 3 per cent, two-thirds is locked up in relatively inaccessible icecaps and glaciers. That leaves a mere 1 per cent or 14 million cu. km. Again, half of this is groundwater and most of it lies too far underground. About 200,000 cu. km can be found in rivers and lakes and 14,000 cu. km in the atmosphere.

How do we get fresh water every year? A key element of the water cycle (see Chapter 2 in this book) helps us get a regular supply of fresh water. Annually, about 430,000 cu. km of water evaporates from ocean and only about 390,000 cu. km returns as precipitation. The remaining 40,000 cu. km moves from the ocean and falls on the land as rain and snow. The first advantage is that we get this supply every year. Second, as the water evaporates from the ocean and falls on land it loses its salt content and thus becomes potable.

How much water does a person need? We need water for domestic purposes (drinking, bathing, and sanitation), agriculture, industrial processes, and energy production. The water needed for each activity varies with climate, lifestyle, culture, tradition, diet, technology, and wealth. Also, the more accessible the water, the more it is used.

The absolute minimum requirement for domestic use is 50 litres per person per day, though 100–200 litres is often recommended. Adding the needs of agriculture, industry, and the energy sector, the recommended minimum annual per capita requirement, according to the World Health Organization (WHO), is about 1700 cu. m.

Is the annual supply of fresh water sufficient for our needs? The annual supply of fresh water is 40,000 cu. km and this amounts to about 6600 cu. m per person per year (for a population of 6 billion). This appears to be plentiful, since the minimum annual needs of a person are only about 1700 cu. m. By this token, we should be comfortable with regard to the world water situation. Yet, there is worldwide scarcity of water.

How do we measure water scarcity?

If a country has about 1700 cu.m of fresh water per person per year, it will experience only occasional or local water problems. If the availability falls below this threshold value, the country will begin to experience periodic or regular *water stress*. If the availability of water further falls below 1000 cu. m per person per year, the country will suffer from chronic *water scarcity.* The lack of water will then begin to adversely affect human health and well-being as well as economic development. If the annual per capita supply of water falls below 500 cu. m, the country will reach the stage of *absolute scarcity.*

In India, the per capita water availability was 5177 cu. m in 1951, but the figure dropped to 2464 cu.m in 1990 and further down to 1820 cu.m in 2001. India will be the next major country to move into the water-stressed category.

In 1990, 28 countries, with populations totalling 335 million, experienced water stress or scarcity. By 2025, about 50 countries (including India) will fall into one of the categories mentioned above, with water scarcity affecting more than 3 billion people worldwide. China, the most populous nation in the world today, will only narrowly miss the water stress benchmark in 2025. In the North China Plain, however, water shortages are already acute, with demand outstripping supply. By the middle of the twenty-first century, only a few countries will escape a major water crisis.

Even now, in the developing world, about 30 per cent of the population or 1.2 billion people are without a safe and reliable supply of water. It is also estimated that 80 per cent of all illnesses in developing countries can be traced to contaminated water.

Why is water becoming scarce?

There are a number of reasons behind this increasing water scarcity. First, distribution of available water is uneven over space and time. Some areas of the world get too much water while other places get too little. Moreover, transporting water over long distances is impractical.

Second, in some regions rainfall occurs over a short period in the year, leading to floods during the monsoon and drought in summer. The run-off, because of deforestation, flows quickly into the ocean before it can be used. Much of the rainfall is in remote and inaccessible places. Some of the rain is also needed to maintain special ecosystems, such as wetlands, lakes, and deltas. Additionally, global warming is now changing rainfall patterns.

The rapidly increasing pollution of rivers, lakes, and groundwater is also reducing the usable supply of water. In many places, available water is polluted and has become unfit for use. Incredibly small amounts of substances like oil can pollute huge amounts of water.

In many places the rates of extraction of groundwater for irrigation are so high that aquifers are getting depleted. As a result, water tables are falling, notably in India, China, and the US (which together produce half the world's food). The major rivers of the world are drained dry before they reach the sea. The Yellow River in China, the Ganga in India, the Amu Darya in Central Asia, the Nile in Egypt, and the Colorado in the US are some examples.

There is gross inequity in allocation of water resources. In general, the richer groups corner a large portion of the available water. Again, urban areas draw a great deal of water from the surrounding areas, depriving the poorer rural people.

Why do our cities face severe scarcity every summer?

The megacities of the world, many of them in the poorer countries, need a lot of water. This water is often drawn from the neighbouring villages and far off rivers and lakes. For example, during summer, the water for Chennai comes from many surrounding agricultural villages. The farmers find it more profitable to sell the water than to cultivate the fields. This clearly constitutes an unwelcome shift in water use that will ultimately threaten our food security.

There are also plans to bring water to Chennai from distant sources like the Krishna River and the Veeranam Lake. Bangalore gets its water from 90 km away. Such practices, apart from increasing the cost of supply of water, also deprive the rural areas of much-needed water.

Even if a city gets good rain, the water is not retained in the area. Buildings, paving, and roads cover most of the land and the rainwater does not percolate into the ground. Chennai, for example, has an average rainfall of 1290 mm per year, which is more than the national average. Yet, 90 per cent of this rain is lost as run-off, through evaporation, or flows into the sewage system.

Megacities also pollute the water. They release large amounts of wastewater into the rivers and the ocean. Worldwide, less than 20 per cent of urban wastewater is at present treated. Another issue is that the urban poor often pay a greater part of their income for water than the rich do.

Leaking taps in cities (and other places) waste enormous amounts of water. In many cities, more than half the available supply is lost through leaks and rotting pipelines. There is an urgent need to devise water-efficient toilets, taps, washing machines, dishwashers, etc.

How is the accessible water used in the world?

Human beings use about 54 per cent of all accessible freshwater supplies. By 2025, this share will increase to 70 per cent. This will have serious implications for all other life forms, including plants. Table 7.1 shows the current pattern of water use in the world.

Table 7.1 Pattern of water use, showing the percentage of total water consumed by each sector

Sector	Global use percentage	Percentage use in industrialized countries	Percentage use in developing countries	Percentage use in India
Agriculture	70	30	82	80
Industry	22	59	10	15
Domestic	8	11	8	5

It is significant that domestic consumption accounts for only 8 per cent of available water and yet there is worldwide scarcity of drinking water. Agriculture continues to consume the major portion of available freshwater. However, there is a clear difference in water use between developing countries and industrialized countries. While agriculture consumes the bulk of available water in the former, industry consumes more than agriculture in the latter. This shows that there will be a severe problem in developing countries once they industrialize to the level of rich, industrialized countries.

How much water do Indian industries consume? We have no reliable figures on water consumption by Indian industry. The estimates range between 40 billion cu. m and 67 billion cu. m per annum. We are sure, however, that this requirement is increasing rapidly. It is estimated that the demand for water by Indian industry will triple over the next 20 years. The same will be true of China and many other developing countries.

Where will all the additional water for industry come from? Since, the total available water supply is limited, an increase of supply to one user means taking away from another user. In the Indian case, irrigation and drinking water for the local community will both be affected (read Box 7.2).

In India, the thermal power plants are the water guzzlers, accounting for about 87 per cent of total industrial consumption of water. Across the board, the efficiency of water use in India is way behind international standards. There is tremendous scope for reduction of consumption by the Indian industry.

There is no incentive for Indian industry to reduce consumption, since it gets water at ridiculously low prices. In order to attract industrial units to an area, state governments often offer assured water supply at low rates. The pricing in most cases ignores the opportunity cost of water (that is, the benefit from possible alternate use) as well as the damage caused, such as pollution of water supplies by the discharge of wastewater.

The amounts paid by the industry to the pollution control boards under the Water Cess (Prevention and Control of Pollution) Act of 1977 are also meagre. In contrast, industries in Europe pay much more for water. In addition, some European countries have introduced the principle of 'Polluter Pays', where industries are charged for the amount of pollution contained in their wastewater discharge.

This leads us to another important aspect of the use of water by industry—the pollution of water bodies by wastewater discharge. This issue is discussed in Chapter 12 of this book.

Box 7.2 Conflict: Who owns the water?

If a soft drink industry owns a piece of land, can it exploit the groundwater without any limit? This is the core issue in what has come to be known as the Plachimada case in Kerala.

In January 2000, the Perumatty Panchayat gave permission to Hindustan Coca-Cola Beverages to set up a bottling plant in Plachimada. The company began drawing groundwater and producing cola. The panchayat received an annual income of about Rs 600,000 in the deal. In addition, the plant provided employment for about 400 people.

By early 2002, the villagers started noticing changes in the water quality in the area. They felt that the plant was drawing too much water from

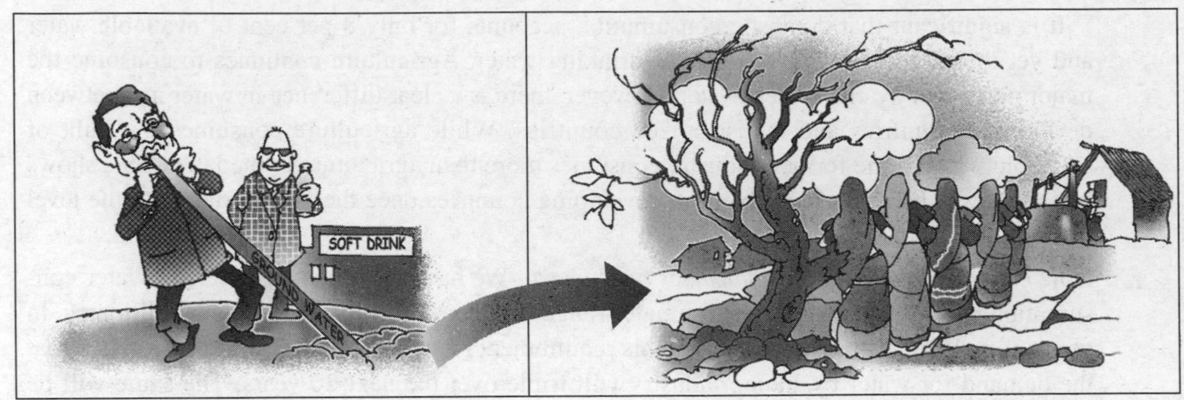

Overuse of groundwater by the soft drink industry

The fields and wells become dry

the ground and that this was affecting the harvest of rice and coconuts. In April 2002, the villagers started an agitation against the plant.

The protest in front of the plant's entrance had been going on for nearly a year, when the panchayat issued a notice to the company in April 2003 cancelling its licence. The company took the matter to the Kerala High Court. In July 2003, a BBC Radio inquiry found that the sludge distributed by the company to the villagers as fertilizer actually contained toxic material.

While the case was going on, the company approached the government, which questioned the panchayat's action and wanted an expert committee to study the issue. The panchayat moved the court against the government order. The company's stand was that it was not drawing excessive amounts of water. Meanwhile the villagers continued their protest, even as the employees started an agitation in favour of the plant's continuance.

In December 2003, the court ordered the plant not to draw water from the ground and to find alternate sources of water. The judge said that the company had no right to extract excessive natural wealth and that the panchayat and the government were bound to prevent it. The groundwater was a national resource, which belonged to society in general and not just to the company, even though it owned the land. The judge quoted the decisions of the Supreme Court of India in this regard. He also directed the panchayat to renew the plant's licence and restrained it from interfering with the functioning of the factory.

We do not know how the story will ultimately end, but the case has become the focal point for the debate on the people's right to water and a company's right to exploit the water under its land. Plachimada was the venue of a World Water Conference in January 2004 after the conclusion of the World Social Forum in Mumbai.

Source: Krishna Kumar (2004) and Vijayalakshmi (2003)

Table 7.2 Personal water use by an Indian urban resident

Use	Litres / Person / Day
Drinking	3
Cooking	4
Bathing	20
Flushing	40
Washing clothes	25
Washing utensils	20
Gardening	25
Total	**137**

How much does a domestic consumer use? Table 7.2 shows the pattern of domestic consumption by urban residents in India. The amount of water consumed can even reach 300 litres per person per day. Rural inhabitants manage with much less water.

The increasing use of flush toilets in urban areas results in enormous wastage of potable water. Leaking taps and pipelines also contribute to the huge loss of water.

Will there be enough water for irrigation?

There has been a five-fold increase in the use of water for irrigation over the course of the twentieth century. The situation in India has worsened since the adoption of the Green Revolution package. In 1992, only 16 per cent of cropland was irrigated, but it produced 36 per cent of the food harvest. In many countries 50 per cent or more of the food production comes from irrigated land. These include countries such as India, China, Egypt, Indonesia, Israel, Japan, Korea, Pakistan, and Peru. By 2025, up to four billion people will live in countries that will have insufficient water to produce their own food.

Even now, inefficient and outdated irrigation practices continue, leading to much wastage of water. When fields are flooded with water, often only half of it gets to the root zone helping the growth of the plant. The other half is wasted as it percolates, evaporates, or runs off. Drip irrigation, the lining of canals, and precision sprinkling must be introduced immediately.

Most new land brought under agriculture uses groundwater resources that are fast depleting. Meanwhile, available agricultural land is also lost due to salinization, reservoir siltation, shift of water use, etc. Salinization is often the result of a rising water table, which can result form the practice of flooding the fields. As the water evaporates it leaves behind a deposit of salt on the topsoil. The siltation of reservoirs in big dams has often cut their lives to a mere 20 or 25 years. Finally, a shift in water use occurs when cultivation is stopped in some places in favour of selling water to a nearby city. As a result of these cropland losses, the total irrigated area may in fact be decreasing.

The agricultural fields in countries like India are often over-irrigated. This is primarily because the farmers get free electricity and subsidized pumps and thus do not pay the real economic price of water. Moreover, the Green Revolution has doubled the amount of water used for irrigation.

Globalization and industrialization have led to an increased demand for water, from industry and urban areas, in different parts of the world. The rapid increase in population also means that more water is needed for drinking and for agriculture.

In many places, irrigation depends on water from dams. The world over, large dams and canal systems are going out of fashion due to spiralling costs, environmental concerns, and displacement of people. Protest movement in many places have stalled the construction of dams.

All these facts seem to indicate that there is not much scope for increasing irrigation at the same rapid rate of the last century. It is also a sobering thought that all irrigation-based societies have failed in the past (Postel 1999).

Why are there so many conflicts over water?

The next world war may well be fought over scarce resources such as oil and water. In fact, there is a long history of conflicts over water. Such conflicts may arise from the desire to possess or control another nation's water resources, thus making the possession of water systems and resources a political or military goal.

Inequitable distribution and use of water resources, sometimes arising from development projects, may lead to disputes. Conflicts may also arise when water systems are used as instruments of war, either as targets or as tools. Worldwide, there are conflicts brewing over many rivers and river basins, with more than 200 water bodies, around the world, being shared by two or more countries.

Strife over water is erupting throughout the Middle East, from the watersheds of the Nile to the Euphrates and Tigris Rivers (read Box 7.3). Likewise, there is a problem in South Asia's Ganga-Brahmaputra Basin, where Bangladesh, India, and Nepal are disputing the use of water. India and Nepal want to exploit the basin's huge hydroelectric power-generating potential whereas Bangladesh wants the water managed in a way that will minimize flooding during the monsoon season and water shortages during the dry season.

Of equal concern are the water conflicts between states in India that share river basins, such as Karnataka and Tamil Nadu, which share the Cauvery River. There are water shortages during the dry season in every major city in South Asia. During the dry season, urban water pipes are often empty, creating situations where water must be delivered by truck to desperate people.

The third category of conflicts is between industries and local communities on several issues: excessive water consumption by industry resulting in reduced availability for irrigation; preferential treatment for industry in pricing; and pollution of groundwater, rivers, and local water bodies by wastewater discharges.

Agitations and public interest litigation on water issues are increasing and more conflicts at all levels are expected in the future (also see Box 7.2).

Box 7.3 Conflict: Water wars?

It is the cradle of civilization, the riverbank where the Egyptian pharaohs and the pyramids flourished. Today it is the scene of conflict and a potential theatre for a war over water.

There are nine countries in the Nile basin: Egypt, Sudan, Ethiopia, Burundi, Congo, Kenya, Rwanda, Tanzania, and Eritrea. Egypt lies at the mouth of the river, yet it uses the maximum water. Egypt gets very little rain and has no groundwater and is thus totally dependent on the Nile. This fact is the source of the conflict and perhaps the trigger for a future war.

The Nile has two major tributaries, the Blue Nile and the White Nile. While the Blue Nile originates in Ethiopia, the White Nile has its source in Lake Victoria in Tanzania. The two tributaries join in Sudan, before flowing into Egypt.

In 1970, Egypt completed building the Aswan High Dam across the Nile and created a huge reservoir called Lake Nasser. The dam prevented flooding of the lower basin, increased the area of cropland through irrigation, and improved transportation. It saved crops during droughts and averted mass famines. The dam also generates a third of Egypt's electricity.

The dam has also created problems. It is now clogged by millions of tons of silt that the Nile brings from the Ethiopian highlands. In due course, the silt will make the dam useless. Previously this silt used to enrich the soil of the Nile delta. Now Egypt has to compensate for it with heavy fertilizer inputs in agriculture. Lack of natural flooding has increased salinization and reduced land productivity.

In 1959, Egypt signed an agreement with Sudan giving itself the right to 'full utilization of Nile waters'. Of the 84 billion cu. m passing through Sudan, Egypt is supposed to take 55.5 billion cu. m. Often it uses up a part of Sudan's share too.

The agreement excluded all the other states of the basin including Ethiopia, the source of the Blue Nile. Ethiopia, which has faced many droughts, now wants to use the Nile waters to grow more food. So does Sudan. Egypt, of course, does not want any cut in the inflow due to economic projects upstream. Its agriculture and energy production are at stake. Of the water flowing into Egypt, 85 per cent comes from the Blue Nile. If Ethiopia tries to build a dam on the river, a war with Egypt could erupt.

In 1999, a Nile Basin Initiative was launched to resolve all the issues over sharing the waters. A resolution to the conflict acceptable to all the countries has proved elusive.

Source: James (2003) and Miller (2004)

What is the water situation in India?

India and China are counted among the water hotspots, primarily because of their large populations that have to be provided with food and drinking water. In India more than 60,000 villages did not have a source of drinking water in 1994 and by now this number must have increased further. Diarrhoea, caused by consuming contaminated water, claims the lives of one million children every year. In addition 45 million people are annually affected by poor quality water.

While the increasing population is one cause of India's water crisis, the problem has been compounded by the steady deterioration, disuse, and disappearance of traditional tanks and ponds. A large number of such local water sources had been created by ancient kings all over India. Local communities exercised control over these resources and there were set procedures for the maintenance of such tanks through community participation.

With the arrival of the British, however, the Public Works Department took over the control of water resources and this started a steady decline in their condition. This has continued even after India's independence leading to the disuse of thousands of such sources of water. These tanks and ponds were very effective in retaining rainwater and recharging the groundwater. Deforestation of the hills coupled with the absence of tanks has increased the run-off into the sea.

In recent years the extraction of groundwater using bore wells operated by electric or diesel pumps has reached gigantic proportions. One estimate is that the extraction rate is already twice the recharge rate. In many states water tables have been dropping at alarming rates.

What do the optimists say?

Even in this grim situation, there are optimists who believe that there will be dramatic solutions that will solve the water problem once-and-for-all. Such dreams include:

- Transport of water in huge bags by sea: American and Norwegian companies are developing huge bags for this purpose.
- Towing of icebergs to places needing water: There are formidable technological problems involved in putting this into practice.
- Discovery of new, deep aquifers full of water: This is similar to the hope of finding huge oil wells and is an unlikely event.
- A new, cheap energy source that will make desalination affordable: This is likely to remain a dream.

Most experts, however, believe that there are real limits to the availability of water and that we are fast reaching them. Perhaps a realistic solution would be to find ways of using water more efficiently.

Is there a way out of the water crisis?

The following is a list of possible solutions to the water crisis in India and elsewhere:

- Reduce demand:
  - Educate people to use less water.
  - Install water-saving devices like self-closing taps, dual-flush toilets, spray taps in sinks, etc.
  - Use decentralized wastewater recycling systems in homes, apartment blocks, campuses, and industries, using natural methods like planted filters.
  - Adopt composting toilets to save water and also to minimise the sewage disposal problem.
  - Adopt agricultural practices that require less water:
    - Replace water-hungry crops by those that require less water.
    - Promote crops that can tolerate salty water.
    - Return to indigenous species that can withstand drought.
    - Switch to organic and natural farming.
    - Get more crop per drop: Use drip irrigation, precision sprinkler method, etc.
  - Persuade people to change to a vegetarian diet that would require less water to produce: According to a recent study, 7000 litres of water are used to produce just 100 grams of beef.
  - Reduce industrial consumption through recycling, reuse, and new water-efficient technologies
- Catch the rain where it falls:
  - Retain water on land as long as possible through check dams and contour bunds. This allows it to percolate into the ground (zero run-off in other words).
  - Implement rainwater harvesting in urban and rural areas.
  - Restore traditional practices (such as the system of ponds and lakes) that were effective.
  - Collect water from dew and fog using large nets: This technique has been developed in Israel and is being tried out in many places.
- Adopt decentralized systems of water supply and sanitation:
  - Plan many small local catchments in place of large ones.
  - Implement small-scale schemes.
- Adopt fairer policies:
  - Give communities control over local water sources.
  - Price water properly.
  - Remove inequities in access to water.
  - Treat water as a common resource and work towards a water ethic.

Will the plan to interlink Indian rivers solve our problems?

The idea of linking rivers has been around for a long time. In 1972, Dr K.L. Rao, the then Minister for Irrigation, proposed a plan to link the Ganga and the Cauvery. In 1974, Capt. Dastur came up with a plan for a garland canal. The government set up the National Water Development Agency (NWDA) in 1982 to carry out studies on the question.

There was not much progress till 2002, when the Supreme Court of India, acting on a public interest petition, directed the government to interlink rivers within 10 years. In response, the government set up a Task Force to build national consensus, work out the detailed plans, and complete the entire work by 2016.

Three major advantages are being cited in favour of the scheme: droughts will never occur, floods in the Ganga and Brahmaputra will cease to be a problem, and an additional 30,000 MW of hydropower will be generated. However, the NWDA has not conducted any detailed assessment of the financial feasibility, technological capability, ecological sustainability, and political feasibility of the project.

The budget for the project ranges from Rs 5600 to 10,000 billion. Even the lower estimate equals 25 per cent of our GDP, 2.5 times our tax collection, and double our present foreign exchange reserves. Obviously, funds of this magnitude will have to be raised from international sources. Even if we succeed in raising the funds, the annual interest alone would amount to Rs 200–300 billion. The water will have to be priced high to meet such a burden. Will the farmers and other users be ready to pay the charges?

It is doubtful if a political consensus can ever be achieved on the project. Would the states be willing to freely share all the river waters? Even the current inter-state water problems, such as the construction of the Sutlej-Jamuna Link Canal, the sharing of Cauvery waters between Karnataka and Tamil Nadu, and the height of the Sardar Sarovar Dam have all defied an amicable solution. It is not even clear whether rivers such as the Ganga, the Brahmaputra, the Mahanadi, and the Godavari are water-surplus. The sources of such rivers are drying up and the rivers themselves are choked with silt.

The likely ecological consequences of building over 200 reservoirs and a network of criss-crossing canals cannot even be assessed now. Judging from the experience of the Narmada Dam and other large projects, the consequences could be disastrous. Environmentalists and social activists fear large-scale submergence of habitats, forests, and fertile land; extensive destruction of wildlife habitats and biodiversity; displacement of large populations; and so on. In some places enormous amounts of energy will have to be spent pumping water uphill.

In the past, every single irrigation project in India has resulted in heavy cost and time overruns. The project to interlink rivers is likely to meet a similar fate.

What is the good news from India on water conservation and management?

In recent times a number of initiatives have been taken by individuals, communities, non-governmental organizations (NGOs), and even by government agencies to implement measures for water conservation and management. Some of the well known cases are:

- Auroville: The spectacular results of the efforts of this international community near Pondicherry over the past 35 years in afforestation, water conservation, wastewater treatment, appropriate technologies, etc., have made it a model to follow. Auroville is now working with local communities and the government to rejuvenate the traditional ponds and let the people manage them.
- Tarun Bharat Sangh, Alwar, Rajasthan: The building of check dams in villages (read Box 7.4).
- Ralegaon Siddhi, Maharashtra: Through the efforts of Anna Hazare, the people have constructed bunds and percolation tanks and brought back greenery to the area.

Rainwater harvesting can solve household water problems

- Rajiv Gandhi National Drinking Water Mission, Madhya Pradesh: This is one of the few successful government schemes in water conservation.
- Gram Gaurav Pratishtan: In an area in Maharashtra, the late Vilas Salunke introduced the idea of a Pani Panchayat for the fair-sharing of water.
- Tamil Nadu: The city of Chennai is blessed with a freshwater coastal aquifer, which was being over-exploited. In the early 1990s, the state government introduced controls and as a result the water table did go up.

  Rainwater harvesting was first made compulsory in Chennai and was extended to the entire state of Tamil Nadu in 2003. It is estimated that rainwater is harvested in at least 30 per cent of the structures in the state. Rainwater harvesting is actively promoted by voluntary organizations like Akash Ganga, which has set up a 'Rain Centre' in Chennai to demonstrate the methods of harvesting rainwater.
- Efficient water use by industry: Faced with water scarcity, increasing costs, and protests by local communities, several industries, such as Arvind Mills, Chennai Petroleum Corporation, J K Papers, and Natural Sugar and Allied Industry have implemented water conservation measures and have benefited by them.

Box 7.4 Inspiration: The rivers that came back to life

On Mahatma Gandhi's birthday in 1985, five young men led by Rajendra Singh got off the bus at Kishori Village in Alwar, Rajasthan. They belonged to the Tarun Bharat Sangh (TBS), founded by Rajendra Singh. They had come to the village 'wanting to fight injustice against all people', but had no clue where to start. Advice came when they met Mangu Lal, an old man in Gopalpura: 'Build *johads* and you will get results', he told them. A *johad* is a small earthen check dam that captures and conserves rainwater. A typical *johad* would be about 1400 feet long, 20 feet high, and 50 feet wide.

Taking this advice seriously, TBS built the first *johad* in Gopalpura. The next year, the wells in the village had water even in summer. Following this success, another *johad* was built in Bhaonta in the Aravali Hills. In this case, the result was even more dramatic, since the Arvari River, which had gone dry, now came back to life. Mangu Lal had been proved right and the mission of TBS was clear.

In village after village in this drought-stricken part of Rajasthan, TBS built *johads* with the active participation of the villagers. To date, 4500 *johads* have been built in 800 villages, bringing back to life more rivers: Ruparel, Bhagwani, Sarsa, and Jahajwali. As the rivers returned, so did the men who had gone away from the villages looking for work in the cities. Agricultural yields and milk production increased and diesel consumption by pumps went down drastically. The green cover in the adjoining Sariska Sanctuary increased by 40 per cent.

What is the secret of Rajendra Singh's success? It is the complete involvement of the villagers in the task. A TBS team would visit a village only if invited. Meetings would be held and work would start only when the villagers were convinced of the feasibility of the project and were ready to contribute at least a third of the cost. Their contribution was in the form of their own labour and their knowledge of traditional methods.

A detailed study of the work (by Prof. G.D. Agarwal, formerly of IIT, Kanpur) talks about the low cost of the structures and the perfection with which they were built, even though no engineering calculations had been made. Further, the structures withstood the heavy rainfall in 1995 and 1996. Agarwal calls it 'the largest-ever mobilization of people in the cause of environmental regeneration'.

The remarkable work of TBS has not been without impediments. Back in 1985, the villagers first thought that the young men were terrorists and child-lifters. Later, when the villagers were with TBS, the fight was with mine owners and the government. TBS helped the people take possession of *nullahs* and rainwater drains from the owner of the marble mines inside the Sariska Sanctuary. Next, the government filed a case against TBS under an obscure Act, dating from colonial times, which actually makes it illegal for villagers to construct check dams!

In June 2001, TBS received a notice from the Irrigation Department that the earthen groundwater-recharge structure, which it had helped to build in the village Lava ka Baas was both technically unsafe and illegal. TBS was given 15 days to remove the structure. The village, with just one hand pump, was desperate for water and the people had contributed three lakhs of rupees out of their meagre resources to build the *johad*.

The Centre for Science and Environment, New Delhi and other groups took up the cause of TBS and the villagers. Eventually, the Rajasthan Chief Minister intervened and the matter was dropped. However, the final sad twist in the story occurred in 2003, when heavy rains washed away the *johad* in Lava Ka Baas.

International recognition and honour came to Rajendra Singh in 2001: the Ramon Magsaysay Award for Community Leadership. Singh responded that it was the villagers who really deserved the honour. They taught him the value of water. He only helped them revive their 'dying wisdom'.

What lies ahead?

It is clear that water scarcity will soon hit most countries, unless immediate remedial measures are taken. Yet there is still no evidence that governments across the world are alive to the seriousness of the problem.

Box 7.5 Make a difference: Observe World Water Day

The international observance of World Water Day grew out of the 1992 United Nations Conference on Environment and Development (UNCED) in Rio de Janeiro. In 1993, the UN General Assembly designated March 22 of each year as World Water Day. Each year, one of the UN agencies is in charge of celebrating and promoting a new theme. While the theme for 2004 was, 'Water and Disasters', that for 2005 is 'Water for Life'.

The objective of World Water Day is to focus attention on the need to:

- address the problems relating to drinking water supplies; and
- increase public awareness on the importance of conservation, preservation, and protection of water resources and drinking water supplies.

World Water Day on March 22 of each year is a unique occasion to remind everybody of the extreme importance of water for maintaining the environment and increasing development in human societies.

To observe World Water Day, you could:

- promote mass media education programmes;
- focus on school children and the youth;
- publish and disseminate documentaries;

- organize conferences, round tables, seminars, and expositions related to the conservation and development of water resources; and
- promote community and self-help programmes
- increase public and private sector support for water conservation.

For more information, access the websites *www.worldwaterday.org* and *www.unesco.org/water/*

Box 7.6 Make a difference: Join the Youth Water Action Team

The Youth Water Action Team (YWAT) was born at the Youth World Water Forum that was held in The Netherlands in 2001. YWAT is a non-governmental, global organization of students and young professionals who aim to increase awareness, participation, and commitment among young people on water-related issues.

YWAT wants to establish a global movement of young people who are interested in water, who either participate in and/or initiate local community initiatives and/or form local YWAT units. Another objective is to influence decision-making processes in the field of water-related issues.

YWAT wants to make young people aware of water problems and water solutions. It also wants to build a worldwide network of students and young water professionals who are involved in local or regional awareness-raising youth water initiatives.

To join YWAT you have to be between 18 and 30 years old and become active in awareness-raising activities. YWAT membership is voluntary.

YWAT activities include:

- Organizing Regional Workshops
- Arranging city walks (These might take place in different cities at the same time)
- 'Infiltrating' water-related organizations, to influence decision-making processes and to get funding/sponsors for activities
- Publishing a local newsletter about awareness-raising activities of young people
- Organizing cultural/lifestyle activities related to water
- Collecting proverbs, stories, cultural views on water

YWAT invites new members. For more information, access the website *www.ywat.org.*

A quick recap

- Water is essential for life and is in most cases irreplaceable.
- The water cycle helps us get a regular annual supply of fresh water.
- There is only a finite amount of water in the world.
- The paradox of the water situation is that there is 'scarcity amidst plenty', primarily due to the impact of human activities.
- By the middle of the century, most countries in the world (and cities in particular) will be facing water scarcity. India will face severe problems.
- Even places with heavy rainfall can experience water scarcity in other seasons if water management is poor.
- The world's food security may be threatened due to a shortage of irrigation water.
- There could be future wars over water issues.
- There are effective ways of conserving and reusing water, which can be followed by anyone.
- There are many examples of successful water conservation efforts in India.
- The attempts to interlink Indian rivers will have financial, social, and ecological consequences.

What have I learned?

Essay questions

1. What are the reasons for the worsening water situation in so many countries? What is likely to happen in another 40 years or so?
2. What negative role do cities play in the demand and supply of water? How can this be corrected?
3. Why do conflicts occur over water? Describe the various aspects of water conflict through the example of the Nile River.

Short answer questions

1. Why does Cherrapunji, which receives an average of 11.5 metres of rainfall annually, suffer from water scarcity?
2. Why is water a unique natural resource?
3. What percentage of the total water in the world is actually accessible to us and how much is it in cu. km?
4. What is the natural process that provides us with fresh water every year?
5. The annual per capita requirement of water is 1700 cu.m, while the annual fresh water supply is at least three times more. Why then do we have water scarcity?
6. What are the levels of water availability that indicate water stress and water scarcity?
7. Which is the next major country that is expected to join the water-stressed category and how is per capita availability of water changing in that country?
8. What are the special problems that big cities face on the water issue?
9. Describe the global pattern of water use across sectors.
10. What are the main differences between the industrialized countries and the developing countries with regard to water use?
11. What are the implications of the statement, 'Any increase of water supply to one user means taking it away from another user'?
12. What is the link between water scarcity and food security?
13. Give two examples of major conflicts over water.
14. What are the main reasons for water scarcity in India?
15. How can we reduce the total demand for water?
16. What is the importance of rainwater harvesting?
17. What are the pros and cons of the plan to interlink Indian rivers?
18. Give two examples of successful Indian attempts in water conservation.
19. What are the lessons we can learn from the efforts of the Tarun Bharat Sangh?

How can I learn by doing?

1. Choose a village, town, or city as a unit for study. Examine the water situation in the place by finding answers to the following questions and write a report:
   - Where does the water the people use come from? What is its ecological footprint?
   - How much does it cost the users?
   - How much does it cost to supply the water?

- What is the pattern of water use by the people?
- Where does the wastewater go?

2. Examine the water usage of a small community (a village, a locality in a city, or an apartment block) and write a report on the action that could be taken to conserve water:
   - How can the demand for water be reduced?
   - How can the ecological footprint of the water they get be reduced?
   - What are the other environment-friendly ways of getting water that could be implemented?
   - How can the quality of the water be protected?
   - How can the people reuse and recycle the water?
   - How can you educate the users about the proper use of water?

What can I do to make a difference?

1. Start with yourself: Measure the daily water that you use and try to reduce the amount
2. Examine the usage pattern of water in your family or in any one family. Educate them on using water with more care and efficiency
3. Check all the public taps and pipelines in your area on a regular basis, and arrange for any leaks to be plugged.
4. Take the initiative to implement rainwater harvesting in your house, apartment block, or on your college campus
5. Observe World Water Day on March 22 (Box 7.5)
6. Join the World Youth Water Team (Box 7.6)

Where are the related topics?

- Chapter 2 has a description of the water cycle.
- Chapter 12 deals with the pollution of water, soil, and the ocean.
- Chapter 18 includes an account of the impact of dams on people.

What are the deeper issues that go beyond the book?

1. There is very little control in India over the use of water by industry. Should the government impose stricter regulations on the use of water and the discharge of wastewater? Should the government also ensure that industry pays a fair price for the water? Will such control increase costs for industry and push up prices?
2. If more water is taken by industry, less water will be available for agriculture. If industry does not get enough water, economic growth will suffer; if agriculture is deprived of water, food security will be hit. What is the way out of this dilemma?
3. Should water be considered a common resource or a private one? Do all citizens have a right to adequate access to water? Should water be privatized and fair prices charged for it?
4. Is it ethical to exploit the groundwater on your land without any restriction? Remember that the groundwater may have seeped in from the surrounding land. A neighbour may take the trouble to recharge the ground water and the geology of the area may be such that all that water collects under your land.

Where can I find more information?

- Postel (1992, 1999) and De Villiers (1999) give a detailed account of the water crisis.
- A number of water-related websites are given in Appendix A2.

What does the Muse say?

A poem by Lovemere Dhoba of the Youth Water Action Team:

Water Laments

Mother of all lives,
A root that gives life
Life serving creature
You used to dedicate all these names to me.

But look at what your sons,
Your daughters and grandchildren are doing to me
Tossing, my innocent soul,
As if I am of their age,
Sewage dumped in me
As if I am of no significance to their lives

Their industrial effluent,
Domestic garbage,
All left for me to swallow

Sanctions, very soon sanctions
I am going to lift up sanctions
Against you cowards

CHAPTER 8

Energy Resources

Human beings and the natural world are on a collision course.
We must move away from fossil fuels to cut greenhouse gas emissions.

World Scientists' Warning to Humanity (1992)

What will I learn?

After studying this chapter, the reader should be able to:

- recall the sources of commercial energy and the proportion of each source in the total energy produced;
- describe the global consumption pattern in energy;
- distinguish between renewable and non-renewable fuels;
- list the problems caused by the excessive use of fossil fuels;
- understand why there is concern about the future of oil as the main energy source;
- explain the status of sources like coal and natural gas;
- appreciate the problems with nuclear power and hydropower;
- describe the status, advantages, and problems with solar and wind power;
- describe the progress toward the hydrogen economy; and
- appreciate the importance of energy efficiency.

Box 8.1 Crisis: Nomads with homes

Basumati Tirkey, a 35-year-old resident of Bangamunda village in Orissa, walks 9 km every day, carrying a 35 kg load of fuelwood, to earn Rs 15. And she has spent a lifetime doing so.

Basumati is just one of the 11 million desperate women in India who eke out a living collecting and selling fuelwood. Seventy per cent of women living in and around forest areas are engaged in this work. It is estimated that, in an average village in a semi-arid region, a woman walks more than 1000 km a year to collect firewood alone. No wonder they are called 'nomads with homes'.

The UN Food and Agricultural Organization (FAO) estimates that the fuelwood business in India has a turnover of US$ 60 billion. It must be the most valuable non-timber forest activity and yet it is the last resort of the poorest of the poor. It is generally regarded as being a threat to the forests and an illegal activity. In fact, a substantial portion of the huge turnover ends up as bribes to forest guards or as profits to middlemen.

Fuelwood is still the most preferred (and often the only available) energy source in the rural areas and for the urban poor. It is estimated that 47 per cent of the total energy consumed by households in India is from firewood, 17 per cent from animal dung, and 12 per cent from crop residues. This leaves just 24 per cent for commercial energy.

While many millions like Basumati make their living by collecting and selling fuelwood, millions more collect wood just for their daily energy needs. Because of low incomes these poor people are unable to shift to commercial fuels like kerosene. All this collection of wood puts tremendous pressure on the environment and leads to deforestation and loss of green cover.

With the increasing population, the demand for fuelwood keeps increasing, even as availability falls. How will we provide sufficient energy to the poor without degrading the environment?

Source: Mahapatra (2003) and Das and Dayal (1998)

Is there a global energy crisis?

The answer depends on what one means by a crisis and also on who answers the question. Different groups have different perceptions of an energy crisis. In industrialised countries like the US, any hint of a shortage of oil supplies or even of a small increase in the price of oil is considered a crisis. In poorer countries like India, the shortage or increasing price of firewood in a village could be a crisis.

There are also widely varying forecasts about energy reserves. Some believe that there will be enough energy for a long time to come; others are sure that a severe shortage is only a few years away.

From where do we get all our energy?

Ninety-nine per cent of our energy comes from the Sun. The commercial energy we pay for is just one per cent of the energy we use. Without the Sun, life on Earth would not exist, since the average temperature would go down to –240ºC. It is this solar energy that gets stored in plants as biomass. Plants use this energy to produce food through photosynthesis.

We can also make use of solar energy directly to heat water and buildings or to generate electric power. Wind energy and hydropower are indirect forms of solar energy. The Sun creates wind patterns and also makes water flow.

Where does the world's commercial energy come from? Commercial energy mostly comes from fossil fuels like oil, coal, and natural gas (Table 8.1).

Table 8.1 Share of different sources in total energy use

Energy source	Percentage of total energy	Sub-total Percentage
Non-renewable sources		
Oil	32	
Coal	21	
Natural gas	23	
Nuclear	6	
Non-renewable total		**82**
Renewable sources		
Biomass (mainly wood)	11	
Solar, wind, hydro and geothermal power	7	
Renewable total		**18**
Total		**100**

What is the global energy consumption pattern?

About 24 per cent of energy consumed globally is for transportation, 40 per cent for industry, 30 per cent for domestic and commercial purposes, and the remaining 6 per cent for other uses, including agriculture. About 30 per cent of energy goes into the production of electricity, which in turn is used by different sectors. Transportation accounts for a substantial amount of oil use.

A third of the world's population, that is, about two billion people, lack access to adequate energy supplies. At least three billion people depend on fuelwood, dung, coal, charcoal, and kerosene for cooking and heating. On the other hand, industrialized countries, with only 25 per cent of the global population, account for 70 per cent of commercial energy consumption. Most of it is oil that is used for transportation and is available cheap because of subsidies. These countries are interested in maintaining and even increasing their current levels of energy consumption. The US is the largest energy consumer in the world.

What is the energy consumption pattern in India and other developing countries? In India, about one-third of the energy comes from non-commercial sources. The rural population depends heavily on fuelwood, dung, and animal waste. In urban areas, there are large numbers of non-motorized vehicles like bicycles, rickshaws, handcarts, and animal carts. Such human and animal energy inputs into the system are not accounted for properly.

Fuelwood provides about 15 per cent of the energy needs in developing countries. According to the UN FAO, 2.7 billion people in 77 developing countries go without adequate supplies of fuelwood. They are also forced to consume wood faster than it can be replenished by nature.

Why do we often hear about energy consumption in the US? With just 4.6 per cent of the world's population, the US consumes 24 per cent of the total commercial energy produced. This exceeds the total amount used by the next four countries, namely, Japan, Germany, Russia and China. India, with 16 per cent of the population, accounts for just 3 per cent of total energy consumed.

A comparison of per capita consumption of energy for transportation shows the stark differences between countries. For every 100 units of energy consumed by a US citizen for transportation a Danish citizen uses 45 units, a Japanese 30 units, and an Indian just two units! Seventy six per cent of US citizens drive to work in a car, while just 5 per cent use public transportation.

Why should we worry about what the US does? First, we should note that 92 per cent of the energy used in the US comes from non-renewable fossil fuels that release huge volumes of emissions. The US has 3 per cent of the world's oil, but consumes 26 per cent of the crude oil extracted in the world. They also waste tremendous amounts of energy. Clearly, their ecological footprint is very large.

Now, many countries, including India and China, are striving to reach the same level of prosperity as the US. If we start consuming energy at the same rate as the US, the world will run out of fossil fuels in a few years (see Box 8.2)!

Box 8.2 Consume and be merry: The Chinese model

An important threshold was reached in early 2004: China exceeded Japan in oil imports, becoming the second largest importer of oil. It is also the fastest growing oil consumer in the world. Two million cars were put on the road in China in 2003, which was 70 per cent more than in 2002. Today, there are 10 million cars in China and the number is rapidly growing.

If China starts consuming at US levels, it will need 80 million barrels of oil per day, which is 10 million more than the entire world production in 1997. Even at current levels of growth, China would need at least 10 million barrels a day by 2025. Where will that oil come from and how much will it cost?

China, however, has huge coal reserves, enough to last the country 300 years. The environmental cost of burning all that coal would surely be very heavy.

What exactly are fossil fuels and why are they non-renewable?

Fossil fuels (coal, oil, and natural gas) are the remains of organisms that lived 200–500 million years ago. During that stage of the Earth's evolution, large amounts of dead organic matter had collected. Over millions of years, this matter was buried under layers of sediment and converted by heat and pressure into coal, oil, and natural gas.

Once we discovered these fuels, we began consuming them at an ever increasing rate. From 1859 to 1969, total oil production was 227 billion barrels. (In the oil industry, the barrel is the preferred unit and one barrel is equal to 159 litres.) Fifty per cent of this total was extracted during the first 100 years, while the next 50 per cent was extracted in just 10 years!

Today, our consumption rate is far in excess of the rate of formation of fossil fuels. We consume in one day what the Earth took one thousand years to form! That is why fossil fuels are considered non-renewable.

Where does all the oil go? The automobile may well be humankind's second worst invention (after the nuclear bomb). Millions of cars, trucks, and buses whizzing along the world's highways consume enormous amounts of oil and, at the same time, release large volumes of carbon dioxide. Oil is easy to carry and hence it is the preferred fuel for transport.

Why is there worldwide concern about oil?

Extraction of crude oil from the bowels of the Earth is not a simple process. When a new oil field is tapped, oil gushes out, but not for long. We then have to wait for the oil to slowly seep into the well from the surrounding rocks. For this and other reasons, the maximum annual production cannot exceed 10 per cent of remaining reserves.

There is also a second constraint on the production, characterized by the Hubbert Curve (Figure 8.1). This curve, proposed by the geophysicist M. King Hubbert, describes the pattern of oil availability in a field over time. In any oil field, the annual production increases until about 50 per cent of the reserve has been extracted. Production peaks at this point and thereafter declines, as it becomes increasingly more difficult and expensive to extract the remaining oil.

At some point, the energy expended in extraction exceeds the energy yield from the extracted oil. Then the field is abandoned. As the price of oil increases, however, it may become economical to remove more and more of the oil. The net energy obtained will, however, be low.

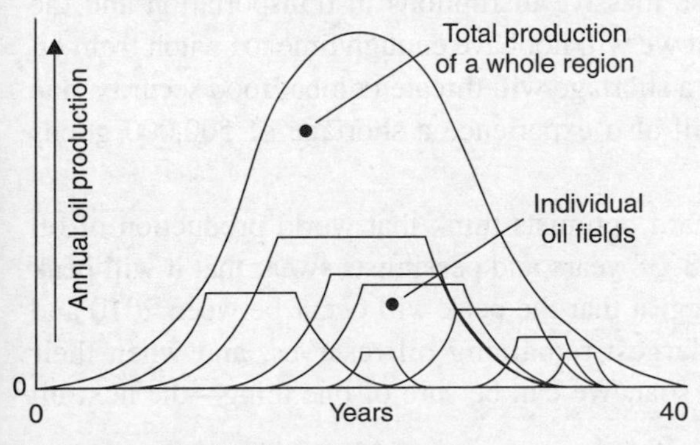

Figure 8.1 The Hubbert Curve for oil

The Hubbert Curve is also applicable to countries and regions. Oil production in US peaked as far back as 1970 and has been declining ever since. It is oil from Alaska that keeps things going. The world peak has not been reached, but it is expected to occur fairly soon.

Oil prices have fluctuated based on political factors, global demand, and the production policies of the Organization of Petroleum Exporting Countries (OPEC). This organization was formed in 1960 with the purpose of jointly regulating the production and price of the exported oil. OPEC currently has 11 members: Algeria, Indonesia, Iran, Iraq, Kuwait, Libya, Nigeria, Qatar, Saudi Arabia, the United Arab Emirates, and Venezuela.

OPEC decisions have a major impact on world oil prices. The price per barrel of oil went up from US$ 4 in 1973 to US$ 38 in 1982. The price dropped due to increased production, but it went up again during the first Gulf War. After the war OPEC increased production and the prices fell again to US$ 10 per barrel.

Since 1999, however, the price has been steadily going up with increasing demand. In mid-2004 the price had reached US$ 44, 40 per cent more than it was a year earlier, and in late September 2004 it crossed US$ 50. Such increases have an immediate effect on transportation costs and lead to inflation in many countries, including India.

The environmental costs of oil extraction and use have been horrendous: air pollution, damage to ecosystems, carbon dioxide emissions, global warming, the list goes on. The world's seas are today criss-crossed by hundreds of huge oil tankers. Major and minor oil spills have caused untold damage to the ocean, coastal zones, and marine life. In addition, the ballast water from the tankers has carried deadly organisms and toxic substances to far-off lands.

How much oil is still left and how long will it last?

There is no agreement on this question. We do not even know how much oil is still left. The estimated recoverable reserves are 1.4 to 2.1 trillion barrels. About 70 per cent of the world's crude oil reserves are with the OPEC countries. Saudi Arabia alone accounts for 25 per cent of the total. China and the US have 3 per cent each.

The OPEC countries often give inflated figures, because the amount they are allowed to produce is a function of the declared reserves. Even large oil companies are known to falsify figures.

Currently, there are 1500 major oil fields in operation, of which the 400 large ones account for 60–70 per cent of production. The discovery of new, big oilfields peaked in 1962 and since 1980 just 40 new fields have been found. Geologists agree that there are no big fields left to discover.

The current world demand is about 24 billion barrels per year and rising. Current new discoveries, however, amount to just 12 billion barrels per year and this is declining!

What are these doomsday forecasts about oil? Several research groups have suggested that an oil shortage is imminent. According to them, by 2010 oil from the current wells will be insufficient to meet demand and there will be massive disruptions in transportation and the world economy. What is more important is that we will not have enough time to switch from oil to other available sources. The impact of such a shortage will threaten global food security, due to a lack of fertilizers and chemicals. We will also experience a shortage of 500,000 goods currently derived from petroleum.

When will oil production really peak? Diehard optimists think that world production of oil (leaving aside the Middle East) will peak in 28–38 years and pessimists swear that it will peak in just 8–18 years. Most computer studies suggest that the peak will occur between 2010 and 2020. The Middle East will then have the largest remaining oil reserves, and when their production peaks, at a later stage, prices will soar. We can be sure of one thing—the next oil crisis will be permanent!

Don't we have huge amounts of coal?

At current rates of use, the world's coal reserves will probably last another 200 years. There are, however, many problems with coal use. Coal has to be mined from underground or from the surface. Underground mines, besides being dangerous, also cause lung disease in miners.

Among the fossil fuels, coal is the most harmful to the environment. The mines create major land disturbances and the burning of coal causes severe air pollution. Coal is currently responsible for 36 per cent of carbon dioxide emissions in the world. In addition, it releases huge amounts of radioactive particles into the atmosphere, more than a properly operating nuclear power plant. Every year, air pollution from coal kills thousands of people and causes respiratory diseases in thousands more.

Is natural gas a good option?

Natural gas, a mixture of methane, butane, ethane, and propane, is found above most oil reserves. While propane and butane are liquefied and removed as LPG, methane is cleaned and pumped into pipelines.

About 40 per cent of the total natural gas is in Russia and Kazakhstan. The available reserves are expected to last another 200-300 years. Its abundance, low production cost, and low pollution make it the ideal fuel during the transition from fossil fuels to renewable sources.

What are the pros and cons of nuclear power?

In a nuclear reactor, neutrons split the nuclei of elements like uranium and plutonium, releasing energy in the form of heat. This heat is used to produce steam, which runs electric turbines.

In 1953, the US President Eisenhower announced that nuclear power would be 'too cheap to meter'. At that time it was predicted that, by the end of the century, nuclear power would be generating most of the world's electricity. This source seemed the ideal answer to the world energy problem. The fuel supply was unlimited, the environmental impact was thought to be very low, and safety was assured.

Things, however, turned out very different. Today, nuclear power accounts for just 6 per cent of total commercial energy. Across the world, very few new plants are being set up. Many orders were cancelled, construction terminated in some cases even after heavy investment, and existing plants are being retired. One plant in New York was abandoned after just 36 hours of power generation! Germany and Sweden have decided to phase out nuclear power within the next 20–30 years. France, Japan, and India, however, are still committed to this energy source.

What went wrong with nuclear power? Many of the calculations and predictions were wrong. The plants cost much more than estimated, the operating costs were high, the technical problems were more intractable than expected, and the economic feasibility came to be questioned. But worse was to come.

The plants generated large amounts of deadly radioactive waste. The low-level waste must be stored safely for 100–500 years, while the high-level waste remains radioactive for a mind-boggling 240,000 years! Fifty years after the advent of nuclear power, we have no satisfactory way of storing the waste. Meanwhile more wastes are piling up in the plants. In addition, plutonium (a byproduct of certain nuclear reactions), removed from old nuclear warheads, must also be stored.

Nuclear power plants and nuclear tests have created thousands of contaminated sites, particularly in the US and the former USSR. The cost of cleaning them up will run into millions of dollars. Similarly, decommissioning an old plant costs more than the original construction cost! Of the large plants in the US, 228 are due for retirement by 2012. There is a move by the US Government to extend the life of all the plants to 50–60 years, but this could be risky.

The biggest problem, however, has been the loss of public confidence in the safety of the plants. Several minor accidents were followed by the major disaster at Chernobyl, which released radioactive dust over thousands of square kilometres and may ultimately cause between 100,000 and 475,000 cancer deaths.

Why can't we move to safer renewable sources? An estimated US$ 200–250 billion is invested in energy-related infrastructure every year and another US$ 1.5 trillion is spent on energy consumption, with nearly all of this investment being for conventional energy. Thus, the world has been locked into indefinite dependence on unhealthy, unsustainable, and insecure energy structures. Things are improving now, with increasing investment in renewables like solar and wind power.

What kind of energy do we get from the Sun?

We receive from the Sun a pure, non-polluting, and inexhaustible form of energy. Solar energy comes from the thermonuclear fusion reaction constantly taking place in the Sun. All the radioactive and polluting byproducts of the reaction are safely left behind in the Sun, 150 million km away.

An enormous amount of solar energy falls on this tiny planet Earth. What we get from the Sun in one month is more than the energy stored in all the fossil fuels we have. Also, using this vast amount of energy does not pollute the biosphere in any way.

Why then are we not using solar energy for all our energy needs? There is also some bad news: solar energy is a diffuse source falling evenly over a vast area and the first problem is collecting it efficiently. The second problem is converting it into a usable form such as electricity.

You can guess what the third problem is: What do we do when it is cloudy and the Sun does not shine? We must have an efficient way of storing the energy. All research in solar energy is about finding cost-effective ways of collection, conversion, and storage.

How has the solar water heater become so popular in some Indian cities? This water heater is in fact a simple and successful use of solar energy. It consists of a flat-plate collector, with a black bottom, a glass top, and water tubes in between. The collector is placed at a suitable angle to catch the Sun's radiation. The bottom gets hot and the heat, which cannot escape through the glass, warms up the water in the tubes.

The insulated storage tank is placed above the collector, the cool water moves down into the tubes and the hot water moves into the tank by natural convection. An electric heater element is also provided as a backup for cloudy days. The design of the solar water heater is being improved all the time, increasing its efficiency.

Solar panels being used to pump water

Since electric water geysers are energy-guzzlers, solar heaters are an energy-efficient alternative. Soft bank loans are now available to buy such heaters. Some cities in India have made it mandatory for all new houses to install solar water heaters.

How do we convert solar energy directly into electricity? This conversion is done by a photovoltaic cell (or PV cell), which consists of two layers of silicon. The lower layer has electrons that are easily lost and the upper one readily gains electrons. When light energy strikes the cell, it dislodges electrons from the lower layer. This sets up an electric current through the circuit carrying the electrons to the upper layer. Each cell generates only a small amount of power, but many cells, placed together on a panel, generate enough power to run an appliance.

The power from a solar panel is usually stored in a battery, to which we connect the appliance. Thus the energy is generated and stored when the Sun shines on the panel and is used whenever needed. The direct current from the battery can be converted into alternating current through an inverter. You can then run normal appliances like a fan or a television.

PV cells are used today in watches, pocket calculators, toys, etc. Larger solar panels can light up a house, run an irrigation pump, operate traffic lights, and so on. Solar power is a viable alternative in remote areas, where power lines cannot be taken because of the high cost involved or the inaccessibility of these areas. For example, in Ladakh in the Himalayas electricity is generated using solar panels, and this has brought about a tremendous change in people's lives.

There are several companies in India that offer solar home-lighting systems. One 100 X 50 cm solar panel can easily power eight lamps. We can also integrate solar panels in buildings, and on roofs, walls, and windows. The initial cost of a solar energy system is about US$ 30 per watt, but it pays for itself in 7 to 10 years. The panel prices are also steadily falling (read Box 8.3).

Box 8.3 Success: The magic lantern

It is evening at the Besantnagar Beach in Chennai. There are hundreds of people on the beach taking in the fresh air, an escape from the humid and polluted air of the city. Inevitably there are also hordes of hawkers selling groundnuts, snacks, and what have you. This story is about them.

The hawkers have always carried petromax lamps that run on kerosene. Pumping them up is quite a job initially. Once lit, however, they give bright light. Since 2001, however, there is an option: A solar lantern that can be rented for the evening.

The hawkers, who have each paid a refundable deposit of Rs 100, collect a lantern in the evening and return it after use. They pay Rs15 for a 4-hour rental and Rs 5 for each extra hour. During weekends, when the demand is high, the rent increases to Rs 20.

The young men who run the enterprise have a small office near the beach, where they charge the lanterns during the day using solar panels. When the lanterns are returned after use, they are immediately cleaned, checked, and repaired, if necessary, and connected to the panels for recharging for the next day.

Why do the hawkers prefer the solar lanterns? Because they spend the same amount as before, with some advantages. They can just switch on and use the lamp without worrying about its maintenance. With the petromax lamp, the flame would go out of control, the mantle would break, or the glass would get heated. Pumping the lamps up was also difficult, especially for women.

The project is the brainchild of Hemant Lamba of Auroville Renewable Energy (AuroRE) at the international community near Pondicherry. AuroRE has been promoting renewable energy and their projects have included installing solar lamps in Ladakh and solar pumps in Punjab. Hemant had been looking for ideas that promote the use of solar energy in a sustainable way, not depending on government subsidies. The idea of providing energy as a service appealed to him.

Hemant's idea was enthusiastically taken up by a group of promoters led by Ananth and Poornima in Chennai. They organized a group of urban poor youth to run it as an enterprise and also put in some capital. The boys are all alumni of the Olcott Memorial School, Besantnagar, which caters to the economically weaker sections. The team was trained by AuroRE in maintaining the solar lanterns. AuroRE has also been providing technical support as well as a loan for buying some of the equipment.

Each lantern costs about Rs 3750 and provides light for about five hours. The annual turnover is about Rs 100,000. The panels are likely to last 15 years, but the battery inside the lamp could be the weak link, especially since the lanterns have to operate under tough conditions of sand and wind.

The enterprise started with three lanterns in February 2002 and now has 100 lanterns serving 70 regular clients. A new unit has been started at the larger Marina Beach. Ananth is also taking the idea to Bhopal as an income generation project for the victims of the gas disaster.

This is a unique project that promotes renewable energy, providing at the same time self-employment for the youth among the urban poor.

Source: Padmanabhan (2003) and Herbert (2002)

Why can't we directly focus the Sun's rays on the material we want to heat? Make a parabolic reflector, something like a dish antenna, and place a bowl at its focal point with, say, rice in it. Your lunch will soon be ready. In fact, solar cookers of this type are available in India and some hotels and community kitchens use them.

The international community of Auroville near Pondicherry has a large solar bowl, 15 m in diameter. It concentrates the Sun's energy on a tube that carries a special liquid. The hot liquid in turn heats water to produce steam, which is used for cooking. The solar kitchen serves 2000 meals every day.

For home needs there are small box-type solar cookers. Food cooked in this manner saves fuel, tastes better, and is good for health. If the cooker is placed out in the Sun facing in the right direction, food will be ready in an hour or two depending on the food and the intensity of sunlight.

What about tapping wind energy?

If you travel from Nagercoil to Kanyakumari in South India, you will see hundreds of sleek-looking windmills, slowly rotating and generating electricity. India is now a leading player on the wind energy scene. Globally, wind farms produce about 25,000 MW of energy. Europe produces 70 per cent of the total, with Denmark in the lead (see Box 8.4). The global wind energy industry is worth US$ 7 billion and is growing rapidly. Europe plans to generate 10 per cent of its electricity from wind by 2025.

Box 8.4 Solution: Life after oil in Denmark

In 1998, the 4400 residents of Brundby on a Danish island decided that they would give up fossil fuels in 10 years. There are now more than twenty wind turbines generating as much energy as the island consumes from fossil fuels. Home heating, which is a necessity in Denmark, is through hot water made from burning straw.

Over six years, the island cut its energy consumption by 25 per cent, drastically reducing emissions of nitrous oxide, sulphuric acid, and carbon dioxide. The European Union (EU) plans to replicate their initiative in 100 other communities so that 12 per cent of the total energy in the EU would come from renewable sources by 2010.

Denmark has been making huge investments in developing green technologies. Twenty per cent of its electricity is now generated by renewable sources. It is also a world leader in wind power. Its windmills and generators are exported to many countries including India. The Danish wind power company, Vestas, has set up a manufacturing unit in Chennai.

The path to sustainable energy use, however, is always difficult. The latest news is that the new Danish government has cut its investment on renewable energy!

Wind energy produces electricity at low cost, the capital costs are also moderate, and there are no emissions. Wind farms can be quickly set up and easily expanded.

Where is the catch? Obviously, you need steady winds of a certain velocity and, therefore, not every place is suitable. In any case, you will need some form of back-up for windless days. Moreover, it involves high land use, though attempts have been made to use the space below the windmills for agriculture or grazing. There is some noise pollution and the monotonous view of hundreds of windmills is visual pollution. In addition, there is the fear that windmills could interfere with the flight of migratory birds.

Is hydropower a good alternative?

Twenty per cent of the world's electricity comes from hydropower. In order to get a sizeable amount of power, we need a high dam on a river with a large reservoir. The potential energy of the water falling from a height runs the power generating turbine.

Hydropower has several advantages. The cost of generation is low and there are no emissions. The reservoir can provide water for irrigation round the year and can also be used for fishing and recreation. It also gives drinking water to towns and cities.

What is the bad news? Dams cost a lot of money and take years to build. Most of the suitable rivers of the world have already been dammed and it is now difficult to find new spots. The reservoir drowns large areas of farmland, wildlife habitats, and places of historical and cultural importance. An example is the town of Tehri, which was getting submerged in the rising waters as a dam on the Bhagirathi River neared completion in mid-2004.

Large dams also cause large-scale displacement of local communities. The people lose their lands and become environmental refugees. Often, compensation for the lost land is meagre (and is not even paid on time) and resettlement is never satisfactory. Dams impede the migration of fish along the river and reduce the silt flowing downstream. In fact, the sediments pile up against the dam and reduce its useful life. There is a worldwide movement against the building of large dams.

There are other renewable sources like tidal energy, ocean thermal energy, geothermal energy, and energy from biomass (plant material and animal wastes). They are useful as small-scale alternatives, but cannot yet satisfy the world's enormous appetite for energy. In particular, they are not as yet suitable for transportation.

What about the hydrogen economy?

Many experts believe that, as we run out of fossil fuels, we will move towards using the element hydrogen as the main fuel to run the world's economy. When hydrogen burns and gives us energy, it combines with oxygen to produce water vapour. In this process, there is no air pollution or emission of carbon dioxide.

This is good news again! What are we waiting for? There is some bad news too. Hydrogen is not available in a free state: it is locked in water and in compounds like petrol and methane. We need to use energy and an effective method to split these compounds to get hydrogen. We can split water by heat or by a process called electrolysis to get hydrogen. We can also get hydrogen from fuels like petrol, natural gas, and methanol. Thus the first problem is one of collection.

As in the case of solar energy, we also have the problem of storage. Hydrogen is highly explosive and if it is stored as compressed gas, the tank needs to be large, heavy, and costly. Only large buses and trucks could hold the tanks. If we store it as a liquid, energy will be consumed in maintaining the very low temperatures that will be needed. Another method being tried out is storage as a solid metal hydride. Here again, energy is needed to release the hydrogen when we want it.

What about fuel cells that are often in the news? A fuel cell is so called because it is an electrochemical unit like a battery. The fuel cell burns hydrogen to produce electricity. In the process, hydrogen combines with oxygen to produce water vapour. Thus there is no pollution and it runs continuously as long as there is input.

Unlike a battery, the fuel cell draws its input (hydrogen and oxygen) from outside. Again, a battery requires recharging, but the fuel cell does not. Finally, there is no toxic output when a fuel cell is discarded.

Fuel cells were developed as far back as 1960 for space applications. The progress since then has, however, been slow. There are now experimental buses and cars running on fuel cells, but they are very expensive. Automobile companies like Daimler-Chrysler, Honda, and General Motors have made prototype fuel-cell cars.

What are the problems with using hydrogen? First, it takes energy to produce hydrogen. Obviously, against this input, we should get much more energy from the hydrogen we produce.

Second, if this input energy comes from fossil fuels, there will be environmental effects. Third, the situation is even worse if we produce the hydrogen from fossil fuels themselves. Finally, we have not fully solved the problem of storing hydrogen.

Where is the hydrogen economy then? We must first find cost-effective ways of producing hydrogen from water using renewable energy like solar energy (read Box 8.5). We should solve all the storage problems too.

Box 8.5 Inspiration: World's first hydrogen economy?

Iceland promises to become the world's first hydrogen economy over the next 25–30 years. For this purpose, the Government of Iceland has teamed up with companies like Daimler-Chrysler, Royal Dutch Shell, and Norsk Hydro. The project is the brainchild of the chemist Bragi Arnason.

The project will produce hydrogen from seawater using the country's abundant renewable energy sources: wind, geothermal, and hydropower. Hydrogen will run buses, cars, fishing vessels, and even factories. Royal Dutch Shell will operate hydrogen filling stations.

Iceland's experiment will be a model for other countries to follow.

We must remember, however, that neither hydrogen nor any other wonder fuel, for that matter, is going to save us if we keep increasing our energy use or even maintain the current usage levels. The hydrogen economy will be viable only if we use energy much more efficiently and reduce our consumption levels. As of now, there is no sign of this happening.

Once again we face the same hurdle: There is a natural limit to the amount of resources on this planet. When we overstep that limit, we will inevitably face an insurmountable hurdle.

Given all the limits and difficulties, are we using energy efficiently?

Increasing the efficiency of energy use will be equivalent to finding free sources of energy. There is tremendous scope for increasing energy efficiency in all our activities. The ordinary incandescent bulb has an efficiency of just 5–10 per cent. Most of the input electrical energy disappears as heat. Replacing it with a compact fluorescent lamp (CFL) will reduce consumption by 75 per cent. The initial cost is more, but over time it pays back. At the same time, we are doing a service to society by consuming less energy. Incidentally, the cost of CFLs has been steadily coming down.

The internal combustion engine that runs our automobiles is another device that wastes 90 per cent of the input energy. Driven by the first oil crisis, the fuel efficiency of American cars gradually increased between 1973 and 1985. Since then, however, it has levelled off primarily because of the consumer craze for the so-called Sport Utility Vehicle (SUV), minivan, etc. and because of the availability of cheap oil.

Makers like Toyota and Honda have introduced hybrid electric cars with much higher fuel efficiency. They run on petrol and a battery. The battery is kept charged by the petrol engine and an electric motor provides energy for acceleration and hill climbing. When the car is braked, part of the heat generated is used to charge the battery.

Whenever we use energy, some waste is inevitable. However, there is a large amount of avoidable waste in energy use. One estimate is that we waste more than 40 per cent of the commercial energy that we buy. Examples are vehicles and furnaces that waste fuel and poorly designed buildings that use up huge amounts of energy for heating and cooling. There is also the huge inefficiency of conventionally-fuelled power stations and large transmission losses.

Technology exists today for increasing the efficiency of most appliances by 50 per cent or more. As the demand for energy-efficient appliances increases, the prices will also come down. We must always consider the life cycle cost of devices, that is, the total of the initial cost and the operating costs over the lifetime of the device. Such an approach will make us take better decisions that will save cost and energy.

Can't we end with some good news?

The good news is that wind and solar energy initiatives are growing fast. Over the past decade, the solar energy industry has grown seven-fold and wind power 13-fold. As a result, costs are steadily declining and this in turn is increasing their penetration into the energy market (read Box 8.4).

A quick recap

- There is a serious fuelwood crisis in India and other developing countries.
- The end of cheap oil is not far away. Fairly soon, prices will soar and there will be widespread repercussions.
- More serious than the depletion of oil resources are the environmental consequences of burning fossil fuels at current rates.
- There will be serious environmental consequences and depletion of sources, if all developing countries seek to copy the energy consumption levels of the richer countries.
- Energy efficiency can and must be increased.
- Both the North and the South must direct more research funds into renewable energy and share the results.
- We must use the most sustainable, least polluting sources with high efficiency.
- Almost all sources of energy have their limits and problems.

What important terms did I learn?

Fossil fuel
Fuel cell
Hubbert Curve
Photovoltaic cell
Renewable energy

What have I learned?

Essay questions

1. Describe the fuelwood situation in developing countries, emphasizing the relationship between poverty and the environment.
2. Describe the different scenarios with respect to the availability of oil. Which scenario appears reasonable to you and why?
3. Explain how almost every source of energy has its limits.

Short-answer questions

1. What is the main source of energy for this planet and how does it give us energy?
2. List the various forms of commercial energy and give the proportion of each in the total.

3. Which country consumes the maximum amount of energy and why is this fact relevant?
4. Why are fossil fuels non-renewable?
5. Explain the Hubbert Curve and the concept of peaking.
6. Compare the advantages and disadvantages of oil, coal, and natural gas as energy sources.
7. What are the special problems of nuclear power?
8. How is solar energy converted to electrical power and what are the problems in this regard?
9. What are the prospects of getting more hydropower?
10. What are the hurdles in moving to a hydrogen economy?
11. How can we improve energy efficiency?

How can I learn by doing?

Do an energy audit of your college. Examine the consumption of all forms of energy over three months and find out how much energy each activity consumes. Find out also the cost aspects. Come up with proposals for (*a*) reducing the demand for energy, (*b*) increasing energy efficiency, and (*c*) using renewable energy sources.

What can I do to make a difference?

1. Gradually replace all the bulbs in your home with Compact Fluorescent Lamps (CFL).
2. Install a solar lighting system and use solar power during the day.
3. If you use hot water, install a solar water heater in place of an electric geyser.
4. Buy energy-efficient appliances when you replace old ones or when you need new ones. Always check the specifications for energy consumption figures.
5. Turn off the lights and fans when you leave the room. Install electronic regulators for fans and auto-switch-off devices for areas like staircases.
6. If you are building a new house:
   - Minimize the use of materials like fired bricks, cement, and steel that use up considerable energy during manufacture. Use earth blocks instead of fired bricks, and vaults and domes in place of concrete roofs, and so on.
   - Provide skylights wherever possible to bring in natural light and reduce the use of electricity.
7. Minimize the use of automobiles for your personal transport:
   - Use a bicycle for local work like shopping.
   - Use public transport whenever possible.
   - Arrange car pools.
   - Live near your place of study or work, if possible.
8. Keep your vehicles tuned for low consumption of fuel.
9. Check fuel consumption data while buying a new vehicle.
10. Shut off personal computers, television sets, music systems, etc., when not in use. Replace the bulky CRT monitor with the flat and thin LCD monitor that consumes much less energy and is easier on the eyes.
11. Follow the advice given by the Petroleum Conservation Research Association (PCRA) with regard to energy conservation.
12. Observe the National Energy Conservation Day: The Government of India has declared December 14 as the National Energy Conservation Day. On that day, National Energy Conservation Awards are given to industrial units who have taken exceptional initiatives to conserve energy.

Where are the related topics?

- Chapter 9 covers forest resources and their uses.
- Chapter 11 covers air pollution caused by emissions from burning fuel.

- Chapter 14 gives an account of nuclear accidents.
- Chapter 16 includes a discussion on the disposal of nuclear waste.
- Chapter 19 discusses climate change, global warming, and resettlement of people due to the construction of dams.
- Chapter 20 includes an account of how a Supreme Court decision led to the buses and trucks in Delhi converting to the use of CNG.

What is the deeper issue that goes beyond the book?

Given that almost every source of energy has its limitations and problems, what should be our attitude to using energy in this world?

Where can I find more information?

- The website *www.handels.gu.se/econ/EEU/* contains papers on the fuelwood crisis.
- Access the site of The Energy and Resources Institute, *www.teriin.org,* for information on India's energy problems.
- Access the site *www.sunminonthenet.com/* for more information on the Chennai solar lantern enterprise.

What does the Muse say?

A short poem by the Sufi poet Jalaluddin Rumi (1207–73) for you to reflect on:

Every moment
the sunlight is totally empty
and totally full.

CHAPTER 9

Forest Resources

The best time to plant a tree was twenty years ago.
The second best time is now.

Anonymous

What will I learn?

After studying this chapter, the reader should be able to:

- appreciate the conflicts between different aspects of forest conservation, such as sustainable use, rights and livelihood of local communities, government policy, and environmental activism;
- recall the main conclusions of two major global forest assessments;
- distinguish between old-growth, second-growth, and plantation forests;
- list the products and services provided by forests;
- describe the role of paper production in the use of forest resources;
- explain how deforestation occurs and the impact it has on the environment;
- establish a connection between forests and climate change;
- recall the international and national initiatives in forest conservation;
- explain the meaning of sustainable forest management;
- appreciate the importance and ways of involving local communities in forest conservation;
- describe some of the efforts of individuals and groups in conserving trees and forests; and
- list what could be done to save the world's forests.

Box 9.1 Crisis: The last chance or a lost opportunity?

It is a complex case involving a century of forest exploitation, the rights of indigenous people, inward migration, road development, and finally the intervention of the Supreme Court of India. Through it all, the plunder of the forests has continued unabated.

The Andaman and Nicobar Islands contain some of the finest tropical evergreen forests in the world. They are also rich in biodiversity with a variety of known and unknown species of flora and fauna.

The British established a Forest Department on the islands in 1883 and began the extraction of timber using convict labour. (The Andaman jail was famous for housing political prisoners, who had opposed the British government.) Even after Independence, logging has continued without a break.

Many recent studies have shown that legal and illegal logging has led to forest degradation. Further, the resultant soil erosion has led to a heavy flow of sediments into the coastal waters that has smothered and killed a substantial amount of corals. Mangroves and corals have also been adversely affected by extraction. Species like the saltwater crocodile and the Andaman wild pig have become endangered.

The indigenous communities have been seriously affected, with their traditional occupations threatened and their rights violated. Communities like the Great Andamanese, the Onge, the Jarawa, and the Sentinelese have lived and flourished here for at least 20,000 years, but they could soon become extinct.

About 150 years ago the population of the tribal communities was estimated to be at least 5000. Over the years, there has been a substantial government-supported migration of people from other parts of the country into the islands. Today, in a total population of 500,000, the four indigenous communities put together account for a mere 500.

A major factor affecting the forests and the communities is the 340-km-long Andaman Trunk Road (ATR). Originating in Port Blair in South Andaman, the ATR cuts across the islands to reach Diglipur in the North. In some places, it traverses through virgin tracts of forestland in the Jarawa Reserve. The ATR has increased the interaction between the Jarawas and the settlers as well as the tourists and eroded much of the natives' original way of life.

The current population is clearly much more than the size that the islands can support. Already, there is a scarcity of drinking water.

Responding to a petition filed by three voluntary organizations in 1998, the Supreme Court of India passed orders in October 2001 prohibiting the felling of naturally grown trees in the entire Andaman and Nicobar Islands. The Court also appointed an expert commission to study the state of the forests and related matters on the islands. In May 2002, accepting the report of the commission, the Court passed a remarkable set of final orders.

The Court orders included a ban on the commercial exploitation of timber from the islands' forests, a ban on the transport of timber to any other part of the country, removal of encroachments, restrictions on inward migration, phasing out of monoculture plantations, reducing sand mining on the beach, and the closure of the ATR in areas where it passes through the Jarawa Reserve. The Court also specified a time frame for implementing its orders.

One might think that it was a happy ending. The local administration, however, did not implement many of the orders. Several months after the passing of the deadlines, they filed a petition asking for a review of the orders.

Perhaps a great opportunity to implement a model conservation programme was lost. What will be the fate of these islands?

Source: Sekhsaria (2002, 2003)

What does the case of Andaman and Nicobar Islands tell us?

The case of Andaman and Nicobar Islands is typical of what is happening to the world's forests. Massive deforestation through the extraction of timber, opening up of forests through road-building, population pressures, the problems faced by indigenous peoples, and similar issues plague forests everywhere.

What is the state of the world's forests?

It is difficult to assess the extent of the world's forests. Data supplied by countries is often unreliable, though satellite imagery is increasingly being used to verify data from ground surveys. The definition of a forest also varies from one assessment to another.

The highlights of two global forest assessments are presented below, one from the World Commission on Forests and Sustainable Development (WCFSD), and the other from the Food and Agricultural Organization (FAO) of the UN.

In its 1999 report, WCFSD (read Box 9.2) came to the following conclusions about the state of the world's forests:

- The world's remaining forested areas amounted to about 3.6 billion hectares in 1999, down from about 6 billion hectares 8000 years ago.
- Fifty six countries have lost between 90 and 100 per cent of their forests.
- Over the last two decades of the twentieth century, 15 million hectares of forest were lost annually, largely in the tropics.
- About 14 million hectares of tropical forests have been lost each year since 1980 due to conversion into cropland.
- Forest decline threatens the genetic diversity of the world's plants and animals. About 12.5 per cent of the world's plant species, and about 75 per cent of the world's mammal species, are threatened by forest decline.
- In developing countries alone, some US$ 45 billion per year is lost through poor forest management.
- In Europe, forests are declining due to drought, heat, pests, and air pollution. The number of complctcly hcalthy trecs in European forests fell from 69 per cent in 1988 to 39 per cent in 1995.

The WCFSD concluded, 'The decline is relentless. We suspect it could change the very character of the planet and of the human enterprise within a few years unless we make some choices.'

Box 9.2 Initiative: Independent Commission

Following the Earth Summit in 1992 it was agreed that solutions to forest degradation were likely to be more political than technical. Accordingly, the InterAction Council, a group of some 30 former heads of government and state, established an independent World Commission on Forests and Sustainable Development (WCFSD).

The Commission, co-chaired by Emil Salim of Indonesia and Ola Ullsten of Sweden, had 22 members including M.S. Swaminathan and Kamla Chowdhary of India.

The objectives of the Commission were to:

- Increase awareness of the dual function of world forests in preserving the natural environment and contributing to economic development.
- Broaden the consensus on the data, science, and policy aspects of forest conservation and management.
- Build confidence between North and South on forest matters, with emphasis on international cooperation.

The Commission held public hearings in Asia, Africa, Europe, Latin America and the Caribbean, and North America. It met with forest-dwelling and other local communities, farmers, industry executives, etc. It took note of what scientists, economists, forest officers, government officials, and other specialists involved in national and international forest policy had to say. It listened to the views of environmental groups.

The Commission published its first report in 1999 after three years of research and public hearings. The Commission's main findings and recommendations have been listed in this chapter.

Source: The website *www.iisd.org/wcfsd/*

The *Global Forest Resources Assessment 2000* (FRA 2000) of the FAO was primarily based on information provided by countries, supplemented by state-of-the-art technology to verify and analyze the information. More than 200 countries and areas were represented in the assessment. For the definition of a forest, FRA 2000 adopted a threshold of 10 per cent minimum cover, for both natural forests and forest plantations.

The main findings of FRA 2000 were:

- The world's forest cover in 2000 was about 3.9 billion hectares, or approximately 0.6 hectare per capita. About 95 per cent of the forest cover was in natural forest and 5 per cent in forest plantations.
- The distribution of forest area by ecological zone, in terms of percentage, was 47 per cent in the tropics, 33 per cent in the boreal zone, 11 per cent in temperate areas, and 9 per cent in the subtropics.
- Deforestation in the 1990s was estimated at 14.6 million hectares per year. The figure represents the balance of annual losses of natural forests minus the area replaced through reforestation with forest plantations. Expressed in another way, during the 1990s, the world lost 4.2 per cent of its natural forests, but gained 1.8 per cent through reforestation (with plantations), afforestation, and the natural expansion of forests.
- Overall, however, the loss of natural forests is still high in the tropics, and increases in plantation establishment and the natural expansion of forests have not been able to compensate for the losses incurred.
- Despite the high losses of the world's natural forests at the global level, new forest plantation areas are being established at the reported rate of 4.5 million hectares per year. The countries with the largest plantation development are China, India, Russia, the US, Japan, and Indonesia.
- Initiatives to promote sustainable forest management have stimulated many countries into implementing forest management plans. At least 123 million hectares of tropical forests were reportedly subject to management plans. In industrialized countries, the majority of the forests are under some form of protection or sustainable management.
- Of the world's forests, 12.4 percent were estimated to be in protected areas according to the categories defined by the World Conservation Union (IUCN).

Estimates of forest cover in India are given in Chapter 3.

How are forests classified from the point of view of exploitation as a resource?

There are mainly three categories of forests: old-growth, second-growth, and plantations. Old growth or frontier forests are uncut forests that have not been seriously disturbed by human activities or natural disasters for several hundreds of years.

Second-growth forests result from secondary ecological succession (see Chapter 2) that takes place when forests are cleared (by human activities or due to natural disasters) and then left undisturbed for long periods of time.

Plantations are managed forests of commercially valuable trees. They are created mostly by clearing old-growth or second-growth forests. Plantations promote monoculture, concentrating on one or a few species. Being less diverse than natural forests, they are more prone to disease and other disturbances.

There is now an increasing reliance on plantations as the source of industrial wood. This is a new development and half of all the plantations of the world are less than 15 years old. More than 60 per cent of plantations are in Asia, many owned by foreign companies.

What are the products and services provided by forests?

Industrial wood and fuelwood: According to FRA 2000, the global production of wood had reached 3335 million cu. m by 1999. Half of it was fuelwood, mostly produced and consumed in developing countries. The remaining half was industrial wood, 80 per cent of which came from industrialized countries.

Non-wood products: In many poorer countries, especially in Asia, non-wood forest products (NWFPs) such as food, fibre, honey, and medicinal plants form an important source of income and a critical component of food security and well-being. Some forests are also sources of minerals.

Ecosystem services: Forests provide a range of services like soil generation, soil and water conservation, purification of air and water, nutrient recycling, maintenance of biodiversity, providing a habitat for animals, mitigation of climate change, and absorption of carbon. Forests contain about half of the world's biodiversity. Natural forests have the highest species diversity and endemism among all ecosystems.

Other contributions: Forests provide employment and income, recreation, education, scientific study, protection of natural and cultural heritage, aesthetic pleasure, and spiritual solace.

How much of wood goes into making paper?

The paper industry is the world's fastest growing consumer of wood. Worldwide, 40 per cent of the wood is used for making paper and this figure is expected to reach 60 per cent by 2050. The US alone accounts for 30 per cent of world paper use, consuming a billion trees a year and releasing 100 million tons of toxins every year during processing.

India has about 600 paper mills with an annual capacity of 8.5 million tons. Of the raw material they use, 39 per cent comes from wood and bamboo, 31 per cent from agricultural residues, and 30 per cent from waste paper. Even though waste paper is abundant in India, the industry annually imports about 1.2 tons of it costing US$ 116 million.

Reforestation has not kept pace with the world demand for paper. The amount of paper recycled has also been too small to make any difference. Some countries like the Netherlands and Germany recover over 70 per cent of waste paper. India has a recovery rate of just 18 per cent, mainly because we have a poor collection and segregation system.

In industrialized countries, about 50 per cent of the paper is used for packaging, 30 per cent for writing, 12 per cent for newsprint, and 8 per cent for paper tissue and towels. American retailers send out 17 billion catalogues to potential customers and 95 per cent of this paper is discarded unread! This junk mail culture is now spreading to countries like India.

Contrary to popular belief, computers did not usher in the era of the paperless office. In fact, studies have shown that paper consumption in an office increases by about 40 per cent after the introduction of computers and email.

In what ways are forests being destroyed?

Commercial logging methods directly and indirectly lead to deforestation. In many places, to obtain one cubic metre of log wood, two cubic metres of standing trees are destroyed. New extractive technologies can cut trees very quickly. When some species are selected for logging, non-target species are also damaged.

Logging companies create infrastructure, especially roads, in forests to make their task easier. However, roads provide easier access to the interiors of forests and encourage the entry of invasive species, hunters, poachers, tourists, plant collectors, and people in general. This, in turn, leads to further exploitation of the forest resources.

Another area of concern is the depletion of forest-based wildlife due to commercial harvesting and trade of bushmeat. This practice of killing wild animals for meat is prevalent in Africa, where many species of primates and antelopes are threatened.

The construction of dams in forests invariably causes enormous damage. Dam reservoirs inundate and destroy forests and their biodiversity.

What is the impact of deforestation?

Deforestation exposes soils and shade species to wind, sunlight, evaporation, and erosion. Soil fertility goes down due to the rapid leaching of essential mineral nutrients. The topsoil is eroded and this accelerates siltation in dams, rivers, and the coastal zone. The increased sedimentation harms downstream fisheries.

When the forest disappears, there is no regulation of the flow into rivers. As a result, floods and droughts alternate in the affected areas.

Deforestation, degradation, and fragmentation of forests affect many species and lead to the extinction of some. In particular, migratory birds and butterflies suffer due to the loss of their habitat.

Local and global climate changes may also result from deforestation. Studies have shown that about 97 per cent of the water absorbed from the soil by the roots of plants, evaporates and falls back on the land as precipitation. When a large forest is cut down, the regional rainfall pattern may be affected.

Deforestation may also lead to global warming by releasing carbon stored in the trees. If the trees burn, the carbon is released immediately. If the trees are cut and removed, half the carbon remains in the form of branches, twigs, etc. When they decompose, the carbon is slowly released.

Clearing of forests affects local communities, who lose their source of food, fuel, construction materials, medicines, and areas for livestock grazing. What is more, they lose their culture and way of life.

What is the role of forest fires?

Wildfires, usually started by lightning, have an ecological role. The combustion frees the minerals locked in the dry organic matter. The mineral-rich ashes are necessary for the growth of plants. The vegetation usually flourishes after a fire.

Fires remove plant cover and expose the soil, which stimulates the germination of certain types of seeds. They also help control pathogens and harmful insects. Occasional fires burn away some of the dry organic matter and prevent more destructive fires from occurring later on. For all these reasons, prevention of fires is not necessarily good for a forest.

Intentional or accidental human-induced fires do however cause damage. Such fires have become a major problem in large forests, especially in countries like Canada and the US. In a number of developing countries, fires continue to be used for land clearing with adverse consequences.

Climate change also causes more fires than usual. In many countries, there was an increase in wildfires during the 1990s as compared with previous decades. The climate phenomenon El Niño was implicated as a major contributing factor to the severe forest fires in the 1990s (as well as in the 1980s). El Niño is an occasional phenomenon that causes change in climate patterns worldwide (see Chapter 14 for a more detailed account). This phenomenon provokes severe droughts in generally humid or temperate areas, enhancing the propensity for devastating fires.

What is the relationship between forests and climate change?

Forests both influence and are influenced by climate change. They play an important role in the carbon cycle and the way we manage forests could significantly affect global warming.

Forests hold more than 50 per cent of the carbon that is stored in terrestrial vegetation and soil organic matter. Hence, deforestation contributes significantly to net emissions of carbon dioxide into the atmosphere.

If the predicted global warming occurs, the impact on forests is likely to be regionally varied, dramatic, and long-lasting. Even now, we can see how any extreme weather has great impact on forests. For example, the 1999 storms in Europe caused heavy damage to forests and also to trees outside forest areas.

The Kyoto Protocol on Climate Change (Chapter 19) may have a great impact on forest management. Under the Protocol, a country with forests earns emission credits, since its forests absorb carbon dioxide. These credits are tradable, that is, a developing country can sell its credits to an industrialized country that has exceeded its quota of emissions. The latter would invest in afforestation and reforestation projects in the developing country.

What are the international and national initiatives in forest conservation?

Sustainable forest management was first discussed at the international level at the Earth Summit held at Rio de Janerio, in 1992. There were major differences between the industrialized countries and the developing countries and the Summit could only come up with a set of non-binding principles.

The International Tropical Timber Organization (ITTO) was set up in 1983 under the United Nations Commission for Trade and Development (UNCTAD). ITTO brings together the producer and consumer countries and is a major platform for issues concerning sustainable forest management. In 1985, FAO, UNDP, the World Bank, and the World Resources Institute came up with the Tropical Forestry Action Plan, later revamped and renamed as the National Forest Action Programme.

The Kyoto Protocol, Convention on Biological Diversity (Chapter 6), and the Convention to Combat Desertification (Chapter 10) are three of the international agreements that have a bearing on forests. The UN Forum on Forests, created in October 2000, is a permanent high-level intergovernmental body with universal membership.

Declaring forests as protected areas or as biosphere reserves is a measure adopted by most countries. This is also supported by initiatives like the UNESCO Man and the Biosphere Programme. The protected area approach has been discussed in Chapter 6 under biodiversity conservation.

What is sustainable forest management?

Sustainable forest management (SFM) is the use of the world's forests in such a way that they continue to provide resources in the present, without depriving future generations of their use. One of the principles of SFM is to fully involve local communities in forest management. Implementing this principle is, however, difficult since forest departments are usually very reluctant to lose their control over forest resources.

SFM has also become an element of climate change negotiations. As mentioned earlier, the Kyoto Protocol would compensate countries for the benefits their forests provide to the world. The industrialized countries are ready to support SFM in developing countries so that they can buy the credits and continue to pollute the atmosphere.

By 2000, 149 countries were engaged in nine international initiatives to develop and implement criteria and identify indicators for SFM, covering 85 per cent of the world's forests. There are 140 countries with national programmes in various stages of development.

Certification of timber as coming from sustainable forests is another approach (read Box 9.3).

Box 9.3 Solution: Forest certification

How can we halt deforestation and save the remaining forests? One way is to act as responsible consumers and buy wood only from companies that follow sustainable practices. How do we locate such companies? Forest certification is meant to help us.

Certification is a voluntary market-based approach that enables us to identify forest products backed by high environmental standards. It focuses on the quality of forest management rather than on that of forest products.

Three certification methods are in operation:

1. Accreditation by the Mexico-based Forest Stewardship Council (FSC): Producers have to meet certain principles and standards for good forest stewardship. They can then use the FSC trademark for product labelling. By 1999, FSC had certified as well as managed more than 15 million hectares in 27 countries. Certification is based on sustainability of timber resources, socio-economic benefits provided to local people, and forest ecosystem health, including preservation of wildlife habitat and watershed stability.
2. ISO 14000 Environmental Management System (EMS): This system, operated by the International Standards Organization, is similar to the ISO 9000 Quality Certification. The ISO 14000 certification is given to a producer who has installed an appropriate environmental management system as specified in the standard. It does not guarantee that the products are ecofriendly. It only ensures that the producer has the right intentions and management procedures. ISO 14000 is applicable to any organization and is not confined to the forestry sector.
3. Pan-Europe Certification Scheme and national certification schemes by individual countries.

By the end of 2000, about 2 per cent of global forests had been certified, most of them in Europe,

Canada, and the US. There is a growing demand for certified wood. Environment-conscious consumers have welcomed certification, though some producers consider it a restrictive practice.

The draft National Environment Policy, released by the Government of India in 2004, promises to encourage eco-labelling and other certification schemes.

Source: Miller (2004) and MoEF (2004)

Foresters and local people working together to conserve forests

How can local communities be involved in forest conservation?

Many local communities have lived in or near forests and have used them in a sustainable manner. In the twentieth century, however, there were two developments in this regard. First, due to an increase in population and poverty, new groups migrated into the forest areas and began over-exploiting the forest resources. Second, when governments began protecting forests and declaring them as protected areas, they viewed the local people as enemies of the forests and tried to prevent them from even entering the forest area. The result was an increase in the illegal use of forest resources and conflicts between government representatives and the local people.

It is now increasingly being realized that local people should be regarded as partners in conservation. Most programmes now involve local communities in planning, decision-making, and implementation. In return for controlled access to the forests, the locals can provide labour and help in conservation. They can become excellent guides in ecotourism ventures.

What are extractive reserves? Extractive reserves are protected forests in which local communities are allowed to harvest products, such as fruits, nuts, rubber, oil, fibres, and medicines, in ways that do not harm the forest. The objective is to improve the lives of the people, while conserving biodiversity.

This approach believes that local people would have a greater stake in conservation if they continue to get the benefits that they previously enjoyed. It recognizes the fact that, in many instances, we have to use the land and forests in order to preserve them.

How are communities involved in forest conservation in India?

In India and some other countries, communities living in or near the forests are involved in conservation in three ways: joint forest management, social forestry, and sacred groves.

What is joint forest management? Around the 1980s, the Government of India came to recognize the important role of local communities in forest conservation. They introduced the concept of Joint Forest Management (JFM), with a view to working more closely with local user communities in the protection and management of forest resources.

In JFM, the local communities are involved in planning the conservation programme. They are allowed controlled access to the forest areas and permitted to harvest the resources in a sustainable manner. In return, they become the guardians of the forest.

The Tamil Nadu Afforestation Project (TAP), implemented with Japanese aid, is cited as a successful example of JFM. During the first phase (1995–2003) of implementation, Rs 6.5 billion was spent and one million acres of degraded forestland was upgraded, involving 400,000 people in 1800 villages.

In the TAP villages, 2500 self-help groups comprising 25,000 women were formed. The basic needs of the people of the TAP villages such as drinking water and roads were met by integrating the development schemes of several departments.

What is social forestry? Social forestry refers to the planting of trees, often with the involvement of local communities, in unused and fallow land, degraded government forest areas, in and around agricultural fields, along railway lines, roadsides, river and canal banks, in village common land, government wasteland, and panchayat land.

The term 'social forestry' was first used in India in 1976 and the idea has been adopted in many Asian countries. A major controversy has been the planting of eucalyptus trees under social forestry projects. Eucalyptus was chosen for the majority of social forestry projects because it survives on difficult sites and out-performs indigenous species in growth, producing wood very rapidly. However, it has some adverse ecological impacts on soil nutrients, water hydrology, biodiversity, and wildlife.

The traditional sacred groves of India play an important role in community participation and conservation of biodiversity (read Box 9.4).

Box 9.4 Solution: Conservation through the Fear of God!

It could be just a few trees or a whole forest. No tree or plant is cut here, no animal or bird is killed, and no form of life is harmed. No one would dare, because it is protected by the local deity. It may even have a temple.

The sacred groves of India are a unique traditional institution devoted to the conservation of forests and biodiversity. They are referred to in the ancient texts and thousands of them must have once existed. They were protected by local communities through social traditions and taboos that incorporate spiritual and ecological values. (Recall also the story of the Bishnois of Rajasthan in Chapter 3, Box 3.3.)

Preserved over many generations, sacred groves contain native vegetation in a natural or near-natural state. They are thus rich in biodiversity and are repositories of species and genetic diversity. They often contain species that have disappeared in other places. Rare medicinal plants are also found in the groves.

Many groves have water sources that help local communities. The groves absorb water during the monsoon and release it slowly during the dry period.

Over time, hundreds of these groves have disappeared under the twin pressures of population and development. However, according to one survey, over 13,000 groves still survive as patches across the country. Other estimates put the number at between 100,000 and 150,000. Only groves in remote areas remain undisturbed, while others need protection.

In recent years, there has been an increasing interest in documenting and conserving the sacred groves. In 1999, the Indira Gandhi Rashtriya Manav Sangrahalaya (the National Museum of Mankind) and several other institutions established a museum for sacred groves in Bhopal, with the goal of raising awareness of the important role these ancient forests play in conservation. The museum celebrates the communities, festivals, and rituals associated with the groves of different states. In addition, the museum has organized a travelling exhibition to build greater ties with local communities and organizations throughout the country. It also organizes a Sacred Grove Festival.

Source: DTE (2003)

How can wood be used more efficiently?

Currently, enormous quantities of wood are wasted, primarily in countries like the US, in many ways: inefficient use as construction material, excessive packaging, excessive junk mail, inadequate paper recycling, not reusing wooden shipping containers, etc. There is great scope to reduce such wastage.

Tree-free paper can be made from natural fibres and agricultural residues from wheat, rice, and sugarcane. China plans to make 60 per cent of its paper from tree-free pulp. A number of publishers in the world now use paper from forest-friendly sources.

What can individuals and groups do to conserve trees and forests?

There are remarkable stories from all over the world of individuals and groups going to great lengths to save or plant trees. Some examples are:

- Chipko, the people's movement to save the forests of Tehri Garhwal (Box 9.5).
- The Green Belt Movement started by Wangari Maathai in Kenya (Box 9.6).
- The struggle of Chico Mendes and the rubber tappers against deforestation in Amazonia (Box 9.7).
- The planting of roadside avenue trees in Karnataka by Thimmakka and Chikkanna (Box 9.8).
- The unbelievable two-year tree-sit by Julia Butterfly Hill to save the redwood trees of California (Box 9.9).

Box 9.5 People power: The women who saved the trees

Chipko women hugging the trees

On March 26, 1974, a group of men arrived stealthily in the forest next to Renni Village in the Garhwal District of the Himalayas. They had been sent by a contractor to begin cutting down 2500 trees in the forest. Anticipating resistance from the people, the contractor had ensured that all the men of the village were away on that day.

Word of the arrival of the axe-men spread in the village and the women came out of their houses. About 25 of them, led by Gaura Devi, confronted the contractor's men. They pleaded with the men not to start the felling operations, but the men responded with threats and abuses. As the confrontation continued, more women joined the protest. Ultimately, the men were forced to leave, since the women would not budge.

This small event was a milestone in the Chipko Movement, which became known all over the world as a symbol of people's action in preventing the destruction of the environment. 'Chipko' means 'to cling to' or 'to hug tight'. The vision of women hugging the trees and daring the axe-men to cut them before cutting the trees inspired environmentalists the world over. And the Chipko Movement remains one of the celebrated environmental movements of the world.

It is not clear when and where the movement to save the trees was given the name 'Chipko'. Neither can authenticated accounts of women actually hugging the trees be found. It is a fact, however, that the Chipko Movement spread rapidly across the Himalayan Region in the 1970s led by dedicated activists like Sunderlal Bahuguna and Chandi Prasad Bhat.

The contract system allowed rich contractors from the plains to make huge profits from the felling of trees in the hills. The Chipko Movement was the hill communities' response to the unfair and destructive nature of this system. As Raturi, the folk-poet of the Movement put it,

Embrace the trees in the forests
And save them from being felled!
Save the treasure of the mountains
From being looted away from us!

The Movement also had a catchy slogan:

What do the forests bear?
Soil, water, and pure air!

In 1995, women in the Ryala Region of Tehri Garhwal launched a Chipko-like movement to save giant trees on the hills, above an altitude of 1000 metres. The women tied a sacred thread (*raksha sutra)* around 1000 trees, marked for cutting. There has been other such direct action by people to prevent felling.

The Chipko Movement brought unprecedented energy and direction to the issue of environmental preservation in India. It spawned action in other places: the Appiko Movement in Karanataka, the Narmada Bachao Andolan, and the Chilika Lake Agitation in Orissa.

Yet, the Movement lost its steam before its immense potential could be realized. Exaggeration of its strengths, excessive adulation, ego clashes among the key players—all contributed to the sapping of its energy. However, Chipko ensured that the contract system was abolished and the indiscriminate felling of trees stopped. We can hope for new Chipkos to arise in the future.

Box 9.6 Inspiration: Nobel Prize for noble work

She started a movement that planted 30 million trees in 20 countries. She campaigned for women's rights and greater democracy in her country. She defied a corrupt regime, was vilified and forced to leave her country for some time, and was even assaulted by the police. Today, however, she is a Member of Parliament and Assistant Minister for Environment. In 2004 she was awarded the Nobel Peace Prize, the first African woman to receive this honour.

Wangari Maathai studied in the US and also received a doctorate in biology from the University of Nairobi, the first woman in east and central Africa to do so. She was a Professor at the Nairobi University in Kenya when she launched the Green Belt Movement (GBM) in 1977.

Her objective was to empower the people and to show them that they could choose to destroy or build the environment. GBM encouraged poor women to plant millions of trees to combat deforestation, in return for which they received sufficient fuelwood.

The movement has set up 5000 tree nurseries run by women and disabled persons. Seedlings are given away free to groups and individuals. For every tree that survives for three months, the planter receives a small payment. By 1988, 40,000 people were planting trees and in due course the movement spread to many other African countries.

Wangari came into the limelight when she led a protest against the building of a 62-storey building in the middle of Freedom Park, Nairobi's most popular public space. The then President, Daniel Arap Moi, labelled her and the GBM subversive. Faced with intense persecution, she had to leave the country for a while. Her marriage also broke up.

When Wangari returned to Kenya, she took up the cause of political prisoners. She criticized the

President for allowing deforestation and displacing people. Once, when she and a group of women were on a hunger strike demanding the release of political prisoners, the riot police came in and knocked her unconscious. Through all the persecution, Wangari stood solid and unbowed, like the trees she had planted.

The political climate changed in Kenya in 2002, when Mwai Kibaki came to power. Wangari became a Member of Parliament and the new President appointed her as Assistant Minister for Environment. In October 2004, she was awarded the Nobel Peace Prize.

Source: Tyler (2004)

Box 9.7 Inspiration: Giving one's life for the trees

The world knows him as a martyr, who died while preventing the destruction of the Amazon forests. Born in 1944 in the Brazilian Amazon, Chico Mendes earned his living as a rubber tapper. Besides extracting latex from rubber trees, the tappers also collected and sold minor forest produce, such as nuts, fruits, and native medicines.

As a leader of the tappers' union, Chico campaigned for extractive forest reserves for tappers, who would use and maintain the rainforest in its natural state for generations. He ran into a conflict with powerful groups interested in clearing the forests for raising cattle. Chico argued that this would destroy valuable forests and that overgrazing would reduce soil fertility. Ranching would only bring short-term gains.

In 1988, the Xapuri Rural Workers' Union led by Chico successfully prevented the powerful cattle rancher, Darli Alves, from deforesting an area that the tappers wanted as a reserve. On December 22, 1988, Chico was shot dead. Darli was later convicted of Chico's murder.

Chico's death brought international attention to the question of saving the Amazon forests. His name has become synonymous with the worldwide fight to save forests. In 1990, the Chico Mendes Extractive Reserve was established and so also the Chico Mendes Foundation.

Chico Mendes had this to say about his work: 'First I thought that I was fighting for the rubber tappers, then I thought I was fighting for the Amazon, then I realized I was fighting for humanity.'

Box 9.8 Inspiration: Trees as children

There is something special about the 300 towering avenue trees on a 4 km stretch of the road from Kudur to Hulikal in Karnataka. All of them were planted and cared for by an elderly couple, Thimmakka and Chikkanna.

Thimmakka was born about 80 years ago in the town of Gubbi and was married off at an early age to Chikkanna of Hulikal Village. They were both landless farm labourers. Thimmakka could not have children and the couple were often lonely in the evenings. Chikkanna kept thinking of 'something to do' with their lives.

About 50 years ago they decided to plant trees on the main road to Kudur. The couple thought that the trees would provide shade for the villagers, who often had to walk the hot and dusty road.

They chose the peepul tree (*Ficus religiosa),* created a small nursery, and began planting the saplings. They built thorn guards around the saplings, watered them daily until they took root. They tended each tree until it was 10 years old.

Every year they planted 15 to 20 trees, until they had covered the entire 4 km stretch. They took care of the trees as if they were their children. In fact, Chikkanna quit working so that he could devote himself full time to this task. The trees grew tall and have been providing shade for the users of the road, shelter for many birds and animals, and biomass for the fields.

Chikkanna died in 1990. Since 1995, many honours have come Thimmakka's way: National Citizen's Award, Priyadarshini Vrikshamitra

Award, etc. An Indian organization in the US has even named itself after Thimmakka. The well-known Bharatanatyam dancer, Malavika Sarukkai, has choreographed a performance based on Thimmakka's life and work.

Source: The websites *www.goodnewsindia.com* and *www.thimmakka.org*.

Box 9.9 Inspiration: The woman who wouldn't come down

In December 1997, Julia Butterfly Hill climbed a redwood tree in California and sat down on a platform high up in the tree. And she remained there for two years! Here is her remarkable story.

The coastal redwood trees, which grow in California, are some of the tallest and oldest trees on the planet. They are also unique: they do not grow anywhere else. Yet, they continue to be cut by timber companies and made into furniture.

In the early 1990s, members of an environmental group erected a small platform and began living in one of the huge Pacific redwoods, which had been slated for cutting. They named the tree 'Luna'. The purpose of this 'tree-sit' was, first, to prevent the tree being cut and, second, to draw wider attention to what was happening. For the timber company in this area of north California was not only cutting redwoods, it was also engaged in clear-cutting entire forests of trees. The result of this was that the soil was destabilized, resulting in massive erosion of topsoil and frequent landslips on precipitous hillsides. Clear-cutting also threatened other forms of life, which depended on the forests for their existence.

Most people could only stay up the tree for a few days at a time, and, as winter approached, there were fewer volunteers. Then a young woman named Julia Hill became involved. The first time she went up for five days. After she came down again, she heard that nobody else was available to continue the sit. So she went up again. She did not come down again for two years!

Of course, she was supplied with provisions by a support team on the ground. But she was living on a tiny platform 180 feet above the ground near the top of a tree, which was exposed on the top of a ridge. She survived two of the toughest Californian winters ever, protected by no more than a thin tarpaulin.

The timber company did everything they could to make her come down. They used a helicopter to generate huge updrafts close to the tree, they trained bright lights on the tree at night, and played loud music. They cut other trees surrounding Luna, and they also threatened to cut the tree while Julia was living in it. Finally, they stationed guards around the tree for a while in an attempt to starve her out.

A number of times she was close to giving up. She contracted frostbite, she broke a toe, and she was almost blown out of the tree by storms. Sometimes she didn't sleep for a week. Yet she survived. She stayed. How? She wanted to protect Luna for the thousands of people across the country for whom she had become a symbol of hope, 'a reminder that we can find peaceful, loving ways to solve our conflicts, and that we can take care of our needs without destroying nature to satisfy our greed.' Above all, she built up a remarkable relationship with Luna itself. In the middle of one of the worst storms, when she was in danger of being blown down to her death, she found herself asking Luna what she should do. She felt that Luna told her to imitate the trees, to bend with the wind and not to try to fight it.

Luna became an extension of herself. She kicked off her shoes and climbed barefoot all over the tree, discovering the beauty of its ecology—the animals which inhabited it, and the way in which the top leaves were shaped differently from the lower ones in order to channel the rain to the roots. In fact, Luna became her school and university, her introduction to the inner world of nature.

Julia also had books sent up to her and, later, even a radio and a phone. She became a celebrity, an inspiration to thousands of people, a symbol of what one individual was willing to do to stand up for her beliefs. She was frequently interviewed on radio programmes and, on the first anniversary of her tree sit, a celebration was held below which attracted thousands of people. She danced at the top of the tree to the music being played below.

Finally, after protracted negotiations with the timber company who owned the tree, an agreement was signed by which Luna was protected from logging in perpetuity. After two years, on December 18, 1999, Julia came down to a huge reception.

A year after Julia climbed down from Luna, a vandal attacked the tree with a chain saw and made a three-foot-deep cut that went more than half way around her 12-foot circumference. A team of volunteer arbourists, engineers, tree climbers, and biologists came together and designed, manufactured, and installed steel brackets to stabilize the cut. It appeared that Luna would die, but sustained efforts including native healing methods seem to have worked and Luna is still standing.

Julia set up the Circle of Life Foundation to activate people through education, inspiration, and connection to live in a way that honours the diversity and interdependence of all life. She travels to many places and gives more than 250 lectures every year.

Source: Hill (2000)

What should we do to save the world's forests?

The summary recommendations of the WCFSD were:

1. Stop the destruction of the earth's forests. Their material products and ecological services are severely threatened.
2. Use the world's rich forest resources to improve lives of the poor and to benefit forest-dependent communities.
3. Put the public interest first and involve people in decisions about forest use.
4. Get the price of forests right, to reflect their full ecological and social value, and to stop harmful subsidies to lumber companies.
5. Apply SFM approaches to forests so that we may use them without abusing them.
6. Develop new ways of measuring forest capital so that we know whether the situation is improving or worsening.
7. Plan for the use and protection of whole landscapes, not the forest in isolation.
8. Make better use of knowledge about forests, and greatly expand this information base.
9. Accelerate research and training so that SFM can quickly become a reality.
10. Take bold political decisions and develop new civil society institutions to improve governance and accountability regarding forest use.

There is no sign of these recommendations being implemented. If we do not act soon, the whole world may go the way of the Easter Island and Madagascar (read Box 9.10).

Box 9.10 Crisis: History repeating itself?

On Easter Suday, 1722, Dutch explorers arrived at a small, isolated island in the South Pacific and named it the Easter Island. They found 2000 inhabitants struggling to live on the barren island. The explorers pieced together the strange story of the island.

The island was first colonized 2500 years ago by the Polynesians. The settlers were almost completely dependent on the island's towering palm trees. They used the trees for shelter, tools, boats, fuel, food, rope, and clothing.

Around 1400 A.D. the population peaked to between 7000 and 20,000. They made and moved large stone structures and sculpted a large number of big statues, many of which are still standing.

The Easter Islanders, however, made a fatal blunder. They did not conserve the palm trees. They cut and used up the trees so fast that soon the last tree disappeared.

The islanders could no longer build boats to catch fish. Without any forests to absorb and retain water, springs and streams dried up, soil began to erode, food crops dwindled, and famine set in. The starving people started fighting among themselves and were perhaps driven to cannibalism. The civilization collapsed.

Easter Island is a stark reminder of what happens if we do not care for trees or other natural resources. Humankind, however, does not learn its lessons. A very similar story is being enacted on the much bigger island of Madagascar, off the coast of East Africa.

Madagascar, the fourth largest island in the world, had an astonishing biodiversity, because it developed for 40 million years in near isolation. What is more, 85 per cent of the species were endemic to the island. Now, however, most of the species are endangered.

Since humans came to the island about 1500 years ago, 80 per cent of the tropical seasonal forests and 65 per cent of the rainforests have been destroyed for lumber, fuelwood, and conversion to cropland. With forests gone, soil erosion has become severe. The rapid population growth has led to slash-and-burn agriculture in these poor soils.

Of the 31 primates inhabiting Madagascar, 16 face extinction. There is also rampant illegal smuggling and export of endangered frogs, chameleons, and lizards. It is expected that half of all the plant and animal species in Madagascar will disappear by 2025.

Some international efforts are being made to save Madagascar's biodiversity, but chances of success are slim. Madagascar is perhaps another Easter Island in the making.

Source: Miller (2004)

What is the future of world's forests?

Forests are being destroyed due to a variety of factors and it is going to be very difficult to save them. None of the remaining forests of the world are free from human intervention. The loss of forests is, however, only a symptom of deeper and possibly unstoppable degradation of the Earth's environment.

A quick recap

- Forests provide invaluable products and services.
- Forest cover is depleting rapidly, especially in the tropics, due to many reasons and this will have serious consequences for the environment.
- In order to conserve forests, we have to reconcile the diverse, and often conflicting, interests of the environmentalists, government officials, and local communities.
- A number of national and international measures have been introduced for sustainable forest management, but implementation has been poor.
- Local communities are increasingly being involved in forest conservation.
- On the whole, the future of the world's forests appears to be bleak.

What important terms did I learn?

Extractive reserve
Joint forest management
Old-growth forest
Sacred grove
Second-growth forest
Social forestry

Have I learnt the stuff?

Essay questions

1. Study the full report of the World Commission on Forests and Sustainable Development (available on the website *www.iisd.org/wcfsd/*) and write a critical review of the same.
2. Write an essay on the international initiatives in forest conservation.

Short-answer questions

1. Describe the efforts of any two individuals who have worked for the conservation of trees or forests.
2. Explain in your own words the problems facing the forests of the Andaman and Nicobar Islands.
3. What is the state of the world's forests according to global assessments?
4. What is the difference between old-growth and second-growth forests?
5. What are the ecosystem services provided by forests?
6. What is the impact of deforestation on the environment?
7. Are natural fires good or bad for the forests? Explain.
8. How can local communities be involved in forest conservation?
9. How can we save the remaining forests of the world?

How can I learn by doing?

1. Most states have programmes to promote Joint Forest Management (JFM). With the permission of the concerned forest department, study any one programme. Spend a few days or weeks living with the forest communities. Interview local people as well as foresters. Write a report evaluating the JFM. Focus on the situation before and after the implementation of JFM.
2. Study any movement against a specific development or conservation project in a forest area, like the construction of a dam or the establishment of a protected area. Write a report giving your assessment of the situation. Suggest alternatives that may be acceptable to the government and the people.

What can I do to make a difference?

1. Plant trees wherever you can—in your compound, neighbourhood, park, streets, under power lines, on the denuded slopes of a hill, etc., but take care to choose an appropriate tree: for example, banyans next to buildings is not a good idea, as their strong root systems may damage the foundation!
2. Contribute to organizations that promote tree planting and have 'adopt a tree' programmes.
3. Join voluntary groups in your area or city that work to save existing trees and plant new ones.
4. Save the forests by saving paper:
   (a) Buy recycled, handmade, or tree-free paper.
   (b) Use both sides of the paper, one-sided paper for notes, etc.
   (c) Reuse paper envelopes.
   (d) Write to companies to take your address off their mailing list.

5. Observe March 21 as World Forestry Day: The day is celebrated in forests around the world to encourage people to appreciate the benefits of forests to the community, such as catchment protection, providing a habitat for animals and plants, areas for recreation, education and scientific study, and as a source of many products including timber and honey. World Forestry Day also aims at providing opportunities for people to learn how forests can be managed and used sustainably for these purposes.

Where are the related topics?

- Chapter 3 has covered forests as ecosystems.
- Chapter 18 discusses how communities could contribute to forests conservation.

What are the deeper issues that go beyond the book?

1. Should we ban the international trade in tropical timber?
2. How would you react to the destruction of tropical forests, if you were:
   (a) A poor landless farmer in Brazil
   (b) The owner of a biotechnology company
   (c) An environmentalist

Where can I find more information?

- For more information on Wangari Maathai and the Green Belt Movement:
  - Maathai (1988)
  - *www.greenbeltmovement.org*
  - *www.nobelprize.org/peace/laureates/2004/index.html*
- More on sacred groves from the following websites:
  - *www.edugreen.teri.res.in/explore/forestry/groves.html*
  - *www.sacredland.org/world_sites_pages/Sacred_Groves.html*
- For the full report of the World Commission on Forests and Sustainable Development (WCFSD), access the website, *www.iisd.org/wcfsd/*.
- Read Hill (2000) and access *www.circleoflifefoundation.org* for more information on Julia Butterfly Hill and her current work.

What does the Muse say?

Here is a well-known piece by the American poet Robert Frost (1874–1963). Incidentally, this poem was found on Jawaharlal Nehru's desk after his death.

Stopping by the Woods on a Snowy Evening

Whose woods these are I think I know.
His house is in the village, though;
He will not see me stopping here
To watch his woods fill up with snow.

My little horse must think it's queer
To stop without a farmhouse near
Between the woods and frozen lake
The darkest evening of the year.

He gives his harness bells a shake
To ask if there's some mistake.
The only other sound's the sweep
Of easy wind and downy flake.

The woods are lovely, dark, and deep,
But I have promises to keep,
And miles to go before I sleep,
And miles to go before I sleep.

CHAPTER 10

Land, Food, and Mineral Resources

There are two spiritual dangers in not owning a farm.
One is the danger of supposing that breakfast comes from the grocery,
and the other that heat comes from the furnace.

Aldo Leopold

What will I learn?

After studying this chapter, the reader should be able to:

- give an account of the degradation of the Earth's land surface including waterlogging, salinization, and desertification;
- recall the sources and availability of food in the world in general and in India in particular;
- evaluate the Green Revolution listing its benefits and the problems it has caused;
- describe different methods of organic farming;
- appreciate the global fisheries crisis, its causes, and possible solutions; and
- describe the environmental impact of mining activities.

Box 10.1 Crisis: Poverty of plenty?

Punjab: India's granary and a success story of the Green Revolution? Farmers raking in fortunes through modern technologies and hard work? A model for the rest of India and the world? That perhaps was the picture some time ago. Today, however, it is a story of degraded soil, depleted water tables, reduced productivity, and farmer suicides. How did things change so quickly?

Later in this chapter, we will learn about the problems facing agriculture and food production the world over. The reasons for Punjab's plight will then become clearer. Here, however, let us listen to some of the farmers through excerpts from two of the many reports that have appeared about the Punjab situation.

The first is from Dasgupta (2001):

As the sun sets on the sepia-coloured horizon, Ram Pal sits alone to tell his story. 'Let the land open wide and swallow us up. My nine acres have become unproductive due to waterlogging. The fields are full of wild grass. I have a family of three to feed and a debt of Rs 50,000 to repay,' laments the 60-year-old farmer from Kalalwala village in Punjab's Bhatinda district. Now Ram Pal goes to the nearby town every day to work as a labourer.

Like many farmers, he has been caught in the vortex of Punjab's agricultural crisis. Today, many farmers living in one of India's richest granaries are in danger of losing their livelihood, as agricultural lands are slowly turning barren due to farming practices aimed at increasing yields to meet demand.

Since the soil has lost its natural capacity to nourish the crops, we have to keep on adding fertilizers. Naturally, the cost of production is going up,' says Jitender Pal Singh, a farmer in Ropar district.

Partap Aggarwal, a pioneer of organic farming in India, visited Punjab in early 2004 and here are some excerpts from his diary:

Four young small farmers who own adjacent plots had come together to meet and talk with us. Here are some of the things we learned from them.

We asked if they thought it was a good idea to grow two heavy crops of cereals on the same land year after year. They knew it was not, and they said so without any doubt in their minds. Then why do they do it? Their answer was that it all came slowly and now they are caught in its vice-like grip. Approximately in the mid-1960s, a whole new package was developed by the agricultural university in Ludhiana and implemented throughout Punjab, with unprecedented energy. It consisted of a high-yielding dwarf wheat seed, a regimen of chemical fertilizer and pesticide use, and most importantly an irresistible bait of a guaranteed lucrative support price.

Their fathers had been subsistence farmers, so they were suspicious of the chemicals but the package as a whole was offered on the basis of take it or leave it all. As the farmers' income rose, they began to stretch the land area under wheat and rice to the limits. As cultivation became more intensive and the gap between crops narrowed, farmers felt the need to buy tractors and other machinery. As their cash income grew, more machinery and chemicals followed.

Soon, with higher incomes, the farmers' lifestyle began to change; large and small brick houses began to replace traditional mud structures; modern gadgets such as refrigerators and motorized two-wheelers as well as cars came into fashion. Clothes became fancier and people more sedentary. Now, almost all Punjab farmers use labour from Bihar, some entirely depend on them. Many more young men and women seek college education than 20 years ago. Food habits have changed. The list is endless. Under these circumstances even if we want to we cannot stop producing the lucrative cereal crops.

The significant thing they mentioned was their concern about the long-term effect of the poisons they spray on the crops and the soil. One effect they had noticed was on the birds. One farmer said, 'They are more or less gone. There are no vultures to eat the dead animals. In fact one does not see or hear any birds except an occasional crow, dove or egret. I think this is a bad omen, but most people do not seem to care.'

Sardul Singh is in his late 60's and has a classic peasant face with his life story written in the lines on it. I asked him if they applied compost to their soil. His answer was, 'Sadly, much less than we should and can. You see, one of my sons raises buffalos and we have that huge hill of dung unappreciated and slowly losing value. I keep reminding my sons to use it up, but they don't. Their indifference to this valuable material is surprising, for they know the good it will do to the soil. In our time we would quickly pick up any amount of compost or raw dung available to us and make the best use of it. But today's youth are used to spreading fertilizers; their noses cannot bear the smell of cow dung. Also, they have grown lazy. We used to work twice as hard and never complained. Times have changed!'

Indeed, times have changed in Punjab! What does the future hold in store for this state?

Source: Dasgupta (2001) and Aggarwal (2004)

What does the Punjab crisis tell us?

Problems like soil degradation, falling water tables, and increasing use of chemicals are not unique to Punjab. Such environmental problems are now common in many parts of India and the world. In this chapter, we will examine the reasons for the crisis.

What is the importance of land as a natural resource?

The land area of the Earth, about 140 million sq km, occupies less than a third of its surface. Yet, it is vital to our existence since it is land that:

- Preserves terrestrial biodiversity and the genetic pool
- Regulates the water and carbon cycles
- Acts as the store of basic resources like groundwater, minerals, and fossil fuels
- Becomes a dump for solid and liquid waste
- Forms the basis for human settlements and transport activities

Even more important, the topsoil, just a few centimetres thick, supports all plant growth and is hence the life support system for all organisms, including humankind.

What is the condition of the world's land surface? UN studies estimate that 23 per cent of all usable land (excluding mountains and deserts, for example) has been degraded to such an extent that its productivity is affected. The main causes of this degradation are deforestation, fuelwood consumption, overgrazing, agricultural mismanagement (planting unsuitable crops, poor crop rotation, poor soil and water management, excessive use of chemicals, frequent use of heavy machinery like tractors, etc.), the establishment of industries, and urbanization.

Soil erosion and degradation, which occur due to loss of green cover, strong winds, chemical pollution, etc., have severe effects on the environment. They affect the soil's ability to act as a buffer and a filter for pollutants, regulator of water and nitrogen cycles, habitat for biodiversity, etc.

How serious are the problems of waterlogging and soil salinity?

When irrigation is not accompanied by proper drainage, waterlogging occurs. This in turn brings salt to the surface of the soil, where it collects at the roots of plants or as a thin crust on the land surface. Rapid evaporation of groundwater also adds salt to the soil.

Pakistan, Egypt, India, and the US are some of the countries worst affected by salinization and waterlogging. In Egypt, 90 per cent of all farmland suffers from waterlogging. In Pakistan, two-thirds of all irrigated land is salinized. In India, between 12 and 25 per cent of land is waterlogged or salinized.

What is the scale of desertification and what is being done about it?

Desertification is land degradation in arid and semi-arid areas caused by human activities and climatic changes. It occurs slowly, as different areas of degraded land spread and merge together. It is progressing slowly but clearly over the planet like a 'skin disease'. Desertification as a problem remains poorly understood, but estimates suggest that a third of the Earth's land area is affected, that is, about 50 million sq km.

A fifth of the world's population is threatened by the impact of global desertification. Its effects can be seen all over the world, in Asia, the African Sahel, Latin America, throughout North America, and along the Mediterranean. Cultivable land per person is shrinking throughout the world, threatening food security, particularly in poor rural areas, and triggering humanitarian and economic crises.

Fertile topsoil takes centuries to form, but it can be washed or blown away in a few seasons. Human activities such as over-cultivation, deforestation, and poor irrigation practices combined with climate change are turning once fertile soils into barren patches of land. When large barren patches merge together, a desert arises.

The international community has long recognized that desertification is a major economic, social, and environmental problem faced by many countries in all regions of the world. The

Earth Summit in 1992 supported a new, integrated approach to the problem, emphasizing action to promote sustainable development at the community level. The UN Convention to Combat Desertification was adopted in 1994 and came into force in 1996. Over 180 countries are now parties to the Convention.

Among the practical measures undertaken to prevent and restore degraded land are:

- The prevention of soil erosion
- Improved early warning systems and water resource management
- Sustainable pasture, forest, and livestock management
- Aero-seeding over shifting sand dunes
- Narrow strip planting, windbreaks, and shelterbelts of live plants
- Agroforestry ecosystems
- Afforestation and reforestation
- Introduction of new species and varieties with a capacity to tolerate salinity and/or aridity
- Environmentally sound human settlements

What is the impact of urbanization and industrialization on land?

Fifty per cent of the world's population lives in urban areas and this figure is expected to go up. Urban areas constantly need more land for settlements, infrastructure, industries, leisure activities, etc., which increase the pressure on land. More and more agricultural land gets converted into urban colonies. Larger cities affect even larger areas surrounding them, thanks to their widening ecological footprint. (See Chapter 1 for a discussion of ecological footprint.)

We see solid waste piled up on many urban streets. Outside cities, there are large dumps of waste brought from the city. It is estimated that about two million hectares of land have been degraded due to waste disposal and landfills.

Urban agriculture has been expanding globally over the past 25 years. In Sao Paulo in Brazil and Havana in Cuba, for example, urban home gardens have been very successful. While urban agriculture provides locally grown food and helps recycle organic matter, it can also cause soil and water pollution if chemicals are used.

What is the future of land as a resource? With an increasing world population, there will be intense pressure on land, particularly in Africa and Asia. More intensive land use will be needed to feed the people. More land will also be brought under agriculture by the conversion of forests and grasslands.

Where does the world's food come from?

Food comes from three sources:

- Croplands that provide 76 per cent of the total, mostly food grains.
- Rangelands that produce meat mostly from grazing livestock, accounting for about 17 per cent of total food.
- Fisheries that supply the remaining 7 per cent.

What is the global food availability? There is enough food in the world to provide at least 2 kg per person a day including grain, beans, nuts, fruits, vegetables, meat, milk and eggs—

enough to make most people fat! The problem is that many people are too poor to buy readily available food. At least 700 million people do not have enough to eat. Every year hunger kills 12 million children worldwide.

India now produces 180–210 million tons of food grains, 5 million tons of meat products, and 6 million tons of fish annually. The increase in grain production since the 1960s is ascribed to the Green Revolution.

What is meant by the Green Revolution and what has been its impact?

The Green Revolution (see also Chapter 5, Box 5.1) refers to the rapid increase in world food production, especially in developing countries, during the second half of the twentieth century, primarily through the use of lab-engineered high-yielding varieties (HYVs) of seeds. It was hailed as a success story of agricultural science and technology. It is now clear, however, that it has also brought in its wake many problems.

In the middle of the twentieth century, there were severe food shortages in many developing countries. With growing populations, countries like India had to import food. The answer to this problem came from Mexico.

In the early 1940s, Mexico's wheat yields were low and the country was importing 50 per cent of its food. The Rockefeller Foundation in the US set up a research programme to increase the production of food grains. An American agricultural scientist, Norman E. Borlaug, who joined the team, developed a high-yielding variety of wheat through new concepts in plant breeding. It was a variety that could respond to high inputs of fertilizer and irrigation. Between 1950 and 1965, Mexican wheat yields increased 400 per cent and a Green Revolution had been born.

By the mid-1960s, the Green Revolution was fully adopted in India, thanks to the efforts of the scientist M.S. Swaminathan and the then Food Minister C. Subramaniam. The new HYVs increased production and made multicropping possible. Ultimately, India achieved self-sufficiency in food. In fact, our warehouses are so full that some of the stored grain is rotting

The Green Revolution was a clear shift from traditional agriculture. It came as a package of HYVs along with high inputs of chemical fertilizer, pesticides, water, and agricultural machinery like tractors. It was an energy-intensive method: apart from the energy that went into the making of inputs, energy was also needed to run the machinery and to pump water. It is estimated that eight per cent of the world's oil goes into Green Revolution agriculture.

Governments had to subsidize many of the inputs to keep the Green Revolution on track. The package also made developing countries dependent on foreign technology.

The farmers found that they had to buy the seeds every year and that, year after year, the inputs had to be increased to maintain productivity levels. On the whole, the Green Revolution has benefited large landowners and not the subsistence farmers.

Farmer dependent on purchased seeds every year

The number of varieties being used by farmers went down drastically. A large number of traditional varieties have disappeared, leaving just a handful of HYVs. The new varieties were more prone to diseases and pest

Organic farming

attacks. The soil has also been degraded and drained of its nutrients through the excessive use of chemicals.

What is the way out for agriculture?

One way out of the crisis is the gradual shift from chemical agriculture to organic farming. Organic farming does not use chemical fertilizers and chemical pesticides. It is in fact a return to traditional methods such as crop rotation, use of animal and green manures, and some forms of biological control of pests.

Organic farming is based on the following principles (Alvares 1996):

- Nature is the best role model for farming since it uses neither chemicals nor poisons and does not demand excessive water.
- Soil is a living system and not an inert bowl in which to dump chemicals.
- Soil's living populations of microbes and other organisms are significant contributors to fertility on a sustained basis and must be protected and nurtured at all costs.
- The total environment of the soil, from soil structure to soil cover, is more important than any nutrients we may wish to pump into it.

Organic farming that blends traditional knowledge with modern scientific ideas can, over a period of time, reverse soil degradation and improve soil health. Conversion to organic farming will mean initial problems and economic losses as the soil recovers, but it will be a better option in the long term. The success of Cuba in shifting to organic farming is a case in point (read Box 10.2).

Box 10.2 Inspiration: From crisis to success story

Imagine a country that suddenly finds itself with no fertilizer or pesticide for its fields and all imports of meat, grains, and processed foods gone. Most countries would not be able to recover from such a crisis, but this is the story of one country that did.

A Cuban home garden

The small island of Cuba has for long been under severe economic sanctions imposed by its neighbour, the US. It had depended on the Soviet Union and the Eastern Bloc, with which it exchanged its sugar for fertilizers, oil, and grains. When the Soviet Union collapsed in 1989, Cuba's economy and food security were seriously threatened. The people were facing starvation.

The Cuban government's answer was a major shift to organic farming. Today, the country is hailed as a success story in organic farming. The strategy was to transform derelict city plots into well-funded vegetable gardens under the supervision of organic farming associations. Today thousands of gardens across Cuba produce organic vegetables and many other crops. Organic farming on small family plots currently provides employment to 326,000 people out of a population of 12 million.

The gardens use organic compost and mulch instead of chemical fertilizers, biological pest control methods instead of chemical pesticides, and other eco-friendly techniques. Giant greenhouses produce vegetables in all seasons. Organic sugar and coffee are now being produced.

The organic gardening associations bring together farmers, farm managers, field experts, researchers, and government officials to develop and promote organic farming methods. Its aim is to convince Cuban farmers and policy-makers that the country's previous high-input farming model was too import-dependent and environmentally damaging to be sustainable, and that the organic alternative has the potential to achieve equally good yields.

In recognition of the remarkable transformation of Cuban agriculture, the apex Cuban Organic Farming Association was awarded the Right Livelihood Award in 1999. The Cuban success in shifting to organic farming points the way to the rest of the world.

Source: Zytaruk (2003)

If you want to shift to organic farming, you will not be alone. Worldwide, there is an active movement towards organic farming and you can get advice and support from many groups. Many young and educated people have taken to this practice. There are organic farmers in India who get as much (if not more) yield in comparison to those engaged in chemical farming (read Box 10.3).

Even coffee growers in India prefer organic manure (read Box 10.4).

Box 10.3 Solution: Small, organic, and profitable!

With agriculture in crisis and small farmers committing suicide, who would believe that one can make more than a million rupees from one hectare of land? Well, Ramesh Chandar Dagar in Sonipat, Haryana can show you how.

Dagar used to wonder whether a small farm could be made viable. Four years ago, he started experimenting on one hectare within his land. His experiment with integrated organic farming has been a success. He now practises it in all the 44 hectares he owns and is busy spreading the message to others.

Integrated organic farming is about more than just avoiding chemicals. It includes many other practices such as bee-keeping, dairy management, biogas production, water harvesting, and composting. The key element in Dagar's experimental plot is the cyclic, zero-waste approach. Paddy waste goes into vermicomposting as well as mushroom production, dung is fed into the *gobar* gas plant with the sludge going into the composting pit, and so on. The excess compost is sold.

A pond collects rainwater that is used to wash the buffaloes and for other purposes. Fish are cultivated in the pond and regularly harvested. Bee-keeping increases crop output through effective pollination and the honey too has a good market. The farm uses solar power for pumping and lighting.

Dagar grows various seasonal vegetables, fruits, paddy, wheat, mushrooms, and flowers. He has begun growing lettuce, baby corn, and strawberry for export. The sale of compost and honey brings in the maximum income. He keeps experimenting with new crops and new ideas.

Dagar has set up the Haryana Kisan Welfare Club and its 5000 members are now busy spreading the message of integrated organic farming. Four thousand farmers gathered at Sonipat in early 2004 to learn about organic farming.

'With a bit of hard work and understanding of nature, any farmer can earn a minimum of Rs one million per annum. I do not understand why everyone is running after a job,' says Dagar. This Haryana farmer is surely a model to follow!

Source: Jamwal (2004)

Box 10.4 Connections: Weather change in Brazil, forest decline in Karnataka!

Brazil is the largest grower of coffee in the world, accounting for 30 per cent of the total production. However, droughts and frosts often destroy Brazil's coffee crops. The frequency of such attacks have increased over the last few decades (an effect of global warming?) and in the mid-1990s, Brazil lost half its output.

World coffee prices shot up and it was an opportunity for countries like India (which produce just 3–4 per cent of the total) to increase their production for export. Three contiguous districts in south India—Kodagu in Karnataka, the Nilgiris in Tamil Nadu, and Waynad in Kerala—account for 57 per cent of India's production. The growers here increased their plantation areas, but needed more manure.

The Indian growers prefer organic manure, which gives the coffee a distinctive taste and value. They wanted huge quantities of dung, but did not want to use the valuable land to rear cattle. The dung had to come from outside.

The ideal source turned out to be a cluster of villages at the periphery of the Bandipur National Park in Karnataka. The farmers here were growing low-yielding millets and pulses. The expenses were high, but the income was low.

The coffee growers had somehow located the right place to get the dung—agriculture was unprofitable, cattle were abundant, and as a bonus free and unlimited fodder was available—from the Bandipur National Park!

Soon, the villages became dung factories. Every day hundreds of cattle were taken into the forest to graze and the dung was collected in the evening. Lorry loads of dung were sent to the coffee estates and the villagers made good money. A whole industry came into being with dung collectors, agents, lorry owners, etc.

Since the demand for dung was insatiable, the villagers bought more cattle. They bought the cattle cheap, since they did not need the milk-yielding varieties, but just any that 'could amble through the forests and defecate'. They did not even retain any dung for their own farming. With the money they earned from the dung, they bought subsidized fertilizers!

The forest periphery was now under a double threat. To the perennial fuelwood collection was now added heavy grazing by the cattle. Tree regeneration was affected, the Park's wildlife had less forage, and degradation set in.

Meanwhile, Brazilian coffee is doing very well again and the Indian coffee industry is in crisis. The dung trade, however, has acquired a momentum of its own. The dung now goes to the ginger, chilli, and tea plantations of Kerala. When the demand goes down, the villagers sell the cattle for meat, again to Kerala.

The poor villagers of Bandipur have found a new livelihood—at the cost of the forest. Should we let them continue? Or should we ban the grazing and take away their incomes? Can we ever implement a ban on grazing? Hard questions with no easy answers!

Source: Sethi (2004)

Organic farming is used as a general label for the following systems of agriculture (Alvares 1996):

- Reliance on the soil's natural fertility, enhanced through composting and vermiculture: While simple composting is the natural conversion of organic matter into manure, vermiculture uses earthworms to speed up the process.
- Natural farming or no-tillage farming: Pioneered by Masanobu Fukuoka of Japan (Fukuoka 1978).
- Biodynamic farming: Exploiting bio- and solar rhythms in farming, based on the ideas of Rudolf Steiner (Tompkins and Bird 1991).
- Biointensive farming: Intensive garden cultivation using deep-dug beds.
- Permaculture: An approach that goes beyond organic farming, developed by Bill Mollison and David Holmgren (read Box 10.5).
- Low External Input Sustainable Agriculture (LEISA): This was developed by Dutch farmers and scientists committed to organic farming.

Box 10.5 Solution: Permanent agriculture?

In the 1960s, environmental issues became part of the public agenda, especially in industrialized countries. People became concerned about problems like the loss of biodiversity, overconsumption of non-renewable resources, pollution of air, water, and soil by chemicals and waste products, etc. In response to these issues, two Australians, Bill Mollison and David Holmgren, developed the concept of Permaculture, short for Permanent Agriculture.

Permaculture is about designing sustainable human settlements. It is a philosophy and an approach to land use, which weaves together microclimate, annual and perennial plants, animals, soils, water management, and human needs into intricately connected, productive communities.

The main features of Permaculture can be summarized as follows:

- It is a system for creating sustainable human settlements by integrating design and ecology.
- It is a synthesis of traditional knowledge and modern science, applicable to both urban and rural situations.
- It takes natural systems as a model and works with nature to design sustainable environments, which will provide basic human needs as well as the social and economic infrastructures, which support them.
- It encourages us to become a conscious part of the solutions to the many problems, which face us, both locally and globally.

Mollison set up the Permaculture Research Institute in Australia and began offering a two-week Permaculture Design Course. Soon, Permaculture spread to other parts of the world, especially to the UK where the movement is still active.

In India, some attempts to propagate the concept were made in Hyderabad. The Annamalai Reforestation Society established a small Permaculture farm in the south Indian town of Tiruvannamalai.

Source: Morrow (1993)

While agriculture is going through a crisis, fisheries, the other major source of food, also have their share of problems.

How important are fish as a food source and where do fish come from?

About two million people, mostly in developing countries, depend on fish as their main source of food. Fifty five per cent of this fish comes from the ocean, 33 per cent from aquaculture, and 12 per cent from inland freshwater fishing in rivers, lakes, reservoirs, and ponds. Aquaculture is the artificial production of fish in ponds and underwater cages. A third of the world's fish harvest is used as animal feed, fishmeal, and to obtain fish oils.

It is a wrong notion that the vast open sea holds an unlimited supply of fish. Most of the global, commercial fish catch (80–90 per cent) comes from coastal waters within 300 km of the shoreline.

Is there a fisheries crisis?

If you visit the world's largest fish market in Tokyo, you will be amazed at the incredible variety of fish on sale coming from every corner of the Earth. The same is true of fish markets and large restaurants in the urban centres of rich countries. This is, however, a misleading picture of abundant fish in the world.

Industrial fishing technology employed in modern trawlers operating in all the seas locate and catch the last remnants of all varieties of fish stocks. The world's fisheries are in fact in deep crisis. An environmental and social catastrophe is in the making, but most consumers of fish do not perceive it.

A massive increase in global fishing began in the 1950s and 1960s with the rapid induction of new technology: factory trawlers, satellite positioning, acoustic fish finders, spotter planes, huge nets, etc. Soon the rate of harvest exceeded the rate of fish population growth.

The first sign of a problem was the collapse of the world's largest fishery, the Peruvian anchovy in 1972. The decline in the North Atlantic started in the mid-1970s and over the next two decades most of the cod stocks in New England and eastern Canada had collapsed. Centuries of fishing tradition came to an end.

The global fish catch increased five-fold between 1950 and 1980. The capacity of the fishing fleet, however, expanded twice as fast as the rise in catch. Now there are too many boats chasing too few fish. Of marine fish stocks, 80 per cent are fully exploited, overexploited, or in a state of depletion. Large-scale fisheries are very likely to collapse within a few decades in most areas.

We are now fishing down the food web. Once the larger fish species are exhausted, the fleets start catching the smaller fish. These being often the prey of the larger fish, there is further decline of the latter. Ultimately, we will be fishing the smallest species and the zooplankton (see Chapter 4 of this book).

Too many boats! Too few fish!

Meanwhile, we are also removing the species living on the ocean floor through bottom trawling, which is like clear-cutting a forest. This removes the species at the base of the food web.

For each species of fish, there is a Maximum Sustainable Yield (MSY). This is the amount that can be harvested annually, leaving enough breeding stock for the population to renew itself. MSY is the amount we can catch every year indefinitely. It is not easy to estimate MSY, but fisheries experts do have some idea of the figure for each species. Implementing any such limit is also a difficult task. What is clear, however, is that we have exceeded the MSY in the case of many important species.

The tragedy is that the massive harvesting of fish was not driven by nutritional needs, but by the demand for luxury foods or livestock feed. The rich countries actively encouraged such unsustainable exploitation through heavy subsidies to their fishing industry. Apart from overexploitation, fisheries are threatened by pollution of water bodies, climate change, and destruction of mangroves and coral reefs.

How far can aquaculture compensate the fishery losses?

As mentioned above, aquaculture is the artificial cultivation of fish under controlled conditions in ponds, lakes, reservoirs, or underwater cages. Aquaculture gives high yields in small volumes of water, does not use much fuel, is very profitable, and brings in valuable foreign exchange for poorer countries.

On the other hand, aquaculture needs large inputs of land, destroys mangroves on the coast, requires special feed and antibiotics, produces toxic effluents, and contaminates water sources. The fish are vulnerable to disease and the tanks become so contaminated that they have to be abandoned after five years or so. Moreover, the increased fishing of small species for feeding large carnivorous aquaculture species could end up destroying both.

Lured by quick and high returns, companies and individuals rushed into the aquaculture business in India and in South East Asia in the 1990s. Epidemics, public protests, and court rulings (see Chapter 20 of this book) have put a brake on coastal aquaculture in India and other places.

Inland aquaculture, however, is still a thriving industry. Production increased five-fold between 1984 and 2001. China is the largest producer, followed by India and Japan. In the long term, however, aquaculture appears to be an unsustainable industry. Apart from other heavy inputs, the fish are fed with large amounts of fishmeal. In fact, the fishmeal consumed is often more in weight than the fish harvested!

What is the state of India's fisheries?

The current annual fish production is about six million tons, with the share being roughly equal for marine and inland fisheries. About 460,000 tons of fish are exported. Ten million people depend directly or indirectly on fishing for their livelihood.

The estimated MSY for India's marine fisheries is about 3.7 million tons. Since we have not reached this figure, we might think that there are still more fish to catch. However, the catch per vessel has been going down in all the coastal states. This is because there are about 35,000 small mechanized boats and two million artisanal crafts competing for the fish.

To add to the problems, the government announced a new deep sea fishing policy in 1991, opening up Indian seas to foreign vessels in the name of joint ventures. When industrial fleets have depleted the huge fisheries of the Atlantic and the Pacific, how long will Indian stocks last?

India has about 1000 freshwater fish species, but this diversity is under threat. Overharvesting, competition from newly-introduced exotics, and pollution are taking their toll.

What is the contribution of aquaculture in India? India has an estimated area of 1.2 million hectares of brackish water, of which over 80,000 hectares are under shrimp aquaculture. As mentioned earlier in this chapter, epidemics, public protests, and a Supreme Court ruling (see Chapter 20) have halted coastal aquaculture, but inland aquaculture continues, primarily for export.

What has been the impact of the UN Convention on the Law of the Sea (UNCLOS) on fisheries?

As we saw in Chapter 4, UNCLOS gives to each maritime country an Exclusive Economic Zone (EEZ) extending 200 nautical miles from its coast. This is an opportunity and a responsibility for the country to manage its fisheries well.

Few countries, however, have used the UNCLOS provisions to conserve and improve their fisheries. On the contrary, lured by hard currency, some have sold away their rights to foreign fleets, which have then plundered the EEZ. This is done under the UNCLOS provision that a country that cannot exploit its EEZ should allow others to do so.

The ocean shelves, which account for less than seven per cent of the ocean area, support about 85 per cent of global fish catch. Most of the ocean shelves are in the EEZ of countries and are now being overfished.

What is the way out for global fisheries?

Some of the recommendations given by experts are:

- Adopt an ecosystem-based approach by considering the food needs of the key fish species.
- Eliminate any fishing gear that destroys the ocean floor or catches non-target species as bycatch.
- Establish marine reserves as no-fishing zones to help populations recover.
- Move away from the notion that the ocean will always give us all that we demand from it.
- Employ traditional aquaculture integrated with agriculture (for example, raising fish in paddy fields at appropriate seasons).

What are minerals and what is the environmental impact of mining?

A mineral is any substance that is naturally present in the Earth's crust and is not formed from animal or vegetable matter. The Earth's geological processes have formed these minerals over millions or billions of years and hence they are non-renewable.

Mining is the process of extracting and processing minerals. Over 100 minerals are mined and these include metals like gold, iron, copper, and aluminium and non-metals such as stone, sand, and salt. Apart from minerals, another major material that is mined is coal.

Underground mining has little direct effect on the environment, but it can cause long-term problems like subsidence and pollution of aquifers. Moreover, workers are at great risk in underground mines. Accidents like flooding and collapses are common and the work itself causes severe health problems like respiratory illnesses.

Surface mining destroys all vegetation in the area and pollutes the landscape with the dust that is thrown up. Once the available material is mined out, large craters are left behind. When hills that act as watersheds are mined away, the water tables go down, as in the case of the Aravallis in Rajasthan (read Box 10.6).

Box 10.6 Crisis: Mining vs *jal, jangal, jameen*

The Rajasamand Lake in Rajasthan had not dried up for at least 300 years. However, this did finally happen in 2001. The likely reason: a decade of marble mining in the Rajnagar area.

The Aravalli Hills, spread across Haryana, Rajasthan, and Gujarat, are the lifeline of the three states as they control the climate and drainage systems of the region. The hills act as a watershed for the region. Unfortunately, the hills are also repositories of immense mineral wealth, including talc, marble, and granite.

Mining and related industries employ about 175,000 workers and 600,000 others are indirectly dependent on mining operations. In Rajasthan alone 9700 industrial units are connected with mining.

Forest cover has been depleted by 90 per cent over the past 20 years since large-scale mining began. When the mines reach below the underground water level, a cone of depression is formed that sucks water from the surrounding areas, drying up wells and affecting agriculture. Several studies have pointed out that the natural drainage system and the groundwater table of the entire region have been badly affected over the years. Pollution levels have also increased.

Studies have shown that labourers are not provided with any health care. Lung diseases like tuberculosis and silicosis are common, making the labourers invalids or even killing them by the age of 40. Child labour constitutes 10–15 per cent of the workforce and women workers 30–40 per cent, and their condition is the worst.

In November 2002, the Supreme Court imposed a blanket ban on mining activities in the Aravallis. The Court based its decision on the report submitted by a committee, which had visited the hill areas to study the impact of the mining activities. The report referred to the devastation caused by the free rein given to unscrupulous mine owners. The basic premise of the ban was the argument that the Aravallis came under the category of forestland.

The Court ruling closed 9900 mines and brought the whole economy to a halt. The closure perhaps affected a large population in the state, which had already been experiencing drought for the fourth year in a row. Finding itself in a fix, the state government filed an appeal and the Court lifted the blanket ban, allowing mining in areas where specific permission had been obtained.

Environmentalists have alleged that the mine owners generally exploit much larger areas than are legally allocated to them. They fear that all the minerals in Rajasthan will be exhausted within 50 years if the mining goes on at the current rate. According to them, mining has affected the water, the forests, and the land. While mining has led to the depletion of water in wells, mining waste has destroyed fertile land.

In fact, it was the Tarun Bharat Sangh, the water conservation NGO of Rajendra Singh (see Chapter 7, Box 7.4) that had first moved the Supreme Court and secured a directive in 1992 against mining in the Aravallis. The Court had at that time imposed a ban on mining inside the Sariska Tiger Sanctuary area.

Rajendra Singh is conscious of the immediate fallout of the closure of mines, such as unemployment and a halt to economic activities in certain areas. 'In the long run it will benefit the poor. Once the displacement stops people would resume farming activities. The wells will be recharged,' he says.

In March 2004, the Supreme Court held that the ban imposed by it in 2002 on the mining activities in the Aravalli Hills would continue. The Court constituted a high-level monitoring committee to suggest ways and means of restoring the overall ecological balance of the hills. It also made environment assessment clearances mandatory before renewal of mining leases.

And so it continues: a typical story of development, environmental effects, industrial interests, confusion in the government, etc.

Source: Sebastian (2003) and the websites *www.hindu.com/2004/03/19/stories/2004031906801500.htm* and *www.hinduonnet.com/thehindu/2002/12/14/stories/2002121407120500.htm*

The processing of the mined material, often done on site, using in many cases mercury, cyanide, and large quantities of water, pollutes rivers and other water bodies. The waste material like slag is often far greater in quantity than what is usable, and is left behind as unsightly, unstable, and dangerous heaps.

The mining of precious metals is today more intense and widespread than in centuries past with far-reaching consequences. Of particular concern is heap-leach gold mining, in which rivers of cyanide are poured over huge piles of low-grade ore to extract the metal.

Heap-leach mining is on the increase and has already caused several serious accidents. Two examples are:

- In 1984 on the Ok Tedi Island in New Guinea, 1000 cu. m of concentrated cyanide were released into a river and the ecosystem was devastated. This gold and copper project, which is tearing down a whole mountain, has already caused extensive environmental damage. It has also destroyed the culture and lifestyle of the native Wopkaimin people.
- In 2000, at the Baia Mare gold mine in Romania, the dam holding the heap-leach waste broke, releasing 80 million litres of cyanide into the Tisza River. The cyanide flowed 500 km into Hungary and Serbia.

Box 10.7 Make a difference: Observe World Food Day and World Fisheries Day

World Food Day (WFD) is a worldwide event designed to increase awareness, understanding, and informed, year-around, long-term action on the complex issues of food security for all. WFD is observed on October 16 every year in recognition of the founding, in 1945, of the Food and Agriculture Organization (FAO) of the UN, the lead agency of the UN system for technical assistance, research, and policy making in world agriculture, fishing, forestry, and rural development.

WFD is a 'tripartite' effort by private voluntary organizations, governments, and the international system that works in many different ways to build public will in the struggle for hunger alleviation and world food security. Each year WFD highlights a particular theme on which to focus activities.

You can observe WFD by increasing awareness, understanding, information, services, support, advocacy, and networking with regard to food issues. You can get more information and help from the FAO website.

The World Forum of Fish Harvesters and Fish Workers (WFF) has declared its foundation day, November 21, as World Fisheries Day. The Forum's main aim is to save the lives of all those who are dependent on the oceans, by promoting sustainable fisheries and by preserving marine ecology. World Fisheries Day is observed with the idea of highlighting preservation of marine ecology, creating and sustaining public awareness, and initiating and strengthening action at various levels.

World Fisheries Day focuses on both positive and negative issues. Achievements are celebrated and current concerns, visions, and aspirations for the future are projected through workshops, rallies, public meetings, symbolic actions, cultural shows, street plays, exhibitions, and art forms like music, dance, poetry, etc., using different modes of communication, such as the print and electronic media. World Fisheries Day is not only for the fish-harvesters and fish-workers, but for all who care for and wish to nurture this planet.

Source: The websites *www.fao.org/wfd/* and *www.wffp.org/*

A quick recap

- The world's land surface is continuously being degraded—the soil is becoming unhealthy, and desertification is increasing.
- There is enough food available to feed everyone in the world, but poverty keeps millions of people hungry.
- While the Green Revolution increased crop production dramatically, it has led to many problems in the long term, leading to a crisis even in states like Punjab.
- The wise option for agriculture seems to be to shift gradually to organic farming, minimizing or altogether avoiding chemical fertilizers and chemical pesticides.
- There are many examples of successful adoption of organic farming.
- There is a global fisheries crisis with most fish stocks in collapse or decline.
- There are too many boats chasing too few fish and curbs on fishing are a must if we want populations to recover.
- Mining has become more intense and widespread and causes many environmental problems.

What important terms did I learn?

Aquaculture
Biodynamic farming
Biointensive farming
Desertification
Green Revolution
Heap-leach mining
Maximum sustainable yield
Natural farming
Organic farming
Permaculture

What have I learned?

Essay question

Write an essay on the Green Revolution answering questions such as: Under what conditions was it introduced? What benefits did it bring? What are the problems now being faced by the farmers who had adopted the Green Revolution package? What is the likely long-term impact of the Green Revolution on the environment?

Short-answer questions

1. In what ways is the world's land surface getting degraded?
2. What are the sources of our food and is there enough food in the world for all?
3. Explain in brief the basic principles of organic farming.
4. Why is there a global fisheries crisis?
5. What are the advantages and disadvantages of aquaculture?
6. What is the environmental impact of mining?

How can I learn by doing?

Spend some time at an organic farm and at a 'normal' farm following the Green Revolution approach. Study the methods used by both the farms and compare them. Write a report detailing the environmental and economic aspects. Describe also how the 'normal' farm can convert itself into an organic one.

What can I do to make a difference?

1. If you have some land or even just a terrace, grow vegetables using organic methods.
2. If you have more land and you cultivate any crops, vegetables, or fruits:
   (a) Plant only indigenous species.
   (b) Prefer polyculture to monoculture.

   (c) Practise composting or vermicomposting and use mulch.
   (d) Shift to organic farming.
   (e) Use drip irrigation.
   (f) Plant appropriate trees wherever possible.
3. If you have no space at all, start a community garden for vegetables and herbs.
4. Buy your food from local markets and small traders and not from supermarkets.
5. Buy only organically-grown food and if it is not available locally, form a consumer group to procure and distribute organic food. Learn from groups and individuals featured on websites such as *www.chennaiorganicfood.com.*
6. If you are a non-vegetarian, find out more about vegetarianism and, if convinced, become a vegetarian.
7. To save minerals and metals:
   (a) Buy durable products that will last long.
   (b) If you are buying a car, buy a small and efficient one.
   (c) Do not buy soft drinks in metal containers.
   (d) Repair and reuse old bicycles.
8. Observe any of the following days:
   (a) October 16 as World Food Day and November 21 as World Fisheries Day (refer Box 10.7).
   (b) June 17 as the World Day to Combat Desertification and Drought . (For information access the website *www.unccd.int.*)
   (c) December 23 as Farmer's Day
   (d) November 21–27 as National Land Resources Conservation Week

Where are the related topics?

- Chapter 3 covers land and desert as ecosystems.
- The Green Revolution has been referred to in Chapter 5 in the context of biodiversity.
- Chapter 20 discusses a public interest case and Supreme Court decisions with reference to aquaculture.

What is the deeper issue that goes beyond the book?

If India has excess food stocks that are rotting, why are millions still going hungry? Why can't we implement a 'Food for Work' Programme and distribute the excess food?

Where can I find more information?

- Alvares (1996) provides more information on organic farming in India.
- For a practical introduction to permaculture, read Morrow (1993)

What does the Muse say?

A poem by the American farmer, environmentalist, and poet, Wendell Berry

Sowing the seed,
My hand is one with the Earth.
Wanting the seed to grow,
My mind is one with the light.
Hoeing the crop,
My hands are one with the rain.

Having cared for the plants,
My mind is one with the air.
Hungry and trusting,
My mind is one with the Earth.
Eating the fruit,
My body is one with the Earth.

MODULE 5

Pollution

Prologue

Asking the right questions about waste

LOIS MARIE GIBBS

Just about everyone knows our environment is in danger. One of the most serious threats is the massive amount of waste we put into the air, water, and ground every year. All across the United States and around the world are thousands of places that have been, and continue to be, polluted by toxic chemicals, radioactive waste, and just plain garbage.

For generations, the main question people have asked is, 'Where do we put all this waste? It's got to go somewhere.' That is the wrong question, as has been shown by a series of experiments in waste disposal and by the simple fact that there is no away in 'throw away'.

We tried dumping our waste in the oceans. That does not work. We've been trying to build landfills that do not leak, but according to the Environmental Protection Agency all landfills eventually leak. We've been trying to get rid of waste by burning it in high-tech incinerators that only produce different types of pollution, such as air pollution and toxic ash. Even recycling, which is a very good thing to do, suffers from the same problem as all the other methods. It addresses waste after it has been produced.

For many years, people have been assuming, 'It's got to go somewhere,' but now many people, especially young people, are starting to ask why.

Why do we produce so much waste? Why do we need products and services that have so many toxic by-products? Why can't industry change the way it makes things so that it stops producing so much waste?

When you start asking these questions, you start getting answers that lead to pollution prevention and waste reduction instead of pollution control and waste management. People, young and old, who care about pollution prevention, are challenging companies to stop making products with gases that reduce ozone in the ozone layer and contribute to the threatening possibility of global warming. They are asking why so many goods are wrapped in excessive throwaway packaging. They are challenging companies that sell pesticides, cleaning fluids, batteries, and other hazardous products to either remove the toxins from those products or take them back for recovery or recycling rather than disposing of them in the environment. They are demanding alternatives to throwaway materials in general.

Waste issues are not simply environmental issues; they are also tied up with our economy, which is geared to producing and then disposing of waste. Somebody is making money from every scrap of waste and has a vested interest in keeping things the way they are...

From my personal experience, I know that decisions made to dump wastes at Love Canal and in thousands of other places were not made purely on the basis of the best available scientific knowledge. The same holds true for decisions about how to manage the wastes we produce today and how to produce less waste.

We live in a world that is shaped by decisions based on money and power. If you really want to understand what's behind any given environmental issue, the first question you should ask is, 'Who stands to profit from this?' Then ask, 'Who is going to pay the price?' You can then identify both sides of the issue and decide whether you want to be part of the problem or part of the solution.

These are excerpts from an essay entitled, 'We Have Been Asking the Wrong Questions About Wastes', by Lois Marie Gibbs in Miller (2004).

The story of Lois Gibbs is an inspirational one. From a housewife with two children near Love Canal, she became an activist when toxic chemicals began leaking into the soil and basements in her neighbourhood. Ultimately, she became a national campaigner for the proper handling of toxic chemicals. (Read about her and about the Love Canal case in Chapter 16.)

Lois Gibbs now runs the Centre for Health, Environment, and Justice (CHEJ) in Washington, D.C., in the US.

What is coming in Module 5?

In this module, we begin with various types of pollution—air, noise, water, soil, and marine pollution. We then discuss solid waste management in depth. The module ends with an account of natural and human-induced disasters as well as those that occur slowly over a long period of time.

CHAPTER 11

Air and Noise Pollution

There's so much pollution in the air now that
if it weren't for our lungs
there'd be no place to put it all.

Robert Orben

Pollution is so bad in New York that
I saw the Statue of Liberty holding her nose.

Jackie 'Moms' Mabley

What will I learn?

After studying this chapter, the reader should be able to:

- describe the primary and secondary pollutants that cause outdoor air pollution;
- trace the sources of the pollutants and describe their effects;
- recognize the major role played by the automobile in causing air pollution and other environmental problems;
- explain the term smog;
- appreciate the impact of air pollution on health, especially among children;
- recall the measures taken to reduce and control outdoor air pollution;
- recount the options available for making cleaner cars and make comparisons between them;
- trace the causes and effects of indoor air pollution;
- describe the state of air pollution in India;
- list the measures taken to regulate vehicular emissions in India;
- list the sources of noise pollution;
- explain how noise is measured and give examples of intensity levels of common sources of sound; and
- describe the effects of noise pollution and control measures against it.

Box 11.1 Crisis: Every breath you take

Eight-year-old Moni caught pneumonia during the winter rains three years ago and she is now an asthmatic, visiting the government hospital every week. Her father, who earns Rs 60 per day, spends Rs 80 per week on her treatment. The family of five cannot buy rations and the house rent is six months overdue.

Moni is not alone. Seven-year-old Umesh Kumar, one-year-old Mushtab, and many more children among the poor of Kanpur city are all asthmatics, condemned to long periods of medical treatment and unaffordable expenses. Anuj Rawal is perhaps among the few happy people in the city, since his medical shop sells Rs 50,000 worth of asthma medicines every month!

Kanpur is one of the most polluted cities in India. It is home to many textile mills and leather units. The most popular form of transport is the rickety three-wheeler tempo, which spews out volumes of smoke from old engines. Dust is thick in the air and the smog never goes away. The hospitals are full of juvenile asthmatics. The story is the same in the shanties of Delhi, the slums of Mumbai, or the colonies of Chennai.

The World Health Organization (WHO) estimates that 10–15 per cent of Indian children in the five–eleven-year-old age group suffer from asthma. It costs an average of Rs 300 per month to buy a child's asthma medicines. Since there is little awareness of asthma among the poor and even among health officials, the disease often remains undiagnosed. Even if it is diagnosed, the child is given steroids because they are cheap. The intake of steroids is bound to have long-term adverse effects on their health.

Several studies have established the link between pollution and asthma. According to a Delhi study, the city's polluted air is responsible for a 40 per cent increase in asthma cases. More people were admitted to the emergency ward on days when the air pollution levels were high.

Moni, Umesh, Mushtab, and all the children in our polluted cities need clean air. How will they ever get it?

Source: Varshney (2004)

What do we learn from the juvenile asthma case?

Indira Gandhi told the 1972 Stockholm Conference on the Human Environment that poverty was the worst form of pollution. The juvenile asthma case shows that when poverty and pollution come together the result is doubly tragic.

Air pollution and noise pollution can have serious effects, and these are the topics of this chapter.

What is air pollution?

Air pollution is said to exist if the levels of harmful gases, solids, or liquids present in the atmosphere are high enough to affect humans, other organisms, buildings, monuments, etc. Pollution may have natural causes, such as a forest fire or a volcanic eruption. In this chapter, however, our primary concern is with human activities, which are today responsible for most of the air pollution. Further, much of the pollution caused by humans is concentrated in thickly populated urban centres (read Box 11.2).

Box 11.2 Hope: Saving a choked city

Just living and breathing the air here is equivalent to smoking two packets of cigarettes a day. It is the second largest city in the world and it also has the second worst levels of air pollution.

Mexico City is located in a bowl-shaped valley with mountains on three sides. It has a population of more than 20 million, which is constantly on the increase. There are more than three million vehicles that emit four million tons of pollutants annually. Many of the vehicles are old, and emit more pollutants than normal.

Emissions from 36,000 industries and leakage of LPG (liquefied petroleum gas) from thousands of containers add to the air pollution. The air also contains particles of dried faecal matter from the millions of tons of sewage dumped on the land near the city.

The Mexican government has taken several steps to reduce the air pollution:

- The hillsides have been reforested to reduce the particulate matter produced by wind erosion—in all, 25 million trees have been planted.
- Some refineries have been closed and others upgraded.
- Cleaner fuel has been imported from the US, unleaded petrol has been introduced and old vehicles have been banned.
- The public transport system is being improved and on one day every week the use of personal cars is prohibited.

With these measures, there has been some improvement, but there is still a long way to go. Meanwhile, Indian megacities like Mumbai and Kolkata are trying to catch up with Mexico in air pollution. Delhi is different (read Box 11.5).

Source: Raven and Berg (2001) and Enger (2004)

Primary air pollutants are harmful chemicals that are directly released from a source into the atmosphere. Secondary air pollutants are also harmful chemicals, but they are produced in the atmosphere from chemical reactions involving primary pollutants.

Primary air pollutants include the following:

- Particulate matter, which includes both solid particles and liquid suspensions. Soil particles, soot, lead, asbestos, and sulphuric acid droplets are examples.
- Oxides of carbon, nitrogen, and sulphur.
- Hydrocarbons like methane and benzene.

Secondary air pollutants include the following:

- Ozone: It is a form of oxygen and is a pollutant in the troposphere or the layer of the atmosphere closest to the Earth's surface. It is a beneficial component of the stratosphere which extends from 10–50 km above the Earth (see Chapter 19 of this book).
- Sulphur trioxide: This is formed when sulphur dioxide reacts with oxygen. In turn, sulphur trioxide combines with water to form sulphuric acid.

What are the causes of outdoor air pollution? The sources of outdoor air pollution are:

- Burning of fossil fuels:
    - in automobiles, domestic cooking, and heating.
    - in power stations and industries (primarily the chemical, metal, and paper industries).
- Mining activities leading to dust as well as fires (read Box 11.3).
- Burning nuclear fuels, biofuels, tropical rainforests, wastes of all kinds.
- Natural emissions from animals and decaying organic matter.

Box 11.3 Crisis: City of fire

In 2000, the town's temple snapped in two and flames leapt out from underneath, spewing noxious gases. The underground fires that had been raging for several decades had reached the residents' doorsteps!

Jharia in the state of Jharkand is now a town surrounded by fires. Raging in the coalmines underground, the fires are now visible on the surface. At night, the smoky haze makes Jharia resemble a cremation ground. Every possible outlet from the mines spits fire. Danger lurks at every step.

Coal fires occur naturally like forest fires. Lightning, forest fires, or frictional heat inside the Earth's crust can trigger the spontaneous combustion of coal. These fires burn away coal, raise the atmospheric temperature, and cause land subsidence. Smoke from these fires contains poisonous gases, such as oxides and dioxides of carbon, nitrogen, and sulphur.

Jharia has a 100-year history of coal fires. Neglected for long and stoked by relentless mining, the fires have consumed 40 million tons of India's best coking coal and rendered 1800 million tons out of bounds. The fires now cover about 17 sq km. There have been many gas eruptions and accidents and many miners have died.

The estimate of the cost of extinguishing the fire ranges from Rs 4–10 billion. It may be worthwhile, because coal worth Rs 30 billion can still be extracted in the area. Bharat Coking Coal Ltd, the public sector company that owns the mines, is said to have spent a billion rupees so far to put out the fires, without much success.

The state government has been considering evacuating the 300,000 people of Jharia, abandoning 100,000 houses. For the residents of Jharia, mere existence now is hell on earth. They live in constant fear of a major subsidence that could cause the entire town to collapse. Yet, they do not want to leave the town.

Will the entire town be evacuated so that mining can go on? Or will the fire be put out incurring high costs? Or more likely, will the precarious existence of the people continue without any action being taken?

Source: Mahapatra (2002)

Temporary, but severe, air pollution can occur due to disasters, such as earthquakes, volcanic eruptions, dust storms, toxic gas leaks (for example, the Bhopal Gas Tragedy), and armed conflicts. Even festivals (Diwali with its crackers, for example) can create air pollution. Dust storms typically occur in desert areas and from there they can spread to places thousands of kilometres away (read Box 11.4).

Box 11.4 Crisis: The fifth season

Schools were closed, many flights had to be cancelled, and hospitals were crowded with patients suffering from breathing difficulties. This was the scene in Seoul, South Korea on April 12, 2002.

The cause was a huge dust storm that had originated in China. It marked the arrival of the 'fifth season', one of severe dust storms that the Koreans and the Chinese have come to fear. Such dust storms have now become a regular feature between winter and spring.

In Seoul, the amount of particulate matter in the air is normally 70 micrograms of dust per cu. m. If this figure reaches 1000 micrograms, many would have breathing problems. On that day, the figure was 2070 micrograms per cu. m.

Apart from affecting the health of many, such dust storms lead to worker absenteeism, lower retail sales, problems in dust-sensitive factories like the ones making computer chips, etc. Hence, they are a threat to the economy. On occasion, the storms from China have even reached the US.

The storms of course wreck havoc in China itself, including the capital Beijing. The real burden of the storms, however, falls on the Chinese farmers. The dust and sand destroy crops, damage trees and orchards, kill animals, fill up water wells, and blow away fodder. Perhaps the worst effect is that they take away the topsoil.

Why do these storms occur? In its efforts to feed a population of 1.3 billion people, China is overgrazing its rangelands, overploughing its croplands, overpumping its aquifers, and overcutting its forests. The land is becoming barren in northern and western China and the strong winds of late winter and early spring generate these huge dust storms. They carry away the fine particles and deposit them in the lungs of bewildered Koreans or even Americans!

Efforts to halt the desertification through afforestation have had only limited success. The key seems to be relieving the pressure on the land created by the grazing of 290 million sheep and goats. How will the Chinese do it?

Source: Brown et al. (2002)

Industries and automobiles are by far the main contributors to air pollution across the world.

What is the role of automobiles in creating air pollution and other environmental problems?

The automobile is the symbol of modern science and technology and is indispensable for transporting people and goods. Environmentalists, however, believe that the automobile is among the worst inventions of humankind, perhaps next only to the nuclear bomb.

The private automobile is one of the most desired items of consumption today and demand for it seems insatiable. From about 50 million in 1950, the number has increased to 530 million in 2002. The automobile industry dominates the economy of many countries. The US alone has 25 per cent of all the cars in the world. Developing countries like India and China actively promote the production and private ownership of cars.

The automobile contributes to a range of environmental problems: increasing air and noise pollution, adding to solid waste, accelerating global warming, taking a heavy toll on human life through accidents, using up natural non-renewable resources like oil and metals, etc.

Lead pollution caused by automobiles has been a serious problem. Lead was added to petrol to prevent 'knocking' in the engine. It is, however, extremely poisonous and tends to accumulate

People in cities wearing gas masks in the future

in most biological systems. Excessive accumulation of lead in the body can result in paralysis, blindness, and even death. It can also affect the mental development of unborn children. The addition of lead to petrol is now banned in most countries.

Automobiles need roads and highways, the construction of which has many adverse environmental effects. They use up land and consume resources like steel and cement, which again require heavy energy inputs to produce and also pollute the environment. Congestion and traffic jams cause enormous loss of man-hours as well as fuel. In cities, automobile emissions are a major cause of smog.

What is smog?

Smog is a form of outdoor pollution and the term was originally used to describe a combination of smoke, fog, and chemical pollutants that poisoned the air in industrialized cities like London. Now the term refers to the effects of air pollution not necessarily associated with smoke particles. It is used to describe air pollution that is localized in urban areas, where it reduces visibility.

What is common now in cities is photochemical smog formed by chemical reactions between sunlight, unburnt hydrocarbons, ozone, and other pollutants. Photochemical smog can be seen as a hazy shroud on warm days in cities like London, Tokyo, Los Angeles, Rome, and in many Indian cities. It is an irritant for humans and very toxic for plants. In developing countries, leaded fuel, old engines, industrial emissions, and the burning of fuelwood combine to create severe smog.

What are the effects of outdoor air pollution?

At low levels air pollutants irritate the eyes and cause inflammation of the respiratory tract. If the person already suffers from a respiratory illness, air pollution may lead to the condition becoming chronic at a later stage. It can also accentuate skin allergies.

Many pollutants also depress the immune system, making the body more prone to infections. Carbon monoxide from automobile emissions can cause headaches at lower levels and mental impairment and even death at higher levels.

Particulate matter can reduce visibility, soil clothes, corrode metals, and erode buildings. On a larger scale, air pollution leads to acid rain, ozone layer depletion, and global warming. All these effects are discussed in Chapter 19.

How can outdoor air pollution be reduced or controlled?

Outdoor air pollution can be reduced by adopting cleaner technologies, reducing pollution at source, implementing laws and regulations to make people pollute less, introducing appropriate transportation policies, etc.

Automobile emissions can be reduced through various measures:

- Making cleaner and fuel-efficient cars (read the discussion below)
- Using lead-free petrol in existing cars
- Introducing policies that encourage the building and use of mass transit systems and discourage the use of personal transport. For example, efficient and low-cost public transport,

congestion charges in city centres, separate lanes for car pools, heavy tax on personal cars, tax incentives on electric cars, etc.

- Shifting from diesel to natural gas (CNG) as a fuel for trucks and buses (read Box 11.5)

Box 11.5 Solution: Supreme Court clears the air

In the early 1990s, Delhi was the fourth most polluted city in the world. There was often a haze over it, and many people had breathing problems. Today, however, the air is much cleaner. How did that happen?

In 1985, responding to a public interest petition filed by M.C. Mehta, the Supreme Court of India issued a series of orders to the government to improve Delhi's environment. These included the mandatory use of the less-polluting CNG (Compressed Natural Gas) by all commercial trucks and other measures to reduce emissions. The government passed many environmental laws, but could not implement them.

Meanwhile, the air quality in Delhi continued to decline. The number of vehicles in Delhi was increasing three times as rapidly as the population. The inefficient two-stroke engines, in particular, were responsible for the heavy pollution. According to a World Bank study, the annual health cost of air pollution in Delhi was about Rs 3.5–14 billion.

In one of its first orders, the Supreme Court asked the government to shift all highly polluting industries out of Delhi. Though the process took several years, the order was implemented by 1997. Other orders included the phasing out of leaded petrol and of all commercial vehicles older than 15 years. The government found it very difficult to implement these orders, especially when elections intervened in 1998. However, the Court was firm and the orders were implemented by December 31, 1998.

Meanwhile the Court had asked the Union Ministry of Environment and Forests to implement its environment plan for Delhi through an Environmental Pollution (Prevention and Control) Authority (EPCA). This Authority, chaired by Bhure Lal, came up with several feasible recommendations that were accepted by the Court in July 1998. The important one was the conversion of all buses to CNG by the end of 2001.

For the next two years, there was considerable debate over this matter and many groups including the bus owners joined issue, blaming the government on many counts: not keeping them informed, accepting an untested technology, not setting up enough CNG stations, not arranging for enough conversion kits, etc. Finally, the long story came to an end in December 2002 with the conversion of all the buses to CNG. Gradually, the air became cleaner.

Many factors contributed to the 'success' in the Delhi case. Though the Supreme Court played a central role, most of the policies had been originated by the government. In fact, the use of CNG had been considered right from 1988. However, successive governments had lacked the political will to take bold action. The Supreme Court was ready to take hard decisions and make them stick. The Court relied a great deal on the EPCA. An active role was also played by M.C. Mehta (who had started it all), NGOs like the Centre for Science and Environment, and the press.

Delhi has now become the model in the region and many neighbouring countries want to try a similar approach. In fact, similar public interest petitions have been filed in Pakistan and Bangladesh.

Sources: Bell et al. (2004) and Brandon et al. (1995)

Particulate matter in the air can be reduced by:

- Fitting smokestacks with electrostatic precipitators, fabric filters, scrubbers, or similar devices
- Sprinkling water on dry soil that is being excavated during road construction.

How can we make cleaner cars?

There are options for making cars that would pollute less than current models do: electric cars, hybrid cars, and hydrogen cars being examples. All three types are under development and prototypes have been made and tested.

The electric car is run by an electrically powered motor. It produces virtually no emissions and its design is simple. The power usually comes from a battery inside the car. You can charge the battery using normal electric power. The battery needs recharging after running a certain distance (current models give about 80 km) and hence long trips are not possible. One can, however, move around within a city, for example.

The electric car seems to be an ideal solution, but there is a catch. If the electricity that charges the batteries comes from a conventional thermal power plant, then we have only transferred the emission problem to the power plant. To produce the extra power needed for charging, the power plant puts out more emissions. Thus, the electric car can be considered clean, only if the electricity comes from a renewable source. The disposal of the discarded battery also poses an environmental problem.

Though there is no net reduction in pollution, the electric car certainly reduces the air and noise pollution in the city. 'Reva', a two-seater electric car made in Bangalore is slowly catching on.

A hybrid car is a petrol-and-electricity driven vehicle. It starts using the petrol engine, but switches automatically to the electric motor at low speeds or while idling. At normal speeds both engines contribute power. When the car slows down, the wheels run the generator to charge the batteries. Japanese carmakers like Honda and Toyota have introduced hybrid models.

A car running on liquid hydrogen would be extremely clean, the only waste being water. In such a car, fuel cells (see Chapter 8) combine hydrogen with oxygen in the air to produce electricity. The hydrogen car has no emissions and gives better performance since there is no waste during idling.

Where is the catch? As we saw in Chapter 8, hydrogen has to be first produced and that process consumes energy. If the hydrogen is made using non-renewable energy, then we are back to square one. The hydrogen car will make sense only if we use renewables like solar energy to obtain the hydrogen.

There have been experiments to run cars on other less-polluting fuels, such as methane and biodiesel. It is, however, unlikely that the zero-emission wonder car running on low-cost, abundant, and renewable energy will arrive soon. Long before this happens, we will be forced to cut down on the number of cars on the road due to oil shortages, the reality of global warming, mounting solid waste, lack of space, and other problems.

What are the causes and effects of indoor air pollution?

We give importance and attention to outdoor air pollution, but we do not realize that indoor pollution can be equally damaging.

Pesticides, mosquito repellents, cleaning agents, etc., used in urban households can cause toxic conditions. Building materials like asbestos, glass fibre, paints, glues, and varnishes are all health hazards. They can cause irritation of the eyes and skin, respiratory ailments, and even cancer.

Air-conditioned rooms and offices cause a broad spectrum of health complaints, because the sealed space accumulates various contaminants. Cigarette smoke affects both smokers and non-smokers (read Box 11.6). The concentration of pollutants indoors may be five times more than it is outdoors.

Box 11.6 Crisis: A grave matter

It is a major cause of suffering and death among adults and it takes about 500 people every single hour to an early grave. What is shocking is that these deaths are completely preventable, if only people would stop smoking.

According to the World Health Organization (WHO), cigarette smoking causes the premature death of four million people every year and this figure is expected to reach 10 million by 2030. Most of this increase will occur in developing countries, China in particular. India is losing 800,000 to 900,000 lives every year to tobacco. The worldwide cost of treating smoking-related illnesses is about US$ 200 billion per year.

Cigarette smoke is a mixture of hydrocarbons, carbon dioxide, carbon monoxide, cyanide, and particulate matter. In addition, it contains a small amount of radioactive material from the fertilizer used for the tobacco plant. All these substances are dangerous when the smoke is inhaled regularly.

The connection between smoking and lung cancer is well-established. Smoking also leads to heart disease and cancer of the bladder, mouth, throat, pancreas, kidney, stomach, voice box, and oesophagus.

Passive smoking, the chronic breathing of smoke from the cigarettes of others, also increases the risk of cancer, besides causing respiratory problems. Passive smoking is particularly harmful for infants, young children, pregnant women, senior citizens, and those with lung disease.

There is the story of the man who read that smoking was bad and promptly gave up—reading! The nicotine in cigarettes is so addictive that only 10 per cent of smokers who want to quit the habit actually succeed in doing so.

Smoking has been declining in the US, Japan, and Europe. And tobacco companies are now aggressively exporting the habit to developing countries. Smoking is on the increase in many of the poorer countries.

The Indian government has taken several steps to curb the advertisement and sale of tobacco products. From January 1, 2005, the sale of tobacco products near all educational institutions was banned. There is also a Supreme Court directive that prohibits smoking in public places.

Source: Miller (2004) and Raven and Berg (2001).

The most common pollutants in urban interiors are cigarette smoke, gases from stoves, formaldehyde (from carpets and furniture), pesticides, cleaning solvents, and ozone (from photocopiers). Organisms like viruses, bacteria, fungi, dust mites, and pollens also thrive in the many ducts found in office buildings.

Urban indoor pollution results in ailments like colds, influenza, and upset stomachs. Since these are common ailments, their connection with indoor pollution is often missed. Indoor pollution can also cause eye irritations, nausea, depression, etc., collectively called the 'Sick Building Syndrome'. Thousands of Indian software engineers, who work for long hours in rooms with artificial lighting and air-conditioning, staring at cathode ray screens, must surely be facing these problems!

What are the levels of air pollution in India?

The air is severely polluted in many Indian cities, with excessive concentrations of suspended particulate matter (SPM), nitrogen oxide, and sulphur dioxide. The World Health Organization (WHO) had consistently ranked Delhi as the fourth most-polluted city in the world. The main cause for this was vehicular emissions followed closely by industrial pollution. (Box 11.5 describes how the air in Delhi has improved in recent times.)

The number of motor vehicles in India increased from 300,000 in 1951 to 37.2 million in 1997. Of these, over 30 per cent are in the 23 metropolitan cities. Delhi alone accounts for 8 per cent of the total.

The exponential growth of vehicles, outdated vehicle technology, bad fuel quality, poor maintenance of vehicles, poor traffic management and planning, etc., all contribute to vehicular pollution. The problem is compounded by the unwillingness on the part of vehicle owners and the auto industry to accept emission norms, and the lack of efficient public transport.

The older vehicles still use leaded petrol although unleaded petrol is now increasingly available. Trucks and buses run on diesel, which has a high sulphur content. The old engines emit vast quantities of SPM, leading to heavy air pollution over many cities.

Polluted air is affecting the peoples' health, particularly that of children. Between 1991 and 1995, the annual number of premature deaths that can be traced to air pollution, increased from 40,000 to 52,000. In Delhi, one out of every 10 school children suffers from asthma (recall the story in Box 11.1).

In rural areas, indoor pollution is taking a toll on the health of women. Traditional stoves that use wood or coal spew out poisons that women inhale directly. This is equivalent to smoking 100 cigarettes a day! Many deaths take place in North India during the winters when doors and windows are closed and poisonous smoke from the stove collects indoors. There have been campaigns to substitute these inefficient and polluting stoves with smokeless chulhas. The effort has, however, failed to make an adequate dent in this huge problem. We will return to this issue in Chapter 17.

How are automobile emissions regulated in India?

The Indian government began regulating automobile emissions in 1991. In the year 2000, the government introduced the Bharat Emission Norms modelled on the basic Euro Norms of the European Union. Since then it has been gradually making these limits more stringent.

Bharat-III norms for passenger cars will come into effect from April 2005 in 11 metros—Bangalore, Delhi, Mumbai, Kolkata, Chennai, Hyderabad, Ahmedabad, Pune, Surat, Kanpur, and Agra. This will account for 40 per cent of the total cars sold in India. The rest of the country will shift to Bharat-II norms.

In the 11 designated cities the new Bharat-III petrol cars would have 28 per cent lower emissions than current Bharat-II cars and 89 per cent lower emissions than cars manufactured in 1991. Similarly, there would be an almost 30 per cent reduction in emissions among Bharat-III diesel cars as compared with the Bharat-II cars. Compared to 1992 diesel cars the reduction in emissions would be 72 per cent.

Lead has been phased out of automobile fuel with effect from February 2000. New stringent norms for petrol and diesel will come into effect from 2005.

The Supreme Court of India has taken a special interest in keeping the air clean, particularly in Delhi. Thanks to the orders of the Court, implemented by the Delhi government, the air quality in Delhi has improved (read Box 11.5).

From a discussion on air pollution let us now move to a discussion on noise pollution.

What is noise pollution and what are its sources?

Noise is defined as unwanted sound and it is an irritant and a source of stress. Most of the noise one hears originates from human activities. The main sources are:

- The transport sector: Aircraft, trains, trucks, tractors, cars, three-wheelers, and motorcycles contribute the maximum noise.
- Industrial and construction machinery: Factory equipment, generators, pile drivers, pneumatic drills, road rollers, and similar machinery also make a lot of noise.
- Special events: High-volume sound from loudspeakers during pop music performances, marriage receptions, religious festivals, public meetings, etc., also contribute to noise pollution.

Noise pollution is increasing in industrial societies and in cities everywhere.

How is sound measured? Sound is measured in decibels (db). It is not a linear, but a logarithmic scale. For example, a change from 40 db to 80 db represents a 10,000-fold increase in loudness. When the sound level reaches 140 db, our ears hurt. However, long exposure to noise even at 85 db can cause hearing loss.

Apart from loudness, the frequency or pitch of the noise also determines whether it is harmful or not. A modified scale called decibel-A (dbA) takes pitch into account. Hearing loss begins if a person is exposed to noise levels of 80–90 dbA, for more than 8 hours a day. A level of 140 dbA is painful and 180 dbA could kill a person.

Table 11.1 gives the typical intensity levels of common sources of sound.

We should also distinguish between plain noise and meaningful noise. We can ignore a constant low-level hum, for example, from a computer, and continue to work. However, we are

Loud music at a marriage reception

Table 11.1 Typical average decibel levels (dbA) of some common sounds*

Source	dbA	Source	dbA
Threshold of hearing	0	Food blender (1 m)	90
Rustling leaves	20	Subway (interior)	94
Quiet whisper (1 m)	30	Diesel truck (10 m)	100
Quiet home	40	Power mower (1 m)	107
Quiet street	50	Pneumatic riveter (1 m)	115
Normal conversation	60	Chainsaw (1 m)	117
Inside a car	70	Amplified Rock music (2 m)	120
Loud singing (1 m)	75	Jet engine (30 m)	130
Automobile (8 m)	80		
Motorcycle (10 m)	88		

Note: *Where necessary, the distance of the source in metres is specified.

much more distracted if we hear a conversation at the same level of intensity. We tend to pay attention to it because it is meaningful. This is only with reference to distraction. In either case, the damage to hearing depends only on the intensity.

What are the effects of noise pollution?

Noise can do physiological and/or psychological damage if the volume is high or if the exposure to it is prolonged. Loud, high-pitched noise damages the fine hair cells in the cochlea of the ear. The vibration of these hair cells in response to sound is transmitted to the brain by the auditory nerve. Since the body does not replace damaged hair cells, permanent hearing impairment is caused by prolonged exposure to loud noise.

Loud noise can also have other ill effects like heart palpitation, pupil dilation, or muscle contraction. Migraine headaches, nausea, dizziness, gastric ulcers, and constriction of blood vessels are some of the other possible outcomes.

Noise can cause serious damage to wildlife, especially in remote regions, where the normal noise level is low. Ways in which animals are adversely affected by noise pollution include:

- Hearing loss, affecting their ability to avoid predators
- Masking, which is the inability to hear important environmental cues and animal signals
- Non-auditory physiological effects such as increased heart rate, respiratory difficulties, and general stress reactions
- Behavioural effects, which could result in the abandonment of territory and loss of ability to reproduce.

What are the control measures against noise pollution? Producing less noise is obviously the best method of reducing this pollution. Almost all machinery can be redesigned to reduce noise. Another way is to provide shields and noise-absorbing material. Ear plugs and earphones can shield the receiver from noise.

What is the state of noise pollution in India? In urban as well as rural areas, noise pollution is on the rise. Public meetings, festivals, marriage receptions, the sound of televisions, etc., have all become louder.

A study conducted in 2003 by the National Physical Laboratory, New Delhi, showed that the noise generated by crackers sold in the market was much higher than the prescribed levels. This was the case with almost all the brands. The noise was more than the sound of aircraft and deleterious to health.

Another study made in 2002 showed that noise levels at three busy traffic junctions in Chennai were above the permissible limits prescribed by the Central Pollution Control Board and the Bureau of Indian Standards. Automobile horns were found to be a major source of noise.

In the 1980s and 1990s, there were several court judgements in India restricting the generation of noise by industries, fire crackers, electric horns, etc. Finally, in 2000, the Indian government notified the Noise Regulation Rules under the Environment (Protection) Act of 1986.

The rules regulate noise levels in industrial (75 db), commercial (65 db), and residential zones (55 db). They also establish zones of silence within a radius of 100 m of schools, courts, hospitals, etc.

The rules specify that no permission could be granted by any authority for use of a public address system in the open after 10.00 p.m. and before 6.00 a.m. Permission may be given for the use of loudspeakers during day time, but the sound levels must be within the limits prescribed in the rules. The rules also fix different noise levels for firecrackers and industrial activities.

What lies ahead?

With increasing urbanization and industrialization coupled with the exponential increase in the number of vehicles on Indian roads, the future does not appear good for air quality in Indian cities, unless, of course, drastic measures are taken.

Box 11.7 Make a difference: Observe Car Free Day

The idea of a Car Free Day originated during the environmental movement of the 1960s and became popular in Germany and the UK in the 1990s.

On September 22, 1998, 66 French cities organized the first 'In Town Without My Car!' Day. On that day, the towns banned vehicular traffic in certain areas and allowed only public transport and bicycles. From 2000, the European Union countries began observing Car Free Day.

Now, September 16–22 has become the European Mobility Week. Over 400 European cities have joined this campaign. European citizens have the opportunity to enjoy a full week of events dedicated to sustainable mobility. Local authorities carry out many initiatives tackling different aspects of urban mobility on each day of European Mobility Week.

Car Free Day on September 22 is the highlight of the European Mobility Week. The challenge is to organize 'In Town Without My Car!' even if it is a working day. The campaign encourages the public to act against pollution, caused by the increase in motorized traffic in the urban environment.

Over 1100 cities across the world observe Car Free Day. There is also the United Nations Car Free Days Programme, which provides information and supports the observance of the Car Free Day in all countries.

In India, it is only a minority that owns cars. However, these cars are already leading to heavy pollution and traffic jams in all the major cities. Further, in the name of development, the government is encouraging the increased production of cars. New models of cars are regularly rolling out of factories.

Try to organize a Car Free Day in your city by bringing together the concerned citizens' groups and organizations. Get the traffic police, the local mayor and some politicians to join. Some possible activities are:

- Get cars banned in busy shopping areas on that day
- Set up stalls on the road to put up posters regarding air pollution, effects of traffic jams, importance of public transport, oil shortage, global warming, etc.
- Demonstrate ecofriendly vehicles like electric cars and scooters
- Launch a campaign for pedestrian zones, better public transport, non-polluting vehicles, etc.
- Arrange a bicycle rally to create awareness of the problems of increasing car use.

Source: The websites *www.22september.org*, *www.mobilityweek-europe.org*, and *www.uncfd.org*

A quick recap

- While there are some natural sources of air pollution, most of the pollution today is caused by human activities.
- Burning fossil fuels in automobiles and industries is a major source of air pollution.
- Air pollution has many adverse effects on human health.
- Technologies are available to reduce automobile and industrial emissions.
- The zero-emission wonder car running on low-cost, abundant, and renewable energy is still a dream.
- The world will be forced to cut down on the number of cars on the road due to a variety of reasons.
- Indoor air pollution can be as damaging as outdoor pollution.
- As the number of automobiles increases rapidly in India, air pollution is becoming a serious problem. In particular, it is taking a heavy toll on the health of the urban poor, especially children.
- Spurred on by the courts, the Indian government is introducing stringent rules to reduce emissions.
- Noise pollution can have very damaging effects and it is best to curb the generation of noise.

What important terms did I learn?

Decibel
Photochemical smog
Primary air pollutant
Secondary air pollutant
Smog

What have I learned?

Essay questions

1. 'The automobile is one of the worst inventions of humankind.' Write an essay in support of this statement. Focus on the environmental impact of the automobile. Comment on the possible impact of the rapid increase in the number of cars in India and China. Review the approaches for making cleaner cars.
2. Describe the state of urban air pollution in India. What is its impact on health, especially on the health of children? What are the measures the government is taking to reduce vehicle emissions? How did the quality of air improve in Delhi?

Short-answer questions

1. What are primary and secondary air pollutants? Give examples.
2. What are the sources of outdoor air pollution?
3. What are the effects of outdoor air pollution?
4. How can outdoor air pollution be reduced or controlled?
5. What are the causes and effects of indoor air pollution?
6. What are the sources and effects of noise pollution?
7. Give an account of the rules framed by the government to control noise pollution.

How can I learn by doing?

Volunteer your services to the Central Pollution Control Board for its programme of monitoring the air quality in the cities. Work with the monitoring unit for two or three weeks and write a report on your experience. Include in your report a description of the measuring equipment, the parameters that are measured, how the parameters changed over the period of your involvement, etc.

What can I do to make a difference?

1. Discourage your friends or neighbours from lighting bonfires. Bonfire smoke is dangerous, containing highly toxic, and possibly carcinogenic, chemicals. Never burn tyres or plastics, which release deadly dioxins and poisonous gases. Join the police and NGOs in discouraging people from lighting such fires during festivals like Bhogi in Tamil Nadu or Lohri in North India.
2. If you must light a bonfire, burn diseased cuttings and branches. If you are at a bonfire, do not inhale the smoke.
3. If any group creates too much noise in your neighbourhood, especially between 10.00 p.m. and 6.00 a.m., try to persuade them to stop or, at least, reduce the volume. If they do not respond, feel free to go the police.
4. Observe Car Free Day on September 22 (see Box 11.7).

Where are the related topics?

- Chapter 10 discusses the use of fossil fuels and the future of hydrogen as a fuel.

What is the deeper issue that goes beyond the book?

What is the psychological basis for the great attraction of the personal car? How can that need be satisfied through public transport? How can cycling be promoted (even as the Chinese are giving up this healthy and ecofriendly habit)?

Where can I find more information?

- Bell et al. (2004) cover in detail the Delhi case (high air pollution, Supreme Court decisions, and the introduction of CNG buses and trucks, etc.).
- The fires and fumes are only one aspect of the Jharia mines tragedy. Read a full account in Mahapatra (2002).

What does the Muse say?

The children's book *The Lorax* by Dr Seuss is a great environmental story and you too will like it for its message and its wacky verse. Here is an extract from the book:

Then again he came back! I was fixing some pipes
when that old nuisance Lorax came back with more gripes.
I am the Lorax, he coughed and he whiffed.
He sneezed and he snuffled. He snarggled. He sniffed.
Once-ler! he cried with a cruffulous croak.
Once-ler! You´re making such smogulous smoke!
My poor Swomee-Swans...why, they can't sing a note!
No one can sing who has smog in his throat.

And so, said the Lorax,
—please pardon my cough—
they cannot live here.
So I´m sending them off.
Where will they go?...
I don't hopefully know.
They may have to fly for a month...or a year...
To escape from the smog you´ve smogged-up around here.

CHAPTER 12

Water, Soil, and Marine Pollution

As we watch the sun go down, evening after evening,
through the smog across the poisoned waters of our native earth,
we must ask ourselves seriously whether we really wish
some future universal historian on another planet to say about us:
'With all their genius and with all their skill,
they ran out of foresight and air and food and water and ideas,' or,
'They went on playing politics until their world collapsed around them.'

U Thant

Man, despite his artistic pretensions,
his sophistication and many accomplishments,
owes the fact of his existence to
a six-inch layer of topsoil and
the fact that it rains.

Anonymous

What will I learn?

After studying this chapter, the reader should be able to:

- appreciate the health impact of excess fluoride in water;
- explain why the contamination of groundwater by substances like fluoride and arsenic has been increasing in India in recent years and why this is causing concern;
- list the main categories of water pollutants and their effects;
- distinguish between point and non-point sources of water pollution;
- describe the process of eutrophication of lakes and marine waters;
- list ways of measuring water quality;
- outline the methods of treating and purifying polluted water and sewage;
- suggest ways of reducing industrial pollution of freshwater;
- describe the condition of Indian rivers;
- appreciate how pesticides have contaminated water in India;
- summarize the status of sanitation facilities in the world and in India;
- explain ecological sanitation and the necessity for the same;
- describe how soil gets polluted and what can be done about it;
- list the sources of marine and coastal pollution; and
- recall the international initiatives for controlling marine pollution.

Box 12.1 Crisis: Neither live nor die!

In Jharana Khurd village, 20 km from Jaipur, Rajasthan, there are no youth. All the 1200 inhabitants, irrespective of age, look old. Their shoulders, hips, and ankles are swollen and ache all the time. All have cracked teeth.

The people of Jharana Khurd are not alone. The story is similar in Baroli Aheer in Uttar Pradesh, Chukru in Jharkand, Kachariadih in Bihar, Annaparti in Andhra Pradesh and thousands of other villages across the country. The people of all these villages suffer from fluorosis, caused by an excess of fluoride in water.

Excess intake of fluoride leads to fluorosis: dental, skeletal, or non-skeletal. Dental fluorosis results in blackened, mottled, or cracked teeth. Skeletal fluorosis means permanent and severe bone and joint deformities. Non-skeletal fluorosis leads to gastro-intestinal and neurological problems.

High fluoride concentration in groundwater occurs naturally in a number of countries. While fluorosis is most severe and widespread in India and China, it is endemic in at least 25 countries across the globe. The Indian geology is such that the bedrock contains minerals with high fluoride content. When the bedrock weathers, the fluoride leaches into the water and soil.

In India, fluorosis has wiped out the economy of whole villages by disabling most of the inhabitants. To women, the disease brings social stigma. Normally, the disease does not lead to death, but to extreme suffering. As someone put it, 'It neither allows a person to live nor to die.'

Why has fluorosis become a problem in recent years? Excessive extraction of groundwater has led to deeper and deeper borewells, which draw water from aquifers containing high fluoride concentrations. There is no fluoride problem if one drinks water from an open well or pond.

Fluorosis is now endemic in 19 Indian states, affecting 65 million people, including six million children. At least 60,000 villages have high fluoride levels. As India digs deeper and deeper to get water, the fluoride problem will become more widespread and more acute.

Until now, state and national level initiatives to combat fluorosis have been weak. There is no regular monitoring and testing of groundwater. Technologies are available for defluoridation, but the plants are expensive and involve significant maintenance costs.

There is a mild hope on the horizon. UNICEF has introduced domestic defluoridation filter units, which can be used in homes to remove excess fluoride from drinking water. Developed in collaboration with the Indian Institute of Technology (IIT), Kanpur, this easy-to-use device, where water poured into the upper chamber is collected and filtered into the lower one, reduces the fluoride content to acceptable levels.

Given the scale of the problem, however, will such attempts ever solve it?

Source: Jamwal and Manisha (2003) and the website *www.fluoridealert.org/fluorosis-india.htm*

What do we learn from the fluorosis story?

Under normal use of water resources, the fluoride problem would not have become such a widespread issue. Fluoride contamination began only when we went deeper and deeper into the ground to extract increasing amounts of water. It is once again a case of exceeding the limits of nature to satisfy our needs and paying a heavy price for it.

In this chapter, we will find out how water, soil, and the ocean are being polluted largely due to human activities.

Which are the pollutants of freshwater and what are the effects of fresh water pollution?

Most of the freshwater sources in the world, such as rivers, lakes, and groundwater are already polluted. This is true even of remote places like the polar regions. Most of the pollution that ultimately reaches the ocean contaminates freshwater sources along the way.

The main categories of water pollutants and their effects are as follows:

- Sediments: Excessive amounts of soil particles carried by flowing water, when there is severe soil erosion. Sediments cloud the water and reduce photosynthesis, clog reservoirs and channels, smother coral reefs, destroy the feeding grounds of fish, and disrupt aquatic food webs.
- Oxygen-demanding organic wastes: Animal manure, plant debris, and sewage; waste from animal feedlots, paper mills, and food processing facilities. Bacteria that decompose these wastes deplete oxygen and cause the death of fish and other aquatic organisms.
- Infectious micro-organisms: Parasitic worms, viruses, and bacteria from infected organisms as well as human and animal wastes. These micro-organisms are responsible for water-borne diseases that kill thousands of adults and children, primarily in developing countries.
- Organic compounds: Synthetic chemicals containing carbon from industrial effluents, surface run-off, and cleaning agents. These chemicals cause many health problems in humans and also harm fish and other wildlife.
- Inorganic nutrients: Substances like nitrogen and phosphorus from animal waste, plant residues, and fertilizer run-off. These nutrients can cause eutrophication (explained below) and can affect infants and unborn babies (blue-baby syndrome).
- Inorganic chemicals: Acids, salts, and heavy metals, such as lead and mercury from industrial effluents, surface run-off, and household cleaning agents. They make water unfit for drinking or irrigation, harm fish and other aquatic organisms, cause many health problems for humans, and lower crop yields.
- Radioactive substances: Wastes from nuclear power plants, nuclear weapons production facilities, and the mining and refining of uranium and other ores. Such substances cause cancers, birth defects, miscarriages, etc.
- Thermal pollution: Hot water from industrial processes. The heat lowers oxygen levels and makes aquatic organisms more vulnerable to disease, parasites, and toxic chemicals. Further, when the hot water is let in, the sudden increase in temperature causes thermal shock in aquatic organisms.

Where do water pollutants come from?

Water pollutants come from point sources and nonpoint sources. Specific places such as sewage treatment plants and factories are point sources. They discharge pollutants into water bodies through pipes, sewers, or ditches. Laws and rules can, however, regulate the discharges from point sources.

When pollutants enter the water body not from a single source but from several points over a large area, it is a case of pollution from nonpoint sources. This is the case when rainwater flows across the soil, picking up pollutants and carrying them into water bodies. Nonpoint sources include surface run-off, mining wastes, municipal wastes, construction sediments, acid rain, and soil erosion. Such nonpoint sources of pollution are difficult to control.

What is eutrophication of lakes?

A lake or a pond that has clear water contains minimal levels of nutrients and supports small populations of aquatic organisms. Eutrophication is the enrichment of such a standing water body by nutrients such as phosphorus and nitrogen. It occurs when sewage and fertilizer run-off bring large amounts of nutrients into the water body.

In eutrophic lakes, there is increased photosynthetic activity. This results in cloudy water covered with a slimy, and smelly, mat of algae and cyanobacteria. When the excessive numbers of algae die, they are deposited on the bottom of the lake and are decomposed. Since this process uses up a lot of the dissolved oxygen, some fish species die. They are replaced by other species that can tolerate lesser amounts of oxygen.

Eutrophication is undesirable, since it changes the species mix in the water body. Reducing phosphorus and nitrogen inputs is the best way of controlling eutrophication.

How does groundwater get polluted?

Around two billion people, approximately one-third of the world's population, depend on groundwater for their daily needs. About 600–700 cu. km of groundwater is withdrawn every year, mostly from shallow aquifers.

It was long believed that water that slowly seeps or infiltrates into the ground would be thoroughly filtered by the soil and hence the groundwater would be free of pollutants. We now know that this is not necessarily true. The filtering capacity of soil varies greatly from place to place. Further, there is also a limit to the amount of pollutants the soil can filter out.

Excessive extraction of groundwater leads to the natural pollution of the same. Examples are the fluoride concentration described at the beginning of this chapter (Box 12.1) and the arsenic contamination discussed below. In coastal areas, when water tables drop due to excessive extraction, there is intrusion of salt water and this is often an irreversible process.

Groundwater receives pollutants from septic tanks, landfills, hazardous waste dumps, and underground tanks containing petrol, oil, chemicals, etc. Substances like paint thinners and motor oil that we pour on the ground very often find their way into the groundwater.

Groundwater pollution lasts long because it neither gets flushed nor does it decompose away. This is because the flow of groundwater is very slow, decomposing bacteria are few in number, and the cold temperatures slow down the decomposing process. As a result, even degradable waste stays in the water for hundreds to thousands of years. Non-degradable waste like lead, arsenic, and fluoride remain in groundwater permanently.

What is arsenic poisoning?

Arsenic contamination of groundwater is now a major problem in Bangladesh and West Bengal. It was first reported in 1983 from the 24 Parganas District of West Bengal. Later, it was found that almost the whole of Bangladesh was affected by this problem.

Arsenic is an element found in combination with oxygen, chlorine, hydrogen, mercury, gold, and iron. The effect of arsenic on body tissues starts showing after two to five years of consuming the contaminated water. The skin develops spots, then hard nodules, leading later to gangrene and cancer. It also causes many other complications, such as blindness, liver and heart problems, diabetes, and goitre.

Arsenic contamination has been steadily worsening in West Bengal and Bangladesh. It has also spread to Bihar. Almost 330 million people may be at risk in India and Bangladesh. Some scientists believe that this is only the tip of the iceberg. Many future generations are at grave risk from this form of contamination, which is perhaps the largest instance of mass poisoning in history.

Where does the arsenic come from? One theory is that it originated in the Himalayan headwaters of the Ganga and the Brahmaputra and had remained below the deltas of the region for thousands of years. Over a long period this arsenic has contaminated deep wells.

The other theory is that arsenic deposits are present in aquifers. When the rate of groundwater extraction is very high, the water level falls below the level of the arsenic deposits, and the poison is released. When the aquifer gets recharged, the arsenic contaminates the water.

In Bangladesh more than 10 million tubewells were installed to provide clean drinking water. Thanks to these water sources, the infant and child mortality rates have come down by 50 per cent over a period of 40 years. Unfortunately, 1.5 million of these tubewells are now heavily contaminated with arsenic.

What is being done to solve the problem? Until now national and international initiatives to tackle the problem have been tentative and ineffective. Testing for arsenic is a complex, difficult, and expensive process, though a new isotope technique has shown promise. There is also a lack of basic knowledge about the movement of groundwater and the location of arsenic in the water sources.

It is a desperate situation. With no other sources of water, the affected people continue to drink the contaminated water.

How do we measure water quality?

Here are three ways in which water quality can be measured:

- Biological Oxygen Demand (BOD): This parameter measures the degree of water pollution from oxygen-demanding wastes and plant nutrients. BOD is the amount of dissolved oxygen needed by the decomposers to break down the organic material present in a certain volume of water, when it is kept in darkness over a five-day incubation period at 20ºC.

  BOD is measured in parts per million (ppm). A BOD level of 1–2 ppm is considered very good. It indicates that there is not much organic waste present in the water supply. A water supply with a BOD level of 3–5 ppm is considered moderately clean. Any water with a BOD level of 6–9 ppm is considered somewhat polluted, with the presence of organic waste and the bacteria that decompose this waste. At BOD levels of 10 ppm or more, the water supply is considered very polluted, containing large quantities of organic waste.
- Presence of disease-causing organisms: The number of colonies of coliform bacteria present in a 100 ml sample of water is another measure of water quality. There should be no coliform colonies in drinking water, while water in a swimming pool could have up to 200 colonies per 100 ml.
- Chemical analysis: The presence of chemicals like pesticides can be measured by analysis. This constitutes another measure of water quality.

How are polluted water and sewage treated?

The first step in water purification is the use of a chemical that makes suspended particles settle down. The water is then filtered and disinfected. The most common disinfectant used is chlorine, though there is concern about the hazards of consuming low doses of chlorine over long periods.

Wastewater, including sewage, goes through several stages of treatment. The primary treatment removes suspended particles by screening and settling. The secondary treatment uses micro-organisms to decompose the organic material in the wastewater. After several hours, the particles and the bacteria are allowed to settle down as secondary sludge. The tertiary treatment is a complex biological and chemical process that removes the remaining pollutants, such as minerals, metals, organic compounds, and viruses.

A major problem in sewage and wastewater treatment is disposal of the sludge. This sludge is often toxic, since people tend to dump all kinds of waste into drains. Treating the sludge is a costly and time-consuming process. Cities in industrialized countries do not know how to get rid of the sludge and so they export it to developing countries (read a discussion on this issue in Chapter 13).

Wastewater including sewage can be treated using natural or artificial wetlands (Box 12.2). This method requires land and an appreciation of the role that wetlands can play. In India, wetlands are rapidly being drained as cities expand (see Chapter 4 of this book).

Box 12.2 Solution: Spend less, save ecology!

How would you like to get a virtually free wastewater treatment system, with a wildlife and bird sanctuary and a natural recreation area thrown in as bonus? Well, this is what a small town in the US managed to set up for itself. This case has become the model for many other cities.

Arcata is a progressive town of 16,000 people on the northern coast of California. Home to the Humboldt State University, it has no chain stores or supermarkets, but only local traders, organic food stores, used bookshops, and weekly markets on the city square. And the town is anti-war!

In 1978, Arcata had to set up a wastewater treatment plant. The first proposal was a US$ 25 million project, which would take the sewage 8 km through pipes and release it directly into the ocean. Arcata citizens questioned the pipeline idea: What if an earthquake occurred? A tsunami? Who was responsible for the clean-up if a ship were to accidentally cut the line? Why should the town choose an energy-intensive and costly option?

The citizens favoured an integrated wastewater treatment process within a natural system. After three years of experiments, they were ready to implement a US$ 7 million low-tech, natural system.

The first stages were the conventional ones: allow the solids to settle, let the dissolved organic wastes degrade, and remove the disease causing agents with chlorine. This does not remove other pollutants like nitrogen and phosphorus, which require expensive treatments. At this stage, Arcata came up with an innovative approach.

The town hired biologists, who worked with the town's engineers to develop a series of six marshes over 60 hectares. Appropriate plants in the marshes absorb and assimilate the pollutants. Algae, fungi, and bacteria in the marsh also feed on the contaminants.

As the water flows through the marshes, it gets purified. When it comes out of the marshes, it goes through a final treatment before being released into the nearby bay.

The marshes constitute a sanctuary that supports a high level of biodiversity with many animals and fish. In addition, thousands of birds reside in the marshes permanently or seasonally. For recreation, the sanctuary includes 7 km of trails, meandering through the marshes.

The project won a national innovation award. More than 800 cities of the world, including large ones like Orlando in Florida, have followed Arcata's example of using wetlands for purification.

What keeps it going still? A citizens' group called Friends of the Arcata Marsh, FOAM for short!

Source: Raven and Berg (2001) and the website of the Humboldt State University, *http://www.humboldt.edu/~ere_dept/marsh/history.html*

What can be done about industrial pollution of freshwater?

Every litre of wastewater discharged by an industry pollutes on an average eight litres of freshwater. The total amount of water polluted in this manner is more than all the water in the largest river basins.

There are ways of reducing industrial pollution of freshwater:

- Improving process technology to reduce water demand.
- Using the same water in series for two or more successive stages in a process.
- Recirculating process water indefinitely, adding only that quantity necessary to make up for unavoidable losses.
- Harvesting rainwater to meet as many of the water requirements as possible.

Governments could take the following steps to regulate the use of water by industries:

- Ensure that they withdraw water from downstream of their own discharge, forcing them to clean their discharges.
- Establish specific and total load-based effluent standards.
- Fix a quota for total freshwater withdrawal for each industry in a watershed.
- Specify mandatory recycling percentage.
- Restrict groundwater withdrawal where levels are low.
- Make rainwater harvesting compulsory.
- Price water properly: Charge high for an industry that can reuse water, charge for groundwater use, and increase unit price with increasing use.
- Introduce pollution charges that are high enough to constitute an incentive for reducing pollution.
- Impose stiff penalties for non-compliance with regulations.

What are the methods of purifying water?

Some of the purification methods are described below:

Reverse Osmosis or the RO method

This process was discovered in the late 1950s. It was slow to catch on but it now enjoys patronage worldwide. In the RO method, water is forced through a semi-permeable membrane. This filters out unwanted substances, producing clear, fresh-tasting drinking water. RO uses no chemicals and can also be used to desalinate seawater. The semi-permeable membranes, however, need to be manufactured and later disposed of.

The UV method

In this method ultraviolet (UV) radiation is directed through clear, pre-filtered, particle-free water. The UV light is extremely effective in killing and eliminating bacteria, viruses, fungi, and certain other harmful organisms. It is mainly used in industry and in hospitals to treat water. This method must be used in conjunction with sediment and carbon filters.

Distillation method

Water is boiled to create steam, which when cooled condenses as clean water. The residual water that contains the contaminants is discarded. When distillation is used in conjunction with other filter mediums such as carbon, we can get very pure water.

Planted filter method

This can be used to purify even sewage water. The wastewater first goes through a septic tank and a baffle reactor, in which all the particulate and organic matter is removed. Next the water is sent through an open, horizontal filter bed containing pebbles and plants like reeds, which absorb many of the impurities. Finally, the water moves through an open polishing pond. The output is good enough for gardening and irrigation.

Let us now look at some aspects of water pollution in India.

What is the state of rivers in India?

While the headwaters of the major rivers retain their pristine quality, they become polluted once they reach the plains. Agricultural run-off, industrial effluents, and domestic sewage find their way into most rivers. The rivers are stressed in their middle stretches, where both extraction rates and pollution levels are high.

Tests of river water in or near cities show that in most cases the water is unfit for drinking. This is the case with the Yamuna near Delhi, the Sabarmati at Ahmedabad, the Gomti at Lucknow, the Adyar at Chennai, the Vaigai at Madurai, and the Sutlej near Ludhiana.

The Central Ganga Authority was established in 1985 to lay down policies for projects to be taken up under the Ganga Action Plan (GAP). In July 1995, this Authority was redesignated as the National River Conservation Directorate (NRCD). The NRCD has drawn up the National River Conservation Plan and coordinates the implementation of the schemes under GAP and the Action Plans for the other rivers.

More than Rs 500 million has been spent on GAP, but pollution levels in the Ganga are as high as ever. Meanwhile, a new National Lake Conservation Plan has come into being.

Tap water being bottled and sold as mineral water

How severe is pesticide contamination of freshwater in India?

Since the beginning of the Green Revolution, large amounts of pesticide have been used (and continue to be used) in agriculture and other sectors. It is not surprising, therefore, that pesticide residues have contaminated many water sources.

We have reached the stage where we cannot be sure of the quality of any water source in the country. Even bottled water samples have shown the presence of pesticide residues (read Box 12.3). This is also the case with soft drinks, which use the same sources of water.

Box 12.3 Danger: Poisons in packages?

When you go to a hotel or travel by train, you are very likely to order a bottle of water as you do not trust the quality of water served by the hotel or the water from the railway tap. You believe that the bottled water is safe. Well, some tests have shown that bottled water could contain pesticide residues above acceptable limits.

A decade ago, bottled water was sold only in expensive hotels. Today, bottles and pouches of water are found everywhere, even in small towns and villages, railway stations, and bus stands. People in cities no longer get enough water in their taps and they do not trust the quality of the little amount that they do get. Their borewells give only brackish water. As a result, most middle class urban residents buy drinking water.

In 2002, the Centre for Science and Environment (CSE) in Delhi tested 34 common brands of bottled water for the presence of pesticides. The tests were conducted in their laboratory using methodology approved by the US Environment Protection Agency (EPA). The results were compared to the norms set by the European Union (EU).

Pesticide residues were found in all the samples except in one imported brand. The residues included deadly pesticides like lindane, DDT, and malathion and the amounts were well above EU norms.

The CSE team traced the pesticides to the sources of water, mostly borewells. The water sources contained even higher levels of pesticides. The bottling plants used chemical and filtration techniques in the purification process, but it seemed that they could remove only a part of the pesticide residues.

There are Indian standards for packaged water and, since March 2001, certification by the Bureau of Indian Standards is mandatory. However, for limits on pesticide presence, the standards refer to the Prevention of Food Adulteration Act, which only says that pesticides should be 'below detectable limits'.

Thousands of bottles of water are consumed in India every day. If they contain deadly pesticides, why are people not dropping dead on the streets? The reason is that, while the levels may be above EU norms, they are not so high as to cause immediate symptoms of poisoning. Over a period, however, the pesticides will have adverse effects on people's health.

Packaged water contains pesticides. Tap water is turbid. Borewell water is brackish. What will you drink?

Source: DTE (2003a)

What is the official response to the presence of pesticides in bottled water and soft drinks? In July 2003, the Union Ministry of Health and Family Welfare notified new standards for pesticide residues in bottled water. In October 2004, the Bureau of Indian Standards decided to treat soft drinks as a separate category for setting final product standards. These standards will cover pesticide residues, caffeine content, and labelling requirements.

Water supply and sanitation are closely connected and it is time to review the global sanitation crisis.

What is the status of world sanitation?

According to the World Health Organization (WHO), at the beginning of 2000, one-sixth (1.1 billion people) of the world's population was without access to proper water supply and two-fifths (2.4 billion people) lacked access to proper sanitation. The majority of these people live in Asia and Africa.

Less than half of all Asians have access to proper sanitation and two out of five Africans lack proper water supply. According to the *Census of India 2001*, 108 million rural households and 14 million urban households have no sanitation facilities.

Other estimates show that 2.7 billion people manage with only basic pit latrines and 0.8 billion with dysfunctional flush sanitation. Only 0.3 billion have proper flush sanitation.

Sanitation coverage in rural areas is less than half that in urban settings, even though 80 per cent of those lacking adequate sanitation (2 billion people) live in rural areas. Of these some 1.3 billion live in China and India alone. These figures are all the more shocking because they reflect the results of at least 20 years of concerted efforts and publicity aimed at improved sanitation coverage. The situation will get worse with the projected steep increase in urban populations in Latin America, Africa, and Asia.

The UN target is to reduce by one-half the proportion of people without access to hygienic sanitation facilities by 2015. To achieve this target in Africa, Asia, Latin America, and the Caribbean alone, an additional 2.2 billion people will need to be given access to sanitation. In effect, this means providing sanitation facilities to 384,000 additional people every day for the next 15 years.

What has been the approach of governments and municipalities to providing sanitation facilities to the population? The cities of the world have tended to set up centralized systems with pipes carrying the waste to sewage treatment plants. Such systems are expensive to build and maintain and they cannot keep pace with the rapid expansion of the area and population of cities. Centralized systems are in any case impractical in rural areas.

The sanitation gap coupled with water scarcity means that a completely different strategy must be adopted.

Why can't we provide everyone with flush toilets? Some environmentalists rate the flush toilet as being one of worst inventions of humankind, comparable to the automobile. In the flush system, water is utilized not just to clean the toilet bowl, but also to transport the excreta. A family of five would need more than 150 thousand litres of water to transport 250 litres of excrement in just one year.

Given the worsening water scarcity, shortage of funds to build large sewage plants, and increasing population, we will never solve the sanitation problem with the current approaches. The only way is to adopt ecological sanitation.

What is ecological sanitation? Ecological sanitation (EcoSan) is a sustainable closed-loop system. It uses dry composting toilets, which are a practical, hygienic, and cost-effective solution to human waste disposal. A key feature of EcoSan is that it regards human excreta as a resource to be recycled rather than as a waste to be disposed of. Recycling of human waste prevents direct pollution caused by sewage, returns it as nutrients to soil and plants, and reduces the need for chemical fertilizers.

The EcoSan toilet does not use water to carry away solid waste. It also separates urine from faecal matter (read Box 12.4 for an account of the EcoSan toilet). If many people use EcoSan, enormous amounts of water will be saved and many more can have piped water.

Box 12.4 Solution: 'But doesn't it smell?'

It is almost dry, crumbly, black stuff that smells light, pleasant, and earthy. Paul Calvert, a British sanitation engineer, describes the smell as 'a walk in the forest—a pleasant, woody, earthy scent'. Six months ago it was fresh human waste. Sounds unbelievable? This is the magic of ecological sanitation or EcoSan for short.

Since 1995, Calvert has set up EcoSan toilets in numerous homes in rural Kerala. These are dry, composting toilets. One model consists of a slab constructed over two vaults. The slab has a hole over each vault for the faeces to drop in and a funnel for the urine to collect. While one hole is used, the other is kept closed.

Before use, the vault is covered with straw to facilitate decomposition. After each use some ash or sawdust is sprinkled on the faeces. The urine is collected and used as fertilizer. After six months, the first hole is closed and the second one is put to use. The faecal matter in the first hole begins to decompose and in six more months the fertilizer is ready for collection.

Calvert's EcoSan work began in 1994 when, in response to the poor sanitary conditions in local fishing communities in Kerala, he began building ecological toilet and wastewater systems. He has built over 250 EcoSan toilets in Kerala and has extended his work to Sri Lanka.

Calvert recalls the first time the not-so-convinced members of the Kerala community he was working with gathered around to see the vaults of the first toilet being opened,

> This was the acid test. Even the slightest imagination of a smell or something unsightly would have ended the compost toilet experiment instantaneously and resulted in the lynching of team and designer. Everybody was looking on in awe as we opened the back wall of the vault. For each brick we removed they retreated another pace, quite convinced we were opening the gates of hell itself!

What greeted them was the pleasant-smelling compost!

Calvert identifies three basic benefits of EcoSan:

> It prevents diseases by removing pathogen-rich excreta from the immediate environment. Besides, it does not contaminate groundwater or use up scarce water resources. It also creates a valuable resource that can be recycled back into the environment. Over time, if stored properly, the excreta are transformed from a harmful product into a productive asset.

He currently spends much of his time in India, especially South India, demonstrating and promoting the ecological sanitation approach among communities, governments, and NGOs.

Calvert says: 'If everyone harvested rainwater, used an ecological toilet, planted a tree and helped their neighbour, India would surely be a land of plenty.' Is anyone listening?

Source: Gobartimes, a supplement to *Down To Earth,* May 1999 and the website *www.eco-solutions.org/*

To change to EcoSan will necessitate a change in mindset. In India, we find anything to do with human faeces disgusting, degrading, and polluting. We end up pretending that it does not exist and in doing so expose ourselves far more to the danger of infection.

The Chinese and Japanese on the other hand have been traditionally faecophilic (faeces-friendly). In those countries, human waste has always been considered a resource. In China raw faeces are used to fertilize fields, though it must be admitted that this exposes the populace to infections.

Let us now take a brief look at soil pollution.

How does soil become polluted?

Soil pollution refers to any physical or chemical change in soil conditions that may adversely affect the growth of plants and other organisms living in or on that soil.

Soil pollution and water pollution are closely connected. Acid rain and excessive use of chemical fertilizers result in the soil's inability to hold nutrients. This in turn allows toxic pesticides or atmospheric fallout to rapidly seep into the groundwater or to run off into rivers and coastal waters. Some of the persistent pollutants remain in the soil and degrade it.

Most soil pollutants are agricultural chemicals, primarily fertilizers and pesticides. It is now known that these chemicals attach themselves to soil particles and persist for long in the soil, continuously releasing contaminants into the surface water, groundwater, and topsoil. Dumping of waste (including garbage, untreated sewage, industrial effluents, nuclear waste, and mining waste) pollutes the soil when dangerous substances from the dumps leak into it.

Salts tend to accumulate in the soils of arid and semi-arid regions. The little precipitation that falls evaporates quickly leaving behind salts. Salinization can also occur in any region due to the continued application of irrigation water containing some salts.

Why are plants unable to tolerate saline soil? Water always moves from an area of higher concentration to one of lower concentration. Normally, plants have a lower concentration of water than the surrounding soil and, therefore, water flows into plants. Saline soil, however, often has a lower concentration of water than plants. Consequently, water starts flowing out of the plant and into the soil.

How can soil condition be restored? Dilution is one way of removing pollutants from the soil. It involves running large quantities of water though the soil to leach out the pollutants. This works only if the soil has good drainage properties. Even then, disposing of the water carrying the pollutants poses a problem. This method also requires lots of water.

In vapour extraction, air is injected into the soil to remove organic compounds that evaporate quickly. Bioremediation cleans up the soil by introducing bacteria and other micro-organisms. Phytoremediation uses plants, whose roots absorb pollutants and store them in their stems and leaves.

From water and soil, we now move on to marine pollution.

How does marine pollution occur?

Since the ocean forms 71 per cent of the Earth's surface, most of the pollutants in the atmosphere fall into it. In recent decades, however, there has been in addition a huge amount of direct marine pollution caused by human activities.

Industrial discharges and agricultural run-off containing pesticides, fertilizers, and various toxic chemicals find their way into the ocean. Treated and untreated human and other domestic waste also ends up in the ocean. In fact, sewage remains the largest source of contamination of the coastal and marine environment.

Ships dumping waste into the sea

The oil industry contributes deadly pollution through leaks, spills, and the cleaning of tankers.

The impact of oil discharges into the ocean, and the remedial measures that can be taken, are discussed in Chapter 14.

There has been a rapid increase of nitrogen inputs into the ocean due to agricultural run-off, atmospheric deposition, and the loss of natural interceptors such as the coastal wetlands, coral reefs, and mangroves. The result is greater marine and coastal eutrophication. Blooms of toxic, or otherwise undesirable, phytoplankton are increasing in frequency, intensity, and geographic distribution. Such blooms or red tides greatly affect fisheries, aquaculture, and tourism (see Box 12.5).

Box 12.5 Mystery: Bloom that kills

It is an explosion of colour on the ocean—orange, red, or brown. Beautiful to look at perhaps, but deadly for humans and animals. It is called a red tide or an algal bloom.

Blooms occur when certain pigmented marine algae experience a population explosion. Some of these algal species produce toxins that attack the nervous system of fishes, leading to massive fish kills. When these fish are eaten by water birds, they also die. The toxins get into the food web and end up killing marine mammals and affecting fish-eating humans.

Even if the bloom is non-toxic, it still causes problems by shading aquatic vegetation and upsetting food webs. Algal blooms are dangerous even after they die. They then sink to the bottom where they reduce the oxygen supply and kill bottom-dwelling organisms, such as crabs, oysters, and clams.

Algal blooms are becoming more common and more severe. What triggers them off? We do not know for sure, but scientists blame coastal pollution. Wastewater and agricultural run-off into coastal waters contain increasing amounts of nitrogen and phosphorus, both of which can stimulate algal growth. A higher ocean temperature due to global warming is another possible cause.

How can we prevent blooms from occurring or end them when they do occur? Again, we do not know, but we are able to predict the conditions that are likely to stimulate them.

Source: Raven and Berg (2001)

Persistent Organic Pollutants (POPs), which spread through the atmosphere, are found everywhere in the ocean. POPs cause reproductive, immunological, and neurological problems in marine organisms and possibly in humans as well. Another concern is the increasing amounts of non-biodegradable waste, for example, plastic articles and floating nets. Large numbers of birds, turtles, and marine mammals are killed when they ingest such waste or become entangled in the nets.

Human activities have changed sediment flows into coastal regions and the ocean. Areas like deltas that need sediments do not get enough, whereas coral reefs are smothered by them.

How bad is coastal pollution?

With the inexorable movement of the world's population towards the coastal areas, the pollution of the ocean nearer the shores has reached alarming proportions. Thousands of tons of sewage and industrial effluents are directly discharged into the ocean in many parts of the world.

Pollution is heavy along the densely populated coasts of India, Pakistan, Bangladesh, Indonesia, Thailand, and the Philippines. The situation is equally critical in some of the industrialized parts of the world. The sewage from most of the cities on the Mediterranean coast ends up untreated in the sea.

A few years ago, shrimp aquaculture was taken up on a large scale on the Indian coast. This industry requires freshwater as well as seawater and uses heavy doses of antibiotics. Within a short time, the effluents from the shrimp farms polluted large areas. The local soil and groundwater were affected. Ultimately, disease outbreaks and court orders (read Box 20.3 in Chapter 20 of this book) put a stop to coastal aquaculture, but the industry flourishes in inland areas.

It is no wonder that marine pollution finds its way into the fish that we catch. For example, three major shrimp species harvested off Mumbai's coastal waters have tested positive for lead and cadmium. The fish catch itself is declining in these waters, probably due to marine pollution.

What are the international initiatives to control marine pollution?

Since 1972, a number of international agreements and programmes have focused on controlling marine and coastal pollution:

- The 1972 London Dumping Convention and its 1996 Protocol: The purpose of the London Convention is to control all sources of marine pollution and prevent pollution of the sea through regulation of dumping of waste materials into the sea. The 1996 Protocol, which is set to replace the Convention, is more restrictive and adopts the 'precautionary principle'. That is, it prohibits all dumping unless explicitly permitted. Further, it prohibits the incineration of wastes at sea and the export of wastes for the purpose of dumping or incineration at sea.
- The 1989 Basel Convention on the Control of Transboundary Movement of Hazardous Wastes and their Disposal: Until 1999, the Convention was principally devoted to setting up a framework for controlling the movement of hazardous wastes across international frontiers. It has now expanded its scope to include the active promotion and use of cleaner technologies and production methods and the prevention and monitoring of illegal traffic.
- The Convention on the Prevention of Pollution from Ships (MARPOL): It is the primary international convention covering prevention of pollution of the marine environment by ships from operational or accidental causes.
- The 1995 Global Programme of Action for the Protection of the Marine Environment from Land-based Activities (GPA-LBA): This programme was adopted in Washington D.C., in the US, by over 100 countries, including those of the European Union (EU). The aim is to control contaminants like sewage, persistent organic pollutants, radioactive substances, heavy metals, oils, nutrients, sediments, and litter that enter the ocean from the land.
- The Regional Seas Programme of the United Nations Environment Programme (UNEP): This programme has fostered regional cooperation on the marine and coastal environment. It has stimulated the creation of various Action Plans for sound environmental management in each region. India, Bangladesh, the Maldives, Pakistan, and Sri Lanka are partners in the South Asian Seas Programme.

India is a party to all the abovementioned agreements and programmes.

What lies ahead?

Water pollution is affecting large populations, soil degradation is endangering food security, and marine pollution will have wide-ranging consequences, some of which we cannot even predict. We must take urgent steps to prevent all these forms of pollution.

A quick recap

- Most of the freshwater sources in the world like rivers, lakes, and groundwater including those in remote regions are already polluted. The pollutants come from a variety of sources, mostly from human activities.
- Excessive extraction of groundwater in India and Bangladesh is leading to increasing contamination with fluoride and arsenic. This has extremely serious effects on the health of the people who consume the water.
- There are several ways of treating and purifying water and sewage. Disposing off the toxic sludge that results is, however, a major problem.
- Most of the Indian rivers are polluted and the National River Conservation Authority has launched Action Plans for several rivers.
- Pesticide residues have been found in bottled water and soft drinks.
- Two-fifths of the world's population, mostly in Asia and Africa, lacks sanitation facilities.
- The adoption of ecological sanitation or composting toilets seems to be the only way to solve the sanitation problem of the world.
- The world's soil is becoming polluted with agricultural chemicals and hence getting degraded.
- Human activities are responsible for heavy pollution of the ocean and the coastal zone. Many international initiatives have been taken to control this pollution.
- The future looks grim with respect to water, soil, and marine pollution and this calls for urgent remedial measures to be taken.

What was new?

Algal bloom
Ecological sanitation
Eutrophication
Fluorosis
Persistent organic pollutants
Reverse osmosis

What have I learned?

Essay questions

1. Write an essay on the various international conventions and agreements concerning water resources, both freshwater and marine waters. Use the information given in this chapter as well as in Chapters 4 and 7.
2. Describe the various methods of treating and purifying water and sewage.

Short-answer questions

1. How did the groundwater in many places in India become contaminated with fluoride and arsenic?
2. How do fluoride and arsenic contamination of water affect peoples' health?
3. Give any five major categories of water pollutants, their sources, and their effects.
4. Explain point and nonpoint sources of water pollution.
5. What is meant by eutrophication of lakes?
6. Describe three measures of water quality.
7. What is the status of sanitation in the world?
8. How can we provide sanitation facilities to the maximum number of people in the world?
9. How does soil become polluted and how can this be remedied?
10. What are the main sources of marine and coastal pollution?

How can I learn by doing?

Approach the National River Conservation Authority and volunteer your services. Join a team that is implementing an Action Plan for a river in your state or in a neighbouring one. Spend at least three weeks working with the team or any ecologist in the team. Write a report describing the state of the river, the sources of pollution, the Action Plan, methods used for field-level implementation, problems encountered, results achieved, etc.

What can I do to make a difference?

1. Many household items like naphthalene balls, drain cleaners, paint thinners, etc., are very toxic. Try to use safer alternatives, such as ammonia, bleaching powder, baking soda, mineral oil, and vinegar.
2. Do not throw unwanted medicines or motor oil down the drain. If the medicines have not expired, donate them to any charitable hospital or voluntary organization that accepts them. There is a thriving industry in cities that recycles motor oil. If there are genuine recyclers in your city, collect all the oil in your neighbourhood and give it to them.
3. Replace lawns with trees and shrubs that need little fertilizer and are drought-resistant.
4. Use fertilizer sparingly, never near a water body.

Where are the related topics?

- Water issues were covered in Chapters 4 and 7.
- Chapter 4 also included a discussion on the ocean.
- Soil degradation was dealt with in Chapter 10.
- The impact of pesticides on health will be taken up in Chapter 16.

What are the deeper issues that go beyond the book?

How can we make people aware of the fact that any pollution that we create ultimately comes back to us? How can we encourage people to avoid polluting water and soil—though regulation or education?

Where can I find more information?

- For a detailed account of arsenic and fluoride poisoning, read Jamwal and Manisha (2003).
- For disturbing news about fluoride poisoning in many parts of India, access the website *www.fluoridealert.org/fluorosis-india.htm.*
- For more information on pesticides in soft drinks, read the editorial and two other articles in *Down To Earth,* vol. 12, no. 6, 15 August 2003.

What does the Muse say?

At the end of Chapter 6, there was a poem by the British poet Gerard Manley Hopkins (1844-1889). Here is another by him:

Wild air, world-mothering air,
Nestling me everywhere,
That each eyelash or hair
Girdles; goes home betwixt
The fleeciest, frailest-flixed
Snowflake; that is fairly mixed
With riddles, and is rife
In every least thing's life,
This needful, never spent,
And nursing element;
My more than meat and drink,
My meal at every wink;
This air, which, by life's law,
My lung must draw and draw
Now but to breathe its praise....

CHAPTER 13

Solid Waste Management

And Man created the plastic bag and the tin and aluminium can
and the cellophane wrapper and the paper plate, and
this was good because Man could then take his automobile and
buy all his food in one place and
He could save that which was good to eat in the refrigerator and
throw away that which had no further use.
And soon the earth was covered with plastic bags and aluminium cans and
paper plates and disposable bottles and
there was nowhere to sit down or walk, and
Man shook his head and cried: 'Look at this Godawful mess.'

Art Buchwald

What will I learn?

After studying this chapter, the reader should be able to:

- understand why waste has become a problem in recent times;
- recall the different categories of solid and liquid wastes and their sources;
- describe how wastes in general are managed;
- appreciate the role that rag pickers play in waste management and the conditions under which they work;
- understand why hazardous and toxic wastes are exported by many industrialized countries and the effects on the receiving countries;
- appreciate the diverse aspects of the ship-breaking industry and the associated problems;
- identify the major polluting industries of India and the associated problems of waste disposal;
- understand the special problems of waste management caused by the widespread use of plastics;
- recall the use of Common Effluent Treatment Plants as a way of helping small industries manage their wastes;
- understand the difficulties in handling municipal waste that is generated in Indian cities with special reference to biomedical waste;
- describe some methods of recycling waste; and
- suggest ways out of the global waste problem.

Box 13.1 Slogan: Clean toxic ships now!

There is a constant clang of metal. Dust and toxic fumes are everywhere. Huge metal monsters are ripped apart and become mountains of scrap. Welcome to Alang, the famous ship-breaking yard!

Alang is located in the Gulf of Kambhat on the Gujarat coast, 56 km south of Bhavnagar city. The unique geographical features of a high tidal range and wide continental shelf, coupled with a mud-free coast, allow very heavy ships to reach the coast easily during high tide. It is ideal for ship-breaking.

The necklace-shaped Alang ship-breaking yard is the second largest in the world, next only to the one in Taiwan. With 184 ship-breaking plots, it dismantles about 300 ships every year, has a turnover of Rs 60 billion, and employs 40,000 people.

The workers, mostly migrants from Orissa, Uttar Pradesh, and Bihar, toil under extremely hazardous conditions in a toxic atmosphere. Given the nature of the job, lack of training, and the absence of protective gear, accidents are common. The workers have no medical facilities, sanitation, housing, or safe drinking water.

They reside in rented shanties in different villages around the yard in relative poverty. The huge influx of migrant labourers, with languages and cultures very different from the local ones, has also created social tensions in the area.

Greenpeace and other organizations have been campaigning for the adoption of an environmentally-safe approach to ship-breaking and for improving the conditions in the ship-breaking yards. In June 2003, a group of activists toured Europe to create awareness among ship-owners and politicians about the conditions in the ship-breaking industry. They rang bells brought form the Alang yard, telling the listeners that the sound of each bell echoed the ship-breakers' call for a cleaner environment and safer work. They secured an assurance from the European Union (EU) that the shipping industry would clean the ships of hazardous material before sending them to Asia for breaking.

In November 2003, the Greenpeace sailing vessel *The Rainbow Warrior* reached Alang on a Corporate Accountability Tour. The activists discovered and exposed the presence of toxic waste on board the UK-owned ship *Genoa Bridge,* which was ready for breaking at Alang. Most of the ships that come to Alang carry toxic waste, which causes environmental and health problems. In addition, the cost of clean-up falls on the Gujarat government.

Thanks to the directives issued by the Supreme Court in 2003 and the Greenpeace campaign, some improvements have been made in the working conditions at Alang. Drinking water and hospital facilities have been planned.

The main problem, however, remains: How do we ensure that the ships are decontaminated before they arrive at Alang or any of the other yards? How do we get the shipping companies to listen to the Greenpeace slogan: *Clean Toxic Ships Now!*

Source: Bavadam (2004) and Chowdhury (2004)

What is the message that the bells of Alang carry for all of us?

Building huge ships may be a technological feat. The bigger problem, however, arises when their lives are over. How do we dispose of the solid waste, the huge hulk of metal with tons of stuff inside? The Alang case demonstrates the complex nature of hazardous waste disposal.

In Chapter 11 and Chapter 12, we discussed the pollution of the air, soil, freshwater, and the ocean. We also discussed noise pollution. What remains of the pollution problem is discussed in this chapter. We will also return to the ship-breaking case. Though the title of this chapter is 'Solid Waste Management', we will consider liquid waste too.

What is waste and why does it require management?

Waste is any material that is not needed by the owner, producer, or processor. Humans, animals, other organisms, and all processes of production and consumption produce waste. It has always been a part of the Earth's ecosystem, but its nature and scale were such that the ecosystem could use this waste in its many cycles. In fact, there is no real waste in nature. The apparent waste from one process becomes an input in another.

It is the exponential growth (see Chapter 1) of human activities that has made waste a problem that needs to be managed. We are simply producing far more waste than nature can handle.

It is, however, better to prevent waste generation than to produce waste and then try to 'manage' it. We cannot simply throw away waste. As they say, 'There is no away in throw away.' What we dispose of remains in the ecosystem and causes some form of pollution. This pollution can have an impact that is far away from the point of generation and far removed in time also.

The composition, quantity, and disposal of waste determine the environmental problems it creates. To minimize the adverse effects of any waste, it has to be recycled, permanently isolated in storage, allowed to decompose and degrade into a harmless state, or treated to remove any toxicity it may have.

Gaseous waste, caused mainly by emissions from vehicles and other sources also carries fine particles of matter, which lead to air pollution and smog (see Chapter 11). Moreover, when gaseous waste is deposited on land as acid rain, it pollutes the soil and water (see Chapter 12).

Most disposable wastes are in the form of solids, liquids, or slurries. The main categories of such wastes are as follows:

- Domestic waste: Sewage, wastewater contaminated by detergents, dirt, or grease, household garbage, and bulky waste including packaging material, appliances, furniture, office equipment, and used cars.
- Factory waste: Solids and effluents from factories of all types; the worst polluters are slaughterhouses, breweries, tanneries, textile mills, paper mills, steel mills, and most chemical industries; power plants discharging heated coolant water, which causes thermal pollution.
- Waste from the oil industry: Oil spills, oil leaks, water used for cleaning tankers, etc.
- E-waste: A new form of waste from discarded computers (read Box 13.2).
- Construction waste: Materials from buildings that are demolished or renovated and materials discarded after completing a building.
- Waste from the extractive industries: Mining, quarrying, and dredging create solid waste (during extraction) and slurries (during processing).
- Agricultural waste: Mostly organic waste from plants and animals; irrigation water from farms containing fertilizers and pesticides.
- Waste from food processing: Organic solid and liquid waste from discarded food material.
- Biomedical waste: Originates mainly in hospitals and clinics and includes blood, diseased organs, poisonous medicines, etc.
- Nuclear waste: Radioactive waste from nuclear power plants and the manufacture of nuclear weapons.

Box 13.2 Solution: Ending e-waste

It is an industry that thrives on obsolescence. New gadgets and new models appear almost daily and the old ones are discarded as junk. The waste is mounting in the electronic and computer industries.

These industries ensure their continued and rapid growth by making their products obsolete as fast as possible. The old ones may still be in good working condition, but they cannot be used because spare parts are not available.

The net result is a rapid increase in e-waste or electronic waste that results from discarded devices like computers, televisions, telephones, music systems, and so on. E-waste contains many hazardous materials like lead, copper, zinc, and aluminium, flame retardants, plastic casings, cables, etc., which can have harmful effects on the environment, if burnt or buried.

Some countries have begun exporting e-waste to developing countries. China and India import such waste and recycle it, with undesirable consequences for health and the environment.

About 1.5 million personal computers (PCs) become obsolete in India every year and this figure will keep increasing. The city of Bangalore alone generates 600 tons of e-waste every year.

Sreenivas Shetty, a former software engineer, has now set up the Indian Computer Crematorium in Bangalore to recycle e-waste. His unit separates different components of the waste and uses the working parts to assemble computers for schools. The remaining waste is recycled into metal and plastic powder for use in other industries.

Many more such crematoria are needed!

Source: Jamwal (2003) and Vyas (2004)

Apart from these regular sources, waste also comes from special events:

- Waste from natural disasters: Rubble from earthquakes, slag and ash from volcanoes, waste left behind by floods, cyclones, and typhoons
- Waste from wars and conflicts: Apart from dead bodies and destroyed buildings, wars leave behind exploded and live shells, landmines, etc. In some cases, deadly material used in weapons of war has effects lasting decades. Agent Orange used in Vietnam (read Box 16.3 in Chapter 16 of this book) and depleted uranium used in the 1991 Gulf War are examples.

It is very difficult to assess the effects of the various types of waste on the environment. We do not know the total amounts, composition, and dispersal of this waste. Nor do we have enough scientific knowledge of the long-term impact of many substances that form a part of waste. What is considered non-hazardous today may be declared dangerous tomorrow. In fact, environmentalists feel that we must follow the precautionary principle and treat every chemical as being potentially harmful unless proved otherwise.

How are wastes managed?

In industrialized countries, household waste is separated into categories such as organic material, paper, glass, other containers, etc. This separation is often done in homes by using different bins for the disposal of different items. In developing countries, waste is not separated, though some cities are trying to persuade the public to separate waste.

The simplest and most common method used in the cities is to collect and dump the waste in a landfill. These landfills are located just outside the city. There are now thousands of landfills in the world with huge piles of waste. In industrialized countries, you can also see separate mountains of used cars and tyres. Many countries and cities have run out of space for landfills.

In the poorer countries, rag pickers sift through the waste, collect the reusable and recyclable material and sell it to the scrap traders. They, in turn, take the material to the recycling units. The rag pickers (the majority of whom are women and children) work in extremely unhygienic conditions and yet provide a great ecological service by manually separating thousands of tons of recyclable waste from the garbage dumps. (Read the story of women scrap collectors in Chapter 17.)

Often, the waste in a landfill is burnt. While this reduces the volume of garbage, it releases deadly dioxins into the atmosphere. Proper incineration of waste needs modern technology and proper management.

How is liquid waste managed? Sewage and industrial effluents are in most cases released directly into water bodies—rivers, lakes, or the ocean. Very often they are not treated before release.

The 1972 London Convention (see Chapter 12) prohibits the dumping of hazardous waste into the ocean. The ocean is, however, still not safe, since the Convention is not observed by all countries. Further, thousands of tons of toxic substances, including nuclear waste, dumped into the ocean before the Convention came into force, are still present in the ocean and are silently polluting the marine environment.

With increasing amounts of waste being generated, its management is becoming difficult and expensive. Industrialized countries have found an easier and less expensive method: exporting the waste to other countries!

Why are hazardous and toxic wastes exported?

Many of the industrialized countries have a waste management problem. Since their economies are based on constant growth, development, and consumption, wastes are piling up and they are running out of suitable spaces for landfills or dumping yards. At the same time, these countries have strict environmental regulations that make waste management expensive.

The most attractive option for the industrialized countries is to export the waste to developing countries, where disposal is cheap and environmental regulations are lax. The latter need the money and the former want to get rid of their waste. Developing countries do have more space for disposal, but their tropical ecosystems are more vulnerable to the damaging effects of the waste. There are many cases of environmental damage caused in the developing countries due to improper management of imported toxic waste.

At any given time, there are a number of ships carrying toxic waste prowling the high seas, ready to dump the waste on an unsuspecting poor country. Alternately, they just dump the cargo somewhere in the middle of the vast ocean (read Box 13.3).

Box 13.3 Believe it or not: Around the world in 16 years

The ship left the US in 1989 loaded with 14,000 tons of toxic flyash from Philadelphia's municipal waste incinerator. And 16 years later, 2500 tons of that very flyash came back to Philadelphia's garbage dump! In between, the original ship travelled to 11 countries in four continents, was sold once, changed its name twice, nearly witnessed a mutiny, had its engineer jailed, was twice turned away from ports at gunpoint, and even disappeared for a time. It is a story fit for a Bollywood movie!

In the mid-1980s, the city of Philadelphia had no space to dump the ash from its trash incinerators. The city signed a contract with a shipping company and the cargo ship *Khian Sea* headed for the Bahamas with the ash.

Even before the ship reached the Bahamas, its government refused to take the cargo. The ship then began a remarkable journey round the world looking for a place to dump the ash. The environmental organization Greenpeace had been alerting countries and no port would allow it to unload the deadly stuff. After two years, it obtained permission from Haiti to unload the cargo as fertilizer.

Hardly had the crew unloaded 4000 tons, when public protests in Haiti forced the ship to leave with the rest of its cargo. It then disappeared from the public gaze, but turned up in Singapore in November 1998—without the ash!

The story did not end there. Spurred on by Greenpeace, the Haitians kept up pressure on the US to take back the 4000 tons of ash that had been dumped on a beach in Haiti. This campaign continued through two changes of government in Haiti. Finally, the US agreed to take the ash back.

In the year 2000, the ash came back to the US and landed in Florida. At least five states and the Cherokee Nation (Native Americans) refused to accept the ash. As the ash waited on a barge in Florida, two full trees grew on it! Finally, in July 2002, what was left of the ash came back to a landfill in Philadelphia.

It was surely a long journey in time and distance. There is, however, an unanswered question: Where are the missing 10,000 tons of the ash? In which part of the ocean were they dumped and with what consequences?

An even more important question: How many other ships with deadly cargoes are on the prowl?

Source: Beans (2002) and the websites *www.globalresponse.org/*, *www.essentialaction.org/return/chron.html*, and *www.monitor.net/rachel/r55.htm*

Waste management has become an international industry and even some industrialized countries like the UK import waste and make money on it. Ninety per cent of the hazardous waste generated in the US is exported to Canada.

The 1989 Basel Convention on the Control of Transboundary Movements of Hazardous Wastes and their Disposal (see Chapter 12) aims at minimizing the creation of hazardous wastes, reducing transboundary movements of such wastes, and prohibiting their shipment to countries lacking the capacity to dispose of them in an environmentally sound manner.

Are hazardous wastes dumped in India?

Ten people were killed and 15 others injured in a blast on September 30, 2004 in a steel factory in Uttar Pradesh. Artillery shells in metal scrap imported from Iran had caused the explosion. The shells were probably smuggled into Iran from Iraq and somehow found their way into the scrap.

According to the Basel Convention as well as India's Hazardous Wastes (Management and Handling) Amendment Rules 2003, scrap metal is hazardous. Import of such waste is not banned in India, but prior permission is needed from the respective state pollution control boards and the Ministry of Environment and Forests.

Huge amounts of wastes are imported into India and the customs authorities do not have the required personnel and equipment needed to check each consignment for explosive or toxic material. They only check if the seals are okay. Another major concern is the hazardous waste that is released and generated during ship-breaking.

What are the environmental consequences of ship-breaking?

Ship-breaking has become a big industry today. The huge vessels, which have served their use and have been decommissioned, are sent to yards where their parts are recycled to the maximum extent possible. India, Taiwan, China, and Bangladesh have large ship-breaking facilities.

Most of the ships sent to the yards contain hazardous material such as asbestos, toxic paints, and fuel residues. However, as we saw in the case of the ship-breaking yard at Alang, at the beginning of this chapter, ship-breaking is not just a solid waste problem. It has many dimensions: the export of hazardous waste to developing countries, the environmental problems of handling toxic waste, the safety and health issues of the workers, the social tensions the enterprise creates, the role of the government, and so on.

Ship-breaking results in the following kinds of pollution:

- Discharge of oil into the ocean, which damages marine organisms and birds and destroys their natural habitats.
- Wastes like blasting residue and paint chips, which contaminate the soil and surface water.
- Improper storage and disposal of scrap metal and other wastes, which causes lead contamination.
- Cutting the metal parts with a blowtorch, which generates smoke, hazardous fumes, and particulates of manganese, nickel, chromium, iron, asbestos, and lead.

Responding to a public interest petition, the Supreme Court of India issued, in October 2003, a number of directives with regard to the ship-breaking industry. Some of the important directives are:

- The owners should properly decontaminate the ship before the breaking operation. They should also submit a complete inventory of hazardous waste on board.
- The Maritime Board and the Pollution Control Board of the concerned state should monitor all ship-breaking operations.
- The industry should be allowed to operate only if it has the facilities for waste disposal in an environmentally sound manner.
- Waste generated during the process should be classified into hazardous and non-hazardous categories and the details made know to the concerned State Maritime Board.

From October 2004, decommissioned ships are also covered under the Basel Convention. Consequently, countries exporting ships for recycling should seek the informed prior consent of the importing countries. This decision would encourage industrialized countries to set up domestic ship recycling facilities. Alternately, they would have to carry out some level of decontamination of ships before sending them abroad for recycling.

In ship-breaking, we are mainly concerned with hazardous waste coming from other countries. What about the waste generated by our own industries?

Which are the major polluting industries of India?

The following are examples of some polluting industries in India and the possible measures to mitigate the problems:

- Around 2500 tanneries discharge about 24 million cu. m of wastewater, containing high levels of dissolved solids and 400,000 tons of hazardous solid wastes, per year. Technologies exist for recycling the wastewater and recovering chrome from the waste. Indian tanneries are, however, reluctant to incur any costs on clean-up operations. (On the positive side, vegetable or EI tanning, which does not use any hazardous chemical like chrome, is coming back into use.)
- About 300 distilleries discharge 26 million kilolitres of spent wash per year containing many pollutants. The wash can be treated and used for making compost, though not much is being done so far in this direction.
- Thermal power plants discharge 100 million tons of flyash and this figure is expected to reach 175 million soon. Flyash contains silicon, aluminium, iron, and calcium oxides and is said to cause silicosis, fibrosis of the lungs, cancer, and bronchitis. Flyash, however, is being utilized in many ways: for making bricks, blending with cement, building roads and embankments, adding micronutrients to soil, etc. The best option is, however, to minimize its generation.

India continues to produce several pesticides banned or restricted in other countries. Examples are DDT, Malathion, and Endosulfan. A sixth of the total pesticides used in India are banned in other countries. The bulk of the production is by small units that have no technology to treat the toxic and non-biodegradable pollutants generated during the manufacturing process. Fifty per cent of the pesticides used in India are for growing cotton.

Industrialized states like Gujarat, Maharashtra, Tamil Nadu, and Andhra Pradesh face major problems of toxic and hazardous waste disposal. For example, the Ahmedabad-Vadodara-Surat industrial belt houses 2000 industrial units and more than 63,000 small-scale units manufacturing chemicals like soda ash, dyes, yarns, and fertilizers. Most of the units dump their wastes in low-lying areas within a radius of two km. A major illegal dump yard has sprung up on the banks of the river Daman Ganga. During the monsoon, the hazardous substances are washed into the river. There are many such cases across the country.

The Ministry of Environment and Forests estimates that 7.2 million tons of hazardous waste is generated annually in India. This will annually require about one sq km of additional landfill area and Rs 16 billion for treatment and disposal. In addition, industries discharge about 150 million tons of high volume-low hazard waste every year, which is mostly dumped on open low-lying land.

Why is pollution by plastic often in the news?

Developed in the 1860s, plastic has become an indispensable part of our lives. Longevity is its main characteristic, but this very quality has also become a major problem. It is the non-biodegradable nature of plastic that is causing a massive environmental problem. It simply does not 'go away' after use.

Plastic bags and bottles are found discarded everywhere. They now form an increasing proportion of municipal waste and cause many environmental problems. They clog sewage lines and canals and litter public places, gardens, wildlife reserves, and forests. Animals and birds consume plastic items and some die as a result. Disposing of plastic by burning it only creates more toxic fumes. The worst offender is the thin carry bag that is easily airborne when thrown away.

The plastic carry bag has become a common item all over the world, both in poor and rich countries. In 2002 alone, between four and five trillion bags were produced, almost all of the non-biodegradable variety. About 100 billion bags are thrown away in the US every year.

Certain plastics can be recycled. Many household articles like water containers and buckets are made of recycled plastic. Non-reusable items include carry bags, cups, plates, and magazine wrappers. Even in such cases, ecofriendly plastics are available as options.

There are strong views both for and against plastic. The plastics industry cites the advantages of convenience, low cost, employment potential, recycling possibilities, etc. Environmentalists point out the problems listed above and also add the difficulty of separating recyclable plastic items in landfills.

Some countries have banned plastic bags. In India, several states and local bodies have enforced bans of various degrees, especially on thin plastic carry bags. The Government of India has banned carry bags made of plastic films of thickness less than 20 micrometers. It is said that manufacturers are hence busy making bags 21 micrometers thick!

What are common effluent treatment plants?

A small-scale industry cannot afford to clean up all its effluents. One solution to this problem is to collect the waste of several units from an industrial estate and treat it at a common location. The various state pollution control boards in India are now compelling specific polluting industries in an area to set up common effluent treatment plants (CETPs).

CETPs are a cost-effective, and perhaps the ideal, solution to controlling pollution. Half the cost is met by the industries and a quarter each by the state and central governments. Each unit also contributes towards the running of the plant.

By 2002, 53 CETPs had been built and 83 more were under construction. The first plants were set up to treat tannery wastes in Tamil Nadu, textile wastes in Rajasthan, and industrial waste in Andhra Pradesh.

CETPs have run into many problems. In many cases, reliable information is not available on the amounts and types of waste expected to reach the CETP. As a result, their design may not be appropriate. Further, a cocktail of different chemicals can be far more toxic than the individual chemicals themselves.

There are also problems of fixing the charges and getting the users to pay them. Questions also remain about the management and proper functioning of the CETPs. For example, who should be held responsible if the treatment is poor and the CETP itself becomes a polluter?

How is municipal waste handled in Indian cities and towns?

According to the Central Pollution Control Board, the daily per capita generation of municipal solid waste in India ranges from 100 g in small towns to 500 g in large towns. (The real figures are likely to be higher.) The recyclable content is said to be between 13 and 20 per cent.

The amount of solid waste is growing faster than the population, at least in the cities. For example, during 1981–91, Mumbai's population increased from 8.2 to 12.3 million (by 49 per cent), while the daily municipal waste generation grew from 3200 to 5355 tons (by 67 per cent). The total amount of solid waste generated annually in Indian cities was 48 million tons in 1997. It is expected to reach 300 million tons by 2050 and will then require about 170 sq km of landfill area.

Seventy per cent of Indian cities do not have adequate waste transportation facilities. Streets piled with garbage, choked drains, and stinking canals are common features of most cities. The garbage that is transported in old, polluting trucks is simply dumped on sites in low-lying areas on the outskirts of cities. The choice of the site is more a matter of availability than suitability.

A dangerous practice is disposing of biomedical waste (from hospitals and clinics) along with municipal waste in dumpsites. The biomedical waste could turn the entire yard infectious. Further, biomedical waste also contains sharp objects like scalpels, needles, broken ampoules, etc., which could injure or infect rag pickers and municipal workers.

The Directorate General of Health Services estimated as far back as 1993 that the total infectious biomedical waste generated in India was 54,000 tons. With the explosive growth in the number of private hospitals in recent years, this amount is likely to have increased tremendously.

In most cities, solid wastes lie unattended in dumpsites and attract birds, rodents, insects, and other organisms. As the waste decays, it releases a bad odour and airborne pathogens. Further, the practice of burning the waste creates toxic fumes and spreads hazardous substances in the air.

Some municipalities in India, like Alandur near Chennai, are selling their organic waste to private companies, which compost the waste into manure. While the Chennai Municipal Corporation is paying a private firm to remove its garbage, Alandur is getting paid for its garbage. There is also a plan in Chennai to convert waste into energy using suitable technology.

How can solid waste be recycled?

A good way of dealing with the solid waste problem is to recycle it. Recycling is the processing of a used item or any waste into usable form. There is a large global recycling industry. In India, we have a thriving unorganized recycling industry, thanks to the itinerant collector, who buys old newspaper, bottles, used clothes, utensils, scrap, motor oil, etc.

Recycling has multiple benefits:

- By taking away some of the waste, it reduces environmental degradation.
- As against expenditure incurred on disposing of the waste, we now make money out of waste material.
- We save energy that would have gone into waste handling and product manufacturing.

Some specific examples of savings through recycling are:

- When aluminium is resmelted, there is considerable saving in cost. The recycling process is, however, energy-intensive.
- Making paper from waste pulp rather than virgin pulp saves 50 per cent energy.
- Every ton of recycled glass saves energy equal to 100 litres of oil.

Safe and profitable technologies for recycling paper, glass, metals, and some forms of plastic are available. Biogas can be produced from landfill waste. Paper factories can also recycle their waste (read Box 13.4).

Box 13.4 Inspiration: Champion of sustainable agriculture

When the young chemical engineer was working for BASF, a German chemical company, he asked the owner of an orchard if he wasn't afraid to eat the apples after they were sprayed with pesticides. The owner replied that he didn't eat the fruit himself but only sold it to other people.

Shocked by this answer, Jose Lutzenberger began a critical examination of modern agriculture. He quit his job and returned to his native Brazil, where in 1971 he helped found Agapan, Brazil's first environmental NGO. Agapan quickly made a name for itself by leading a successful public campaign to stop pollution at a Norwegian wood pulp factory in Pôrto Alegre.

Along with other agricultural scientists, Lutzenberger campaigned for rules to control the indiscriminate use of pesticides. They had to battle against the powerful chemical industry, but managed to create awareness of pesticide problems and regulatory provisions. By the mid-1980s, Lutzenberger had become known as one of the leaders of Brazil's environmental movement. He was also a successful industrialist running a waste management company.

When he became Brazil's Environment Minister in 1990, he managed to persuade President Fernando Collor to recognize and protect the land of the Yanomami people in the northern Amazon. Soon, however, he resigned from the government, unable to withstand the corruption and intrigue.

Lutzenberger devoted the remainder of his life to promoting sustainable agriculture and denouncing industrialized farming methods. 'The modern farmer,' he said, 'is only a tractor driver or a poison sprayer.' He also took a strong stand against the ills of globalization.

For many years, Lutzenberger fought Riocell, a large cellulose and paper factory in southern Brazil, because it polluted the environment. During all those years, however, he remained on speaking terms with the factory's director. Eventually, the director hired Lutzenberger as a consultant—with dramatic results.

Riocell used to bury its waste in huge pits, which polluted and devastated the environment. Now, the factory hands over its total waste to Lutzenberger's company, which converts it into manure and other products for organic farming.

As a result, the environmental degradation has stopped, the factory saves half a million US dollars a year, and the waste brings in money. The processing involves low technology and is thus labour-intensive, supporting 50 full-time jobs.

Lutzenberger created a park right on the Riocell factory site. Instead of waste dumps, there are now fishponds and reeds with an abundance of birds. The whole park is a thriving ecosystem, wrapped right around the factory. Thus, at Riocell, ecological sustainability has become an industrial achievement.

Lutzenberger, environmental hero, winner of the Right Livelihood Award, family farm advocate, and opponent of multinational trade agreements, died on May 14, 2002 at the age of 75.

Source: The websites *www.theecologist.org, http://education.guardian.co.uk/higher/biologicalscience/story/,* and *www.unu.edu/unupress/unupbooks/uu16se/uu16se04.htm*

Recycling is not a solution to managing every kind of waste material. For many items, recycling technologies are unavailable or unsafe. In other cases, the cost of recycling is too high.

What is the way out of the waste management problem?

We have to move from waste management to waste prevention. That is, we should design clean production technologies or zero-discharge systems that use minimum amounts of raw materials, energy, and water and do not generate wastes.

The life cycle of a product should be such that at no stage is any natural ecosystem adversely affected. This should apply to raw material extraction, design, manufacture, material transport, actual use, and disposal. Clean production technologies do exist and they will be cost-effective if the true environmental costs are taken into account.

Even as we move to clean technologies, there is another way out of the industrial waste problem. If the industries in an area cooperate, they could design a system in which the waste from one industry becomes an input for one or more industries in the neighbourhood. The classic example of such a system is the Kalundborg industrial ecosystem in Denmark (read Box 13.5).

Box 13.5 Solution: More profits with zero waste!

Suppose industries in an area begin exchanging their outputs that are normally considered wastes. What is waste for one could be an input or raw material for another. If we are lucky, we could evolve an almost closed system in which matter and energy circulate within the system. The system as a whole would not generate any waste.

Ten units in the small town of Kalundborg in Denmark work together to produce almost zero waste and make more profits in the bargain. The system was not designed to be so. It began as one-to-one exchanges that made economic sense and evolved over 10 years as a zero discharge system.

The Kalundborg system comprises five core partners:

- Asnæs Power Station, Denmark's largest coal-fired power station with a capacity of 1500 MW.
- Statoil Refinery, Denmark's largest, with a capacity of 4.8 million tons per year.
- Gyproc, a plasterboard factory, making 14 million sq m of gypsum wallboard annually.
- Novo Nordisk, an international biotechnological company, with annual sales over US$ 2 billion.
- The City of Kalundborg that supplies heating to its 20,000 residents, as well as water to homes and industries.

The other partners are a cement plant, a producer of sulphuric acid, local farmers, greenhouses, and a fish farm.

The power station supplies its waste heat to the other four partners as well as to the fish farm. The surplus natural gas from Statoil goes to the power plant and to Gyproc. The flyash from the power plant becomes an input in the cement plant.

Statoil first removes the sulphur from the natural gas and sells it to the acid plant. The power plant is required to remove sulphur from its coal smoke and this goes to Gyproc as a cheaper substitute for gypsum. The farmers buy the sludge from the fish farm as well as Novo Nordisk as fertilizer.

This web of recycling and reuse has generated new revenues and cost savings for the companies involved and has reduced air, water, and land pollution in the area. The partners have reduced their water and energy consumption. In ecological terms, Kalundborg exhibits the characteristics of a simple food web: organisms consume each other's waste materials and energy, thereby becoming interdependent.

However, the Kalundborg model, an example of 'industrial symbiosis', will work only if the industries are of the right mix, are located near one another, and share an attitude of openness.

Source: Raven and Berg (2001) and the websites *www.symbiosis.dk* and *www.uneptie.org/pc/ind-estates/casestudies/kalundborg.htm*

What does the discussion on various forms of pollution tell us?

In this module, we have learned about air, water, soil, marine, and noise pollution. It is important to understand that the atmosphere, land, and water are interconnected parts of the Earth's ecosystem. Pollution of any one part affects the entire ecosystem.

Any form of pollution immediately or over a long period affects, either directly or indirectly, all the organisms on the land and in the ocean. Many organisms are very sensitive to pollution and migrate away from the source or simply die. Others like cockroaches, bacteria, and disease organisms continue to survive and even grow in number. The population of humans, who are now the major contributors to pollution levels, is also growing rapidly. It is, however, not clear how long they can continue to pollute and also exist!

What lies ahead?

Consumers, industries, and governments across the world have underestimated the economic and environmental costs of generating and managing waste. With increasing amounts of waste, growing costs of waste management, greater awareness of the damage caused by wastes, tougher controls on transport and disposal, and more determined resistance of communities to the location of disposal sites near their habitats, there will be a pressure on everyone to 'make waste slowly'.

A quick recap

- Waste has always been a part of the earth's ecosystem, but human activities now generate so much waste that it needs proper 'management'.
- Human activities generate many categories of wastes, each with its own characteristics and disposal problems.
- Hazardous and toxic wastes generated in industrialized countries are often exported to developing countries, where environmental regulations are lax, for disposal.
- Hazardous wastes are imported into India and cause many problems.
- Disposing of decommissioned ships is a complex task, since it involves handling large amounts of toxic waste.
- Tanneries, distilleries, and thermal power plants are some of the polluting industries in India.
- The widespread use of plastics is a problem because many plastics are non-biodegradable, persist for long, and cause many environmental problems.
- Indian cities face huge problems in managing municipal waste and sewage, though some amount of recycling is done.
- The only way out of the waste problem is to prevent or minimize the generation of waste.

What important terms did I learn?

Biomedical waste
Common effluent treatment plant
E-waste
Landfill
Ship-breaking

What have I learned?

Essay questions

Write an essay on the export of hazardous and toxic wastes from one country to another. Why are these wastes exported? What are the environmental and social issues involved? Which international agreements cover the case of such movement and disposal of waste? What are the recent developments in this regard?

Short-answer questions

1. What is ship-breaking and why is it a complex task?
2. Why has waste become a major problem in the world?
3. List any five categories of waste and give the source of each.
4. What are the ways of managing waste?
5. List the major polluting industries in India and describe the scale of their pollution.
6. What are the effects of plastic waste?
7. What are common effluent treatment plants (CETPs)?
8. How are Indian cities handling the increasing amounts of municipal wastes?
9. How can solid waste be recycled?

How can I learn by doing?

Trace the path of the waste that you dump in your street bin. Interview conservancy workers, municipal officials, rag pickers, waste dealers, etc. Write a report answering questions such as:

- Where does the truck take the waste?
- Is it treated in any way or is it just dumped on the land?
- What do the rag pickers collect, how much, and where do they sell it? How can we assess the role of this unorganized sector in waste management?
- Does the waste contaminate the soil, water, and air?
- How much of the waste is recycled and what are the processes used? Is there waste from recycling?
- What are the economics of the recycling trade?

What can I do to make a difference?

Examine every item of waste that is generated in your home. Find out where it came from and where it is headed. In each case, try to prevent the generation of that waste item. If it cannot be prevented, can it be reused or recycled in any way? Is there an alternative to throwing it in the garbage bin?

Some practical steps you can take:

- Minimize packaging waste by taking your own cotton bags to the market and the grocery store. Do not buy pre-packaged items. Buy reusable refill containers. Avoid supermarkets and if you do shop in them, take your own bag. Refuse the large plastic carry bag offered by the shop. Even better, support the small store, which uses old newspaper to pack groceries.
- Do not throw toxic materials, such as batteries, thermometers, and insecticides into the garbage. Find out if the manufacturers will take them back or if any one can recycle them. Otherwise, encase them in something like concrete before burying them.
- Avoid buying dangerous substances, such as mosquito repellents, insect sprays, chemical cleaners, detergents, etc. As far as possible try to use natural or organic alternatives.
- Reuse every possible item at home—paper, plastic bags, cards, envelopes, wood, etc. You can make artistic items using waste.
- Donate old clothes, books, etc., to NGOs for redistribution.
- Try to persuade supermarkets in your area to offer a small discount to shoppers who bring their own bags. Tell them that they can benefit by spending less on bags and by advertizing their 'green' concerns.
- Buy recycled products (paper, stationery items, etc.)
- Buy rechargeable batteries if you can afford them.

Where are the related topics?

- The problems of handling nuclear waste and oil spills are covered in Chapter 14.

- Chapter 16 deals with the impact of toxic substances and industrial accidents like the Bhopal gas leak on human health.
- The laws that control industrial and other pollution are discussed in Chapter 20.

What are the deeper issues that go beyond the book?

Can a country shift its environmental problems to another country? Should the export of hazardous wastes be banned? What then will be the consequences in developing countries, where thousands of people are employed in handling foreign waste (in Alang, for example)?

Where can I find more information?

- The Greenpeace website, *www.greenpeace.org,* gives full details of the export of hazardous wastes and the Alang issue.
- Beans (2002) and many websites such as *www.monitor.net/rachel/,* describe the fascinating story of *Khian Sea*.
- For more on Lutzenberger, access *www.unu.edu* and *www.theecologist.org.*
- The website *www.symbiosis.dk* gives details of the Kalundborg case.

What does the Muse say?

A poem for reflection written by Fuad Rifka of Syria:

A Vision

All day long
in the sky this bird,
Where to rest?
'On the tree'
Where is the tree?
'In the field'
Where is the field?
'North at the wood'
Where is the wood?
'East of the mountain'
Where is the mountain?
'There you blind man, don't you see?'
Oh, yes, I see,
I see mountains of refuse, its smoke smells of fields and trees.
(a voice from the end of the 20th century):
'Signs are imminent winds. In the stars,
On that day,
the mother will not recall her infant,
Nor the sky its stars,
And between the seasons
No time for making the crossing'

CHAPTER 14

Disaster Management

It wasn't the Exxon Valdez *captain's driving that caused the Alaskan oil spill. It was yours.*

Greenpeace advertisement

What will I learn?

After studying this chapter, the reader should be able to:

- appreciate how coastal mangroves reduce the damage caused by cyclones;
- understand that natural disasters could also be the result of human activities;
- enumerate the damage caused by natural disasters in the world and in India;
- explain where and why most of the floods and cyclones occur in India;
- recognize that extreme weather events are becoming more common;
- explain how earthquakes are caused and how their intensity is measured;
- describe the impact of earthquakes in general;
- list the main features of the draft National Policy on Disaster Management of 2003;
- list the major nuclear accidents that have occurred and describe their consequences;
- explain why managing nuclear waste is a problem;
- describe the environmental impact of oil spills and list the major spills that have occurred; and
- compare the various ways of coping with oil spills.

Box 14.1 Disaster: When the mangroves are gone ...

Will it or won't it? That was the all-important question in Bhubaneswar, Orissa's capital, on the night of October 28, 1999. Would the severe cyclonic storm that had developed in the Bay of Bengal hit the Orissa coast, or move away towards West Bengal and Bangladesh?

Just 12 days earlier, a cyclone had struck Ganjam District in the state, killing 100 people and destroying the town of Berhampur. As cyclone warnings began sounding again from October 25 onward, the people were more disbelieving than alarmed. How could two cyclones strike the same coast in such quick succession? The administration made only feeble preparations hoping that the cyclone would bypass Orissa. It is said that the Chief Minister consulted his astrologers, who assured him that Orissa would be spared.

The second cyclone, however, beat the odds and struck the coast. Not only that, it struck with a fury that made Ganjam seem like a gentle breeze.

A gale with winds of up to 300 km per hour and tidal waves 8 m high lashed the coast from Puri to Balasore, covering six districts. It lasted for nearly two days.

What was the cyclone's toll? There was heavy damage to life and property in 18,000 villages, with

20 million people affected. More than 10,000 people were killed, 15 million were rendered homeless, half a million cattle were lost, 1.8 million hectares of agricultural land was damaged, 90 million trees were uprooted, besides significant damage to infrastructure. The damage caused was threefold: the physical impact of strong winds; the resultant storm surge and flash floods; and saline inundation.

Previously, forests had formed a five-km wide buffer zone against strong winds and flash floods. However, large tracts of Orissa's mangroves had been cleared to make way for shrimp farms and the coast had lost its natural protective shield. When the cyclone struck, its path was unfettered and it travelled as much as 100 km inland. Wherever mangroves were still intact, however, they absorbed a part of the shock, and the damage was less. An area near Paradeep where the forests were intact was largely saved from the ravages of the cyclone.

The two cyclones uprooted practically all the tree cover in the immediate vicinity of the coast and caused much damage to trees tens of kilometres inland. This continues to affect the microclimate in these areas. The lack of protective forest cover also made it possible to inundate large areas causing much destruction.

What about the future? Supercyclones can hit Orissa or any part of the eastern coast again. Is Orissa better prepared? Is the planting of mangroves being undertaken vigorously?

Source: Banerjee (2001) and the websites *www.nic.in/cycloneorissa and www.oras.net.*

What are the lessons of the Orissa supercyclone?

The important lessons from the Orissa cyclone are: First, we should restore and conserve coastal mangroves. Second, weather patterns are changing and we should expect the unexpected. This became apparent when a totally unexpected tsunami devastated the coasts of Thailand, Sri Lanka, and India in December 2004.

How can disasters be classified?

We could categorize disasters as natural (cyclones, hurricanes, floods, earthquakes, tsunamis, landslides, etc.) and human-induced (nuclear and industrial accidents and oil spills, for example). We should remember, however, that natural disasters could also be due to human activities. Landslides occur when mountainsides are deforested, hurricanes and cyclones are very likely the result of global warming induced by greenhouse gas emissions, an earthquake could occur due to a large reservoir created by the construction of a dam, and so on.

There are also disasters waiting to happen. The mounting nuclear wastes and the accumulation of nuclear and biological weapons are examples. We think of disasters as swift events, but they can also occur slowly over a period of time. Radioactive contamination near nuclear power plants and uranium mines and arsenic poisoning by contaminated groundwater are examples.

What is the natural disaster scene in the world?

Between 1970 and 2000, natural disasters in the world killed at least three million people and affected millions more. The average annual economic losses due to disasters were eight times more than in previous decades. The losses in the 1990s were more than US$ 400 billion.

Ninety per cent of natural disasters and 95 per cent of all deaths in such disasters occur in developing countries. The average annual population affected is highest in China (90 million),

'Today's Weather' replaced by 'Today's Disaster'

followed by India (56.6 million) and Bangladesh (18.5 million). In November 1970, a cyclone claimed 500,000 lives in Bangladesh.

Why is India classified as a disaster-prone country?

India's size, geographical position, and the behaviour of the monsoon make it one of the most disaster-prone countries in the world. The subcontinent is highly vulnerable to droughts, floods, cyclones, and earthquakes. In addition, the Himalayan region experiences landslides, avalanches, and bush fires. However, volcanoes are uncommon in India, with just two active ones in the Andamans.

The number of people affected in earthquakes, cyclones, and floods is the highest, followed by those affected by droughts. The areas prone to different types of natural disasters are as follows:

- Cyclones: The eastern coastline and the islands of Lakshadweep, Andaman and Nicobar.
- Floods: The major river valleys such as those of the Ganga and the Brahmaputra.
- Earthquakes: Fifty six per cent of the land area.
- Droughts: Sixteen per cent of land area spread over 16 states.
- Landslides: The Himalayan region and Western Ghats.
- Fires: Bihar, West Bengal, Orissa, and the North East.

Cyclones occur mainly in the Indian Ocean. They are violent tropical storms in which strong winds move in a circular fashion. Hurricanes and typhoons are violent storms with very strong winds experienced mainly in the western Atlantic Ocean.

How do floods occur in India?

Floods are the result of the peculiar rainfall pattern in most of the country. Of the total annual rainfall, 75 per cent occurs over three to four months. This leads to a very heavy discharge from the rivers, which floods large areas.

Of all the natural disasters that occur in India, the most frequent and devastating are river floods. The Ganga-Brahmaputra-Meghna basin, which carries 60 per cent of the total river flow in India, is most susceptible to floods. The rivers Brahmaputra, Ganga, and their tributaries carry tons of debris and water throughout the year, and during the monsoon the water flow exceeds the capacity of the rivers, breaks the man-made ridges, and floods whole areas.

The states most affected by floods are Assam, Bihar, and Uttar Pradesh. About 40 million hectares of land in the country are flood-prone. Every year, an average of 19 million hectares of land becomes flooded.

What about cyclones?

We use the name cyclone for the strong tropical storms that develop in the Indian Ocean. A cyclone can travel 300–500 km in a day and often brings large amounts of rainfall with it. When the storm approaches the coast, the sea level rises, creating what is called a storm surge.

The Indian Ocean is one of the six major cyclone prone regions of the world. India is exposed to tropical cyclones arising in the Bay of Bengal and the Arabian Sea. On an average, every four out of five cyclones is generated in the Bay of Bengal and has an impact on the eastern coast.

Siltation too plays a role in increasing storm surges and the resultant floods. The major rivers of India and Bangladesh, including the Ganga, Yamuna, and Brahmaputra, flow into the Bay of Bengal. Due to deforestation and soil erosion, the rivers bring vast amounts of silt. When the silt is deposited in the Bay of Bengal, the area around the Gangetic Delta becomes shallower. This creates ideal conditions for high tidal waves and storm surges. Infrastructure development on the coast prevents the seawater from receding, thus aggravating floods.

A UN study reports on a potential danger in the Himalayas. More than 40 lakes in the mountains could burst their banks at any time and flood communities up to 100 km downstream. Very few of the people at risk would get any advance warning.

The lakes are formed as mountain glaciers melt, a process that is probably being accelerated now by global warming. The water is kept in place by ice or piles of sediment called moraines that were deposited long ago. As the lakes grow, however, the moraines are beginning to collapse and every monsoon season the risk of a disaster increases.

Are cyclones and hurricanes increasing in frequency and severity?

Many scientists and environmentalists are certain that global warming (Chapter 19) is occurring and as a result natural disasters like cyclones and hurricanes are increasing in frequency and severity. We cannot as yet conclusively establish a connection between any individual weather event and global warming. However, there is considerable evidence to make us believe that climate change is having an overall impact on the pattern of extreme weather events.

A statistical analysis of 98 years of data (1891–1988) reveals the increasing frequency of pre- and post-monsoon cyclones in the northern Indian Ocean. Storms have become more frequent especially in the past four decades. On the other hand, cyclones have become less frequent during the summer monsoon.

There is evidence of a tendency toward weather extremes in the Bay of Bengal. Maximum summer temperatures in Orissa have shown a rise since the 1950s. At the same time, winter and night temperatures have fallen. Another observed change is that the buffer zone between the warmer surface seawater and the colder water at the bottom of the Bay of Bengal has been lost. This might indicate that even small changes in temperature affect the weather in the area very fast. Computer simulations suggest that if tropical sea surfaces warm by a little more than 2°C, the stronger cyclones will have wind speeds 5–12 per cent higher than before, that is, the worst tropical storms will get even worse.

How are earthquakes caused and how are they measured?

Earthquakes are violent upheavals of the Earth's surface caused by stresses occurring deep below. When a weak point gives way under stress, huge masses of underground rock begin to shift, and energy is released as shock waves. These waves move outward from the earthquake's focus and reach the surface.

The focus of an earthquake is the initial point of shift and its epicentre is the point on the Earth's surface directly above the focus. The severity of an earthquake is measured on the modified Richter scale. It is a measure of the amount of energy released, which is indicated by vibrations on a seismograph.

A magnitude of less than 4.0 on the Richter scale denotes an insignificant event. It is a minor earthquake if it measures between 4.0 and 4.9; moderate if it is between 5.0 and 5.9; and severe if it is between 6.0 and 6.9. A very severe earthquake will range between 7.0 and 7.9 and an extremely severe one will measure 8.0 or above. Each unit on the Richter scale represents an amplitude 10 times greater than the next smaller unit. Thus, a 5.0 quake is 10 times more severe than a 4.0 quake.

Mild earthquakes occur all the time, mostly under the ocean. Every year, about 20 serious earthquakes occur, about half of them in built areas. An earthquake is often followed by aftershocks that gradually fade over several months.

About 95 per cent of all earthquakes in the world occur along two belts. The first is along the Pacific Rim and affects Japan, Alaska, California, and the west coast of South America. The second belt stretches from the East Indies through Indo-China, the Himalayas, the Caucasus, and the Mediterranean.

About 56 per cent of India is vulnerable to seismic activity of varying intensities. Most of the vulnerable areas are generally located in the Himalayan and sub-Himalayan regions and in the Andaman and Nicobar Islands.

Some scientists believe that large reservoirs behind dams exert enormous pressure on rock formations and cause earthquakes. In India, there was a widespread belief that the Koyna Dam caused the 1967 earthquake in that region of Maharashtra.

Accurate prediction of earthquakes is still difficult and people in vulnerable areas depend more on quake-proof structures to protect them.

What is the impact of earthquakes?

The primary effect of an earthquake is the shaking and possible displacement of the ground. This results in damage to buildings, roads, dams, pipelines, etc., and causes loss of life and property. The secondary effects include flooding caused by subsidence of land, fires, epidemics, etc. Coastal areas could be hit by earthquake-generated waves called tsunamis (see the section on tsunamis).

On an average, about 15,000 people are killed every year in earthquakes. They destroy property and cause fires and floods. Earthquakes in the ocean create giant waves and cause coastal or underwater landslides. We should note that damage and fatalities resulting from secondary causes such as fire, tidal waves, flooding, and aftershocks can be as great, if not greater, than the impact of the main earthquake. The death-toll and loss of property are heavy in the poorer countries that do not have quake-resistant structures.

According to historical accounts, the highest number of deaths resulting from a single earthquake occurred in 1556, when 830,000 people died in Shaanxi, China. In the 1976 earthquake in Tangshan, also in China, the official death toll was 225,000—although unofficial estimates put the figure as high as 655,000.

Three earthquakes in recent times have each claimed in excess of 10,000 lives. While the Turkey quake in 1999 killed 17,000, over 20,000 lost their lives in the Gujarat quake in 2001 (read Box 14.2), and 26,000 people were killed in Iran in 2003.

Box 14.2 Disaster: Only the town tower remained!

It was Republic Day, 2001. Nancy Takkar a 12-year-old in Anjar in Gujarat had reached school well before time for the celebrations. A few minutes later, at 8.45 a.m., the earth started shaking violently, tossing the children around the school building. For Nancy and her friends in the Anjar School it was just the beginning. Within a few seconds the school building collapsed with 200 children crushed beneath it. Nancy was the lone survivor.

A severe earthquake with a magnitude of 6.9 on the Richter scale had hit Gujarat with its epicentre 20 km northeast of the town of Bhuj in Kutch. The shock was felt in most parts of the country. The districts of Kutch, Bhavnagar, Surendranagar, Rajkot, and the Ahmedabad districts were devastated.

It was the most severe earthquake in the last 50 years in India, with 20,000–30,000 dead, 150,000 injured, and 15.9 million affected. The total economic loss was estimated at Rs 225 billion.

Bhuj was the worst affected town with about 10,000 people killed. Almost half of its structures had been levelled. Amazingly, its historic tower was still standing. This was the case in other places too, with many old buildings remaining intact, while new buildings had collapsed.

As in the case of the Orissa cyclone (Box 14.1), the state government was grossly unprepared to meet such an emergency. In many places help arrived too late to save people trapped under the debris.

Could the earthquake have been foreseen? About a month earlier, on December 27, 2001, there were indications of heightened seismic activity, when a minor earthquake (4.0 on the Richter scale) occurred in Bhuj. But it was ignored.

Kutch is known to be an earthquake-prone area and between 1845 and 1956, one very severe, five severe, and 66 moderate earthquakes have occurred in this area. There is no evidence of human activity having been responsible for the 2001 earthquake. Are we sure, however, that the heavy extraction of groundwater and oil as well as the construction of big dams are not aggravating tectonic activity, which could result in further disasters similar to Bhuj?

Source: Venkatesan and Swami (2001), Swami, Menon, and Bavadam (2001), and the website *http://gujarat-earthquake.gov.in/.*

Death and destruction are often just the beginning of the hardships that communities face after an earthquake. Quakes may leave thousands injured and permanently disabled, putting enormous strain on medical and health care systems that may already be stretched to the limit. They can also cause intense psychological and social damage. Post-traumatic stress disorder persists long after the event and, in certain cases, interferes with individual's ability to function properly for the rest of his/her life.

If an earthquake ever hits a nuclear reactor, the consequences will be unimaginable. This is an argument against the building of reactors, since no place on Earth is really quake-proof.

What are tsunamis?

Tsunami is a Japanese word meaning 'harbour wave'. A tsunami is not a tidal wave and is not caused by winds or the gravitational pull of the Moon or the Sun. Most tsunamis are caused by undersea earthquakes that set off waves in water.

A tsunami moves silently but rapidly across the ocean and when it hits the coast, it unexpectedly rises as destructive high waves. These waves may last just minutes, but can cause widespread devastation along the coast.

Most tsunamis occur in the Pacific Ocean. During the 1990s, 82 tsunamis occurred worldwide, many more than the historical average of 57 a decade. They are relatively rare in the Indian Ocean but not unprecedented.

The tsunami that hit South East and South Asia on December 26, 2004, was the biggest ever in history. It was triggered by a massive undersea earthquake measuring nearly 9.0 on the Richter scale that occurred in Sumatra. It moved with a speed of about 900 kmph and hit the Andaman and Nicobar Islands barely an hour after the quake occurred.

Nearly 300,000 people died in this disaster and entire coastal villages were wiped out in Thailand, Sri Lanka, and India. Car Nicobar, Cuddalore, and Nagapattinam were the worst affected places in India. Thousands of people, particularly fisherfolk, lost their homes and livelihoods. It is noteworthy that the damage was less in those parts of the coast that had natural barriers like mangroves and casuarina trees.

How can the effects of earthquakes be mitigated?

We can locate fault zones and map them, so that their presence is considered while planning settlements and irrigation projects and when siting nuclear establishments, etc. Building codes have been established for the appropriate design of structures in earthquake-prone areas. In places like Japan and California, shock-proof structures are common.

A disaster management plan should be prepared for all quake-prone areas. This should include creation of public awareness, setting up disaster management cells, citizens' drills, etc.

Is there a general policy on disaster management in India?

In recent years, there has been a shift of focus from post-disaster management to preparedness and mitigation. Forecasting and monitoring systems are now in place for earthquakes, droughts, floods, and cyclones.

The draft National Policy on Disaster Management released in 2003, proposes the following:

- A holistic, and proactive approach towards prevention, mitigation, and preparedness.
- Each ministry and department of the central and state governments should set apart adequate funds for vulnerability reduction and preparedness.
- Mitigation measures should be built into ongoing schemes and programmes.
- Each project in a hazard-prone area should include mitigation measures and vulnerability reduction.
- A national disaster management law should be enacted covering all the existing mechanisms.

Let us now turn to human-induced disasters.

How do nuclear accidents occur?

Nuclear power was expected to be 'too cheap to meter'. In reality, it turned out to be an expensive form of energy. Scientists and administrators have always assured us that nuclear power was safe, given all the foolproof features of reactors. Yet, accidents, big and small, have occurred (and continue to occur) in nuclear establishments all over the world.

The worst nuclear accident to date is the one that occurred in Chernobyl in 1986 (Box 14.3). Here are some of the other reported cases:

- The most serious nuclear accident in the US occurred on March 29, 1979 at the Three Mile Island Power Plant in Pennsylvania. The failure of a cooling system in a reactor was followed by human error. This led to a 50 per cent meltdown of the reactor core. Unknown amounts of radioactivity escaped into the atmosphere, but there were no immediate casualties. Five thousand people were evacuated and 50,000 fled of their own accord. It took 12 years and US$ 1.2 billion to repair and reopen the plant. This accident shook public confidence in the safety of nuclear plants and led to delays in, and cancellations of, new projects.
- In 1957, human error caused a disastrous fire at Sellafield, a major nuclear complex in England. There was heavy leakage of radioactivity, but no warnings were given and nobody was evacuated. The eventual death toll could have been about 1000. Several hundred accidents in Sellafield have been reported over a period of 40 years. There are reports of unsafe exposure of workers and incidence of cancer in the local communities due to radiation leaks.
- Several accidents have taken place at the large Hanford nuclear power complex in the US. In 1973, Hanford released 450,000 litres of radioactive liquid into the Columbia River. There have been persistent reports of higher-than-normal cancer rates among the Hanford workers and possible lack of safety of its waste storage tanks.
- Accidents have also been reported from Japan, which has a large number of nuclear power stations.

Nuclear power stations do have many safety features. The accidents have shown, however, that such features can never be a total defence against human error and unforeseen technical problems.

Box 14.3 Disaster: The giant wheel does not turn

There is a giant wheel meant for the amusement of children in the town of Pripyat, Ukraine. Since 1986, the wheel has not moved. No children come there. In fact, Pripyat is a ghost town and nobody lives there any more.

The giant wheel is a reminder of the life of the city, which was specially built for the workers of the nuclear power station at nearby Chernobyl. In the early hours of April 26, 1986, Reactor No. 4 at the nuclear power complex at Chernobyl exploded with terrible and long-lasting consequences. It remains the biggest nuclear accident in the world.

The Chernobyl plant in the Ukraine (then a part of the USSR) was considered one of the most efficient in the country and had been operating without problems since it was commissioned in 1983. What happened on that fateful day?

The operators were conducting an experiment and, to prevent interruptions, they switched off the reactor's automatic shutdown mechanism. The experiment went wrong and caused runaway chain reactions. There was a massive steam explosion, which blew off the roof. The core of the reactor combined with water to produce hydrogen, which exploded, blowing radioactive gases and debris high into the air. More than 30 fires were set off in the complex.

It took 10 days for hundreds of firemen to control the fires and prevent them from spreading to the adjacent reactor. Helicopters dropped 5000 tons of lead, boron, and other materials on the core to smother the radioactive gases. Over the next six months, the reactor was entombed in a giant steel and concrete building.

Land was contaminated for 1000 sq km around the complex. About 60,000 buildings had to be decontaminated, and 135,000 people along with 80,000 animals were evacuated from the area over the next 10 days.

Radioactive clouds travelled across Europe and deposited the dust at random depending on weather conditions. There was high radioactive fallout on Poland, Sweden, Norway, Finland, Germany, Italy, and France. Eventually radioactivity reached Siberia, Saudi Arabia, and North America.

The radioactive fallout was more than the total fallout from all the atomic tests ever conducted in the world. Some of the brave firemen who worked to put of the fires in the reactor, took the risk knowingly. Some died immediately and others died later on of radiation-related illnesses. But for their self-sacrifice, more reactors would have caught fire. If the radioactivity released had been just three times more, the damage in Europe would have been 200–400 times greater.

The actual damage caused by the radioactivity will never be known due to the lack of monitoring and the long time spans of radioactive decay and cancer growth. The world simply did not have the experience and ability to monitor such a disaster and measure its effects. There was in fact complete confusion in Europe in assessing the scale of damage and responding to it. Many areas in Europe will remain contaminated well into the twenty-second century or even longer.

Chernobyl radically undermined public faith in the safety of nuclear reactors. Public memory, however, is short and there are no voices of protest even as many governments, including those of India and the US, go ahead with more nuclear power projects. In spite of all the assurances given by the nuclear establishment, there will surely be more accidents.

Whose turn is it next?

Source: Pope et al. (1991) and the website *www.chernobyl.info*

What is nuclear waste and why is it a problem?

There are three types of radioactive waste—low-level, intermediate-level, and high-level waste. Hospitals, laboratories, and other buildings where radioactive technology is used emit low-level radiation. This is not a health hazard as far as we know, but any rubble from such buildings is normally buried in trenches or dumped in the ocean.

Intermediate-level waste consists of substances from nuclear power stations like cleaning agents and sludge. This material is bulky and decays slowly. Hence, it is encased in bitumen or concrete before being stored. Until 1980, this kind of waste was just dumped in the ocean. Since then it is being stored in deep landfills because of an international convention against ocean dumping.

High-level nuclear waste is extremely dangerous and must be isolated for thousands of years. This is mostly spent nuclear fuel that comes from nuclear reprocessing plants. There is no totally safe method of storing this waste. Tons of this hazardous waste are piling up all over the world (read Box 14.4). Environmentalists believe that there is urgent need to stop producing more of this waste.

Box 14.4 Controversy: 'Chernobyl would be small potatoes'

Today there are 45,000 tons and by 2035, there will be 100,000 tons of the deadly stuff to be stored and secured. It will ultimately become safe to handle, but it will take just 10,000 years!

This is the story of spent nuclear fuel in the US. More than 100 nuclear reactors produce 20 per cent of the energy in that country. The nuclear waste that the reactors generate needs permanent and safe storage facilities. For years, the US has been looking for such a central site for all the waste.

Every nuclear complex has the facility to store the spent fuel on site. The problem is that local communities object to such storage and the states place limits on the amount of on-site storage. Further, many units are running out of space and some aging reactors have to be closed down.

In 1985, the US Department of Energy chose the Yucca Mountain desert region in Nevada for building a permanent underground storage facility. It was to cost US$ 58 billion, and was to be financed by a tax on nuclear power.

Citizens and scientists oppose the plan for several reasons: The area is a seismic zone, there is an active volcano nearby, rock fractures could allow water to leak in and corrode the storage casks. One geologist has said that if water were to flood the site, the resulting explosion would be so large that 'Chernobyl would be small potatoes'.

To transport all the waste from the power stations to Yucca would need six shipments a day for 30 years, passing through 109 populated cities. At the end, there will still be the same amount of waste in the plants, since they will be producing new wastes as fast as they ship out the old ones.

Despite all the objections, the US Congress approved the plan in July 2002. However, it will be 10 years before the Yucca facility is ready to accept waste. Since the problem is already acute, the power companies have joined together to develop a temporary site.

The native American Goshute tribe has offered to host the temporary facility on its 7500-hectare reservation in the state of Utah. The tribe once numbered 20,000, but is now reduced to less than 500 members, of whom only 25 live on the reservation. The amount of money offered to the tribe is not known, but the facility would create 60 local jobs.

The planned facility would hold 40,000 tons of waste in 4000 special concrete casks designed to last for 40 years. The cost would be more than US$ 3 billion.

The state of Utah, citizens' groups, and even some of the Goshute people are against the project. They do not want any radioactive waste in their backyard and they also fear that the facility would become permanent since the Yucca one would store only 77,000 tons. And so the controversies continue and wastes are piling up!

The next time you see an advertisement proclaiming how clean nuclear power is, remember Yucca, 100,000 tons of waste, and 10,000 years of safekeeping!

Source: Enger and Smith (2004) and Miller (2004)

Though nuclear bombs have been used only once so far (by the US on Hiroshima and Nagasaki in 1945), innumerable nuclear tests have been conducted since then by a number of countries, including India and Pakistan. During the first two decades of the twentieth century, the tests were conducted above ground and thousands of unsuspecting people and many habitations were exposed to dangerous levels of radiation. Some Pacific islands and the desert regions of Australia and the US were the worst affected. Since 1963, the tests are conducted underground, but there is still contamination of air, soil, and water.

Apart from nuclear accidents and nuclear waste, there are other sources of radioactive contamination. Hospitals use nuclear medicine involving radioactive capsules, needles, and so on. In India and in other places, small items such as these are stolen or find their way to the junkyard. Unsuspecting people handle them and are exposed to harmful radiation (read Box 14.5).

Exposure to low-level radiation over long periods can have damaging effects on the human body. Such contamination could occur in nuclear power plants and near uranium mines (read Box 14.6). On the whole, there is a need for constant vigilance.

Box 14.5 Disaster: Deadly blue crystals

It happened in September 1987 in the Brazilian city of Goiania. Two men found a lead capsule in an abandoned radiotherapy unit. They tried to open it before selling it to the local junkyard. Unfortunately, the capsule contained highly radioactive caesium chloride.

The blue crystals inside the capsule were beautiful to look at and were passed around as curiosities. People started getting sick and those who came to visit them became sick too. The local doctors thought that the sickness was due to food allergies and it was a while before the real cause was found. By that time, the contamination had spread over an area of 300 sq km.

Some of the sick died, more than 200 people were evacuated, and several houses demolished. Some 3500 cu. m of radioactive waste were removed to a storage facility outside the city. There were not enough instruments to screen all the people and the actual number of people affected was never known.

There was public panic in Brazil and products from Goiania were boycotted, which seriously affected the local economy. Protesters halted even the funerals of victims, fearing continued contamination from the buried bodies. The World Health Organization (WHO) termed this disaster as second in severity only to the Chernobyl disaster (Box 14.3).

How did the accident happen? Due to lack of resources, the authorities could not monitor the condition of old equipment. In fact, they admitted that 30 other radioactive items had been abandoned across the country. Incidentally, only two-thirds of the abandoned radioactive material in Goiania was ever found!

In India, there have been several cases of missing radioactive items, often stolen from hospitals and laboratories. Where are they and when will they unleash their lethal power?

Source: Pope et al. (1991)

Box 14.6 Controversy: Mining in 'Magic Land'

Once the Lippis and the Peos, beautiful birds with sweet voices, were abundant here. Beautiful girls used to be named after these birds. Now the birds have gone. Mainas, hariyals, and owls have declined, so have monkeys and honeybees.

The people have been falling ill with fatigue, lack of appetite, and respiratory illnesses. A Health Survey showed a unusual variety and prevalence of problems including miscarriages, birth defects, cancer, tuberculosis, thalassemia, and so on.

What black magic has been happening in Jaduguda (literally meaning Magic Land) in Singhbhum District of Jharkhand? Is it the effect of the uranium mining that has been going on here for over 30 years? This is an unresolved controversy.

Since 1964, the public sector Uranium Corporation of India Limited (UCIL) has been extracting uranium ore from three underground mines in Jadugoda. The waste products that come out of the initial processing of the ore are called 'tailings'. In Jadugoda the tailings are stored in three ponds, which are in reality small-sized dams. Additional tailings come to Jadugoda from the Nuclear Fuels Complex in Hyderabad.

The waste in the tailing ponds is radioactive and is supposed to be isolated. In Jadugoda, however, the isolation has never been strict. People move freely in and out of the tailings area with their cattle, and children used to play there. What is worse, the tailings have been used to construct roads, schools, playgrounds, and other tailing dams.

Water in the wells near the tailing dams has turned black and salty. During the monsoon, water from the tailing dams overflows and reaches the local river. On the whole, the tailings have not been well protected.

More than one committee set up by the Atomic Energy Commission, however, has concluded that there was no abnormally high level of radioactivity in the area and that the local disease patterns could not be ascribed to radioactive exposure.

There is, however, considerable circumstantial evidence that things are not all right with the people of Jadugoda. The incidence of tuberculosis is very high among the miners and very few of them have been able to work until the retirement age. Birth defects and deformities are common.

There have been protests by citizens' groups, intervention by the Bihar Assembly, and petitions in court. On April 15, 2004, the Supreme Court of India dismissed a Public Interest Litigation (PIL) on the hazardous impact of uranium waste disposal in Jadugoda. Meanwhile, we need nuclear energy, we need the uranium, and the mining goes on. But the doubts linger.

On the one hand, there is the personal experience and the sense of 'something wrong' on the part of the villagers, and on the other the official denials and committee reports. Will we ever find out the truth? Should we wait until contamination effects are proven or should we use the precautionary principle, listen to the people, and take protective measures?

Source: Bhatia (2001) and the website *www.antenna.nl/wise/uranium/*.

From nuclear contamination, we now move to the problem of oil spills.

What are the environmental effects of oil spills and oil leaks?

Most cases of oil spills and oil leaks pollute the ocean. Oil spills from tanker accidents and blowouts at offshore drilling rigs (when high-pressure oil escapes from a bore-hole on the ocean floor) attract media attention, though they occur only occasionally. In terms of quantity, however, more oil is discharged into the ocean by the normal drilling operation, cleaning of tankers, and leaks from pipelines and storage tanks.

Most of the oil pollution of the ocean actually comes from the land. Waste oil of all kinds from industries, automobile workshops, and homes ultimately reach the ocean.

Some chemicals in oil kill many marine organisms and coral reefs. Other chemicals form a black layer on the surface of the water, which coats the feathers of birds and the fur of marine mammals, which die or drown as a result. The heavier components of the oil sink to the bottom and kill organisms like crabs and mussels or make them unfit for human consumption. When the oil spill reaches the coast, it destroys fishing activities and tourism.

Which are the major oil spills that have occurred?

Table 14.1 gives the top 10 oil spills in terms of the size of the spill.

Table 14.1 Major oil spills

Position	Name of Ship	Year	Location	Spill Size (tons)
1	Atlantic Empress	1979	Off Tobago, West Indies	287,000
2	ABT Summer	1991	700 nautical miles off Angola	260,000
3	Castillo de Bellver	1983	Off Saldanha Bay, South Africa	252,000
4	Amoco Cadiz	1978	Off Brittany, France	223,000
5	Haven	1991	Genoa, Italy	144,000
6	Odyssey	1988	700 nautical miles off Nova Scotia, Canada	132,000
7	Torrey Canyon	1967	Scilly Isles, UK	119,000
8	Sea Star	1972	Gulf of Oman	115,000
9	Irenes Serenade	1980	Navarino Bay, Greece	100,000
10	Urquiola	1976	La Coruna, Spain	100,000

Source: The website *www.itopf.com/stats.html*

In a number of these incidents, despite their large size, the oil did not impact coastlines. That is why some of the names may not sound familiar. They have, however, done great damage to the marine ecosystem.

A well-known case of an oil spill is that of the tanker *Exxon Valdez* that ran aground on Bligh Reef, located in Prince William Sound, Alaska, in March 1989 (read Box 14.7). A more recent example of an oil spill, which devastated bird life, is that of the oil tanker Erika that broke up and sank 70 miles or so off the coast of France in late 1999. Around 10,000 tons of oil escaped into the sea, killing up to 500,000 birds.

Box 14.7 Disaster: How to kill 22 whales and a million birds without trying

It was around 9 p.m. on March 23, 1989 and Gregory Cousins, third mate of the oil tanker *Exxon Valdez*, was tired, perhaps due to an excessive workload. Joe Hazelwood, the captain, and William Murphy, the expert pilot, had left Cousins to manoeuvre the tanker through the icebergs near Prince William Sound in Alaska.

The tanker had been taken out of the shipping lanes to go around some icebergs. Cousins was to steer the tanker back into the shipping lanes at a certain point. For reasons that remain unclear, Cousins and the helmsman Kagan failed to make the turn back into the shipping lanes and the ship ran aground on Bligh Reef at 12:04 a.m.

When the oil tanker hit the reef, it spewed 38,800 tons of crude oil into the sea. The spill contaminated 2000 km of shoreline, four national wildlife refuges, three national parks, and a national forest. The containment work was poor and the spill ended up affecting 4000 km of the coast.

There have been many bigger oil spills since then. However, the *Exxon Valdez* spill is widely considered the number one spill worldwide in terms of damage to the environment. The timing of the spill, the remote and spectacular location, the thousands of miles of rugged and wild shoreline, and the abundance of wildlife in the region combined to make it an environmental disaster well beyond the scope of other spills. Today, the very name *Exxon Valdez* evokes images of what an oil spill can do to the environment.

No one knows how many animals actually died because of the spill. The carcasses of more than 35,000 birds and 1000 sea otters were found after the spill, but since most carcasses sink, this is considered to be a small fraction of the actual death toll. The best estimates are: 250,000 seabirds, 2800 sea otters, 300 harbour seals, 250 bald eagles, up to 22 killer whales, and billions of salmon and herring eggs. Some estimates talk about the death of three million birds.

Exxon, the oil company that owned the tanker, spent US$ 1.8 billion in a controversial clean-up operation using 11,000 people. Some of the methods used by them damaged more wildlife. More than 150 cases were filed against the company.

Exxon agreed to pay US$ 900 million to restore the natural and other resources that had suffered a substantial loss or decline as a result of the oil spill. The Exxon Valdez Oil Spill Trustee Council was formed to oversee restoration of the injured ecosystem through the use of these funds.

The disaster created tremendous awareness of environmental issues in the US. The US Congress set up a billion dollar clean-up fund and blocked several new offshore drilling operations. At the international level, many new regulations came into being for oil transport.

Even 10 years after the accident, many fish and wildlife species injured by the spill had not fully recovered. It is less clear, however, what role oil plays in the inability of some populations to bounce back. The ecosystem is dynamic and continues its natural cycles and fluctuations at the same time that it struggles with the impacts of spilled oil. As time passes, separating natural change from oil-spill impacts becomes more and more difficult.

The *Exxon Valdez* disaster was not the first such oil spill. On March 18, 1967, the tanker *Torrey Canyon* sank between the Isles of Scilly and Lands End. Over the next two weeks, an oil slick covering 650 sq km killed thousands of birds and other marine animals. Eleven years later, the supertanker *Amoco Cadiz* was grounded in Brittany and released 230,000 tons of crude oil into the sea. It killed all marine life in a 15 km radius around the wreck and caused damage to a depth of 60 m. In 1991, there was an oil spill in the Persian Gulf because of the Gulf War. There have been more spills since then.

The *Exxon Valdez* was clearly a case of human error and more such cases have followed. With oil tankers becoming bigger and bigger, the ocean is surely in for more and more oil pollution—until of course the world runs out of oil!

Source: Pope et al. (1991) and the website of the Exxon Valdez Oil Spill Trustee Council, *www.evostc.state.ak.us*

Can the environment be restored after an oil spill?

After an oil spill has occurred, about 12 to 15 per cent of the oil can be removed through mechanical and chemical means. The oil can be sucked up by vacuum pumps in boats or can be absorbed by pads. In addition, floating booms are used to prevent the spill from spreading or reaching more sensitive areas.

Chemical means include using coagulating agents to form oil clumps that can be removed easily or dispersing agents that break up the slick, though the chemical used can also adversely affect the environment. The oil can be burnt off, but that will cause air pollution.

After crude oil has been discharged into the ocean, many marine organisms recover in about three years. If refined oil like petrol is involved, the recovery may take upto 10 years. Beaches covered with oil slicks may take several years before they become clean and usable again.

It has been found that ocean processes in general can absorb oil spills and break down hydrocarbons. More recently, it has been discovered that certain bacteria in the microbial mats or layers of bacteria in salt marshes are specifically involved in performing clean-up operations. The catch, however, is that it takes a long time. Further, the bacteria do not work well on complicated hydrocarbon molecules, preferring the taste of straight-chain hydrocarbons. The best approach is to prevent oil spills from occurring at all.

What will the future bring in the form of disasters and what should we do?

Since we can expect the frequency and severity of natural disasters to increase, we should be better prepared to meet disasters. The public and the government respond when disasters strike, but forget the whole issue after a while. Immediate relief arrives, but funds for long-term projects for better preparedness are rarely received.

During the immediate months after the supercyclone in Orissa, relief worth two billion rupees came from international agencies. The rest of the year brought the same amount for the more expensive long-term measures. Five hundred cyclone shelters had been planned so that every village would have access to one within a distance of 2.5 km. A year later, not a single one had been built. A crucial satellite communication system had been recommended, but not a single phone had been set up. This is a story repeated in many disaster-prone areas. We can only hope that the new National Policy on Disaster Management will help.

Some natural disasters can be prevented, or their effects can be reduced, if we conserve the environment. For example, maintaining forest cover on hills will reduce landslides and conserving mangroves on the coast will reduce the damage caused by cyclones.

The case of man-made disasters is more complex. We do not know where the enemy is lurking—which nuclear power plant, oil tanker, or industrial unit will be the next cause.

You may have noticed that we did not cover industrial disasters like the Bhopal Gas Tragedy in this chapter. We will cover that topic in Chapter 16 under Environmental Health.

A quick recap

- It is vitally important to maintain mangrove cover on the coasts in order to reduce the impact of cyclones.
- Natural disasters in the world are increasing in severity and frequency.
- Most of the natural disasters and maximum damage occur in developing countries.
- India is one of the most disaster-prone countries, with occasional earthquakes and regular floods, cyclones, and droughts.
- Earthquakes have caused major damage in India.
- Nuclear accidents do occur due to human error and unforeseen technical problems.
- While the Chernobyl accident has been the worst so far, several others have occurred in different parts of the world.
- The safe storage of deadly nuclear waste from reactors is posing a major problem.
- Though oil spills attract attention, more oil is discharged into the ocean by normal drilling operations, cleaning of tankers, and leaks.
- Large oil spills cause great damage to the marine and coastal environment
- We cannot easily and quickly clean up a spill and restore the environment.
- With increasing natural and human-induced disasters, greater vigilance and urgent environmental conservation are necessary.

What important terms did I learn?

Richter scale
Seismograph
Cyclone
Hurricane
Tsunami
Typhoon
Tailings

What have I learned?

Essay questions

Write an essay on the environmental health problems created by nuclear power plants through accidents and the generation of radioactive waste. Include an account of the major nuclear accidents and their effects. Why do such accidents occur? What are the issues concerning the storage of nuclear waste? What is the impact of uranium mining?

Short-answer questions

1. What role do mangroves play in the event of a cyclone?
2. List the various types of natural, human-induced, and slow-acting disasters.
3. What is the scale of damage that occurs due to natural disasters in the world and in India?
4. Where do most floods in India occur and why?
5. Why have the cyclones on the eastern coast become more severe in their impact?
6. Give a brief account of the Gujarat earthquake of 2001.
7. List the main features of the draft National Policy on Disaster Management.
8. In what ways does oil find its way into the ocean?
9. List four major oil spills. Describe in brief the Exxon Valdez accident.
10. How is an oil spill cleaned up?

How can I learn by doing?

Visit a place, which has recently been struck by a natural disaster. For example, you could visit Bhuj in Gujarat, Ersama in Orissa, or Nagapattinam in Tamil Nadu. Study the environmental consequences of the disaster and write a report. What were the remedial measures attempted? How far has the environment been restored? What can still be done?

What can I do to make a difference?

1. Observe National Day of Disaster Reduction: October 29, the day the supercyclone hit Orissa in 1999, is observed every year in India as the National Day of Disaster Reduction. The activities are conducted during the fortnight ending October 29. You can organize events to highlight the major disasters that have occurred in India, their impact, mitigation measures, lessons learnt for the future, etc.

   If you live in a disaster-prone area, you should focus on preparedness and explain to the people how they should meet disasters, what kind of protective measures they should take, where to turn to for help if a disaster strikes, etc.
2. Alternately, you can observe the International Day for Natural Disaster Reduction: The UN and its agencies have been observing the International Day for Natural Disaster Reduction on the second Wednesday of October. Each year a theme is also announced. The theme for 2004 was 'Learning To Live With Risk', with the aim of providing practical examples of how educational activities and public awareness can help societies be less vulnerable to natural hazards such as earthquakes, hurricanes, floods, droughts, and landslides.

Where are the related topics?

- Industrial disasters like the Bhopal Gas Tragedy are covered in Chapter 16.
- Global warming, which could cause extreme natural disasters is dealt with in Chapter 19.
- Chapters 11, 12, and 13 discuss in detail all forms of pollution and their effects.

What is the deeper issue that goes beyond the book?

Given the risks of accidents and the problem of storing deadly waste for thousands of years, should we persist with nuclear power? Is it really as clean as its proponents claim?

Where can I find more information?

- For an account of natural disasters in India, refer to UNEP (2001) and the website *http://www.ndmindia.nic.in/* of the Natural Disaster Management Division, Ministry of Home Affairs, Government of India.
- Details of the Gujarat earthquake can be found on the website *http://gujarat-earthquake.gov.in.*
- UNEP (2003) gives an account of the world disaster scene.

What does the Muse say?

Seeking a breath of fresh air in the middle of that fateful night, Lyubov Sirota went out on to her balcony in the city of Pripyat and watched the Chernobyl nuclear reactor explode in front of her. In the days that followed, she and her son fell gravely ill from heavy doses of radioactive contamination. To express her grief and rage, she turned to writing poems. The poems of Lyubov Sirota are available on the website:*http://www.wsu.edu/~brians/chernobyl_poems/chernobyl_index.html.*

The following is the first stanza of one of her poems:

To Pripyat

We can neither expiate nor rectify
the mistakes and misery of that April.
The bowed shoulders of a conscience awakened
must bear the burden of torment for life.
It's impossible, believe me,
to overpower
or overhaul
our pain for the lost home.
Pain will endure in the beating hearts
stamped by the memory of fear.
There,
surrounded by prickly bitterness,
our puzzled town asks:
since it loves us
and forgives everything,
why was it abandoned forever?

Module 6

Human Population and the Environment

Prologue

Children speak up!

Severn Suzuki

Hello, I'm Severn Suzuki, speaking for ECO, the Environmental Children's Organization. We are a group of four 12- and 13-year-olds from Canada trying to make a difference—Vanessa Suttie, Morgan Geisler, Michelle Quigg, and me. We raised all the money ourselves to come 6000 miles to tell you adults you *must* change your ways.

Coming up here today, I have no hidden agenda. I am fighting for my future. Losing my future is not like losing an election or a few points on the stock market.

I am here to speak for all future generations to come. I am here to speak on behalf of all the starving children around the world whose cries go unheard. I am here to speak for the countless animals dying across this planet because they have nowhere left to go.

I am afraid to go out in the sun now, because of the holes in the ozone. I am afraid to breathe the air, because I don't know what chemicals are in it. I used to go fishing in Vancouver, my hometown, with my Dad, until just a few years ago we found the fish full of cancers. And now we hear about animals and plants going extinct every day—vanishing forever.

In my life, I have dreamt of seeing the great herds of wild animals, jungles, and rainforests full of birds and butterflies, but now I wonder if they will even exist for my children to see.

Did you worry about these things when you were my age?

All this is happening before our eyes, and yet we act as if we have all the time we want and all the solutions. I am only a child and I don't have all the solutions, but I want you to realize, neither do you!

You don't know how to fix the holes in our ozone layer. You don't know how to bring the salmon back up a dead stream. You don't know how to bring back an animal now extinct. And you can't bring back the forests that once grew where there is now a desert.

If you don't know how to fix it, please stop breaking it!

Here you may be delegates of your governments, businesspeople, organizers, reporters, or politicians. But really you are mothers and fathers, sisters and brothers, aunts and uncles. And all of you are somebody's child.

I am only a child yet I know we are all part of a family, five billion strong; in fact, 30 million species strong. And borders and governments will never change that.

I am only a child yet I know we are all in this together and should act as one single world towards one single goal.

In my anger, I am not blind, and in my fear, I'm not afraid to tell the world how I feel.

In my country, we make so much waste. We buy and throw away, buy and throw away. And yet Northern countries will not share with the needy. Even when we have more than enough, we are afraid to lose some of our wealth, afraid to let go.

I'm only a child yet I know if all the money spent on war was spent on ending poverty and finding environmental answers, what a wonderful place this Earth would be.

At school, even in kindergarten, you teach us how to behave in the world. You teach us not to fight with others, to work things out, to respect others, to clean up our mess, not to hurt other creatures, to share, not be greedy.

Then why do you go out and do the things you tell us not to do?

Do not forget why you are attending these conferences, who you are doing this for—we are your own children. You are deciding what kind of a world we will grow up in.

Parents should be able to comfort their children by saying, 'Everything is going to be alright. We are doing the best we can. It's not the end of the world.'

But I don't think you can say that to us any more. Are we even on your list of priorities?

My dad always says, 'You are what you do, not what you say.' Well, what you do makes me cry at night.

You grown ups say you love us. I challenge you, *please,* make your actions reflect your words. Thank you for listening.

These are extracts from a speech given by Severn Suzuki, a 13-year-old girl from Canada, on June 11, 1992, at the Earth Summit in Rio de Janeiro. Reprinted from Suzuki (1994).

CHAPTER 15

Population Growth

Too many people brings suffering to the land, and the land returns its suffering to the people.

O.Soemarwato

Everywhere on the ground lay sleeping natives—
hundreds and hundreds.
They lay stretched at full length
and tightly wrapped in blankets, heads and all.
Their attitude and rigidity counterfeited death.

Mark Twain
(on a nocturnal drive through Bombay in 1896)

What will I learn?

After studying this chapter, the reader should be able to:

- appreciate the scale and complexity of the problems facing the governments of the large cities;
- understand the meaning of population explosion in the world and in India;
- explain the Malthusian Theory of Population;
- describe how the population and its growth are distributed in the world;
- appreciate the trends with regard to the populous countries of the world, especially India and China, and their environmental implications;
- list the main features of India's population;
- compare the family planning and welfare programmes of India, China, and Thailand;
- define birth rate, death rate, and population growth rate;
- understand the rate and trends of urbanization in the world and their implications;
- describe how the large cities of the world are growing;
- explain what is meant by ecological architecture; and
- describe the goals of the Cairo International Conference on Population and Development held in 1994 and how far they have been implemented.

Box 15.1 Survival: Deprivation and desperation, yet energy and enterprise!

There are around 5000 small industries in the area with a total turnover of Rs 20 billion. The industries include plastic recycling, garment-making, printing, zari making, leather products, and pottery. Leather products made here are exported to France and Germany. Every day, about 200 tons of snacks like banana wafers and groundnut sweets are produced in over 1000 units. Surely it is a prosperous and developed place with good roads and nice buildings? Well, it happens to be Asia's largest slum and all the people here are 'illegal' occupants!

Spread over 175 hectares and swarming with one million people during the day, Dharavi in Mumbai is (to quote Kalpana Sharma) 'an extraordinary mix of the most unusual people'. They have come from many parts of India, driven to the city from their villages by drought, discrimination, or deprivation. There are potters from Kumbharvada in Saurashtra, dhobis from Gujarat, and tanners form Tamil Nadu. Many began as workers and ended up as owners of small factories. They have found ways of living together and there were hardly any communal clashes here until 1992.

Living in Dharavi is no cakewalk. Within congested Mumbai, Dharavi has the highest density of population, an unbelievable 45,000 persons per hectare. Everywhere there are open drains, piles of uncleared garbage, filth, and pitiful shacks. Water supply and sanitation, once non-existent, are now poor. The people suffer incredible hardships during the monsoon, with flooded lanes and rivers of sewage.

The other part of the city's population would like to believe that Dharavi does not exist. For them, the slums are dirty and the inhabitants are criminals. It would be good to get rid of Dharavi and all the slums. Or at least to hide them from visitors!

Dharavi has changed over the years. As the migrants reclaimed and developed the land, their kin from home joined them. The tanning industry grew into a thriving leather trade and you can now find air-conditioned leather showrooms on the main road, which display every conceivable designer label. Small-time garments manufacturers have expanded their units into export-oriented garment 'factories'. Many of the old timers now live in two-storey structures.

When the migrants were banished to Dharavi years ago, it was a swamp outside the city. Now, it finds itself in the heart of the vastly expanded city in a strategic location between the two main railway lines. And so the land has become extremely valuable!

Over the years, unsuccessful attempts were made 'to develop' Dharavi. In early 2004, however, the state government announced a Rs 56 billion project to completely transform Dharavi. Not to be outdone, the then Prime Minister Atal Behari Vajpayee, announced during an election campaign a grant of Rs 5 billion for the project's infrastructure development.

The new dream project is the brainchild of an NRI businessman. According to the plans, Dharavi will be divided into 10 sectors, allocating space for residential, commercial, industrial and recreational use. Around 15 per cent of the land will be reserved for open spaces such as playgrounds and gardens. Every resident, who had come there before January 1, 1995 will be eligible for a free home of 25 sq m (225 sq ft). The space released will be given to builders to develop the remaining land commercially.

The government plans to create ceramic-training and leather-training institutes as well as special economic zones for the famous potters of Kumbharvada and for other kinds of industry. However, it will not allow polluting industries like leather tanneries or the existing illegal chemical godowns.

'This is a top-down plan made without taking residents into confidence,' says A. Jockin, of the NGO SPARC, which has been working in Dharavi. 'People have to just fit in. Earlier, people had a choice whether they wanted to opt for the Slum Redevelopment Scheme. With this new plan, they won't have an option.'

Will the new scheme work? Where is the space to build transit camps for at least 600,000 people during the construction phase? Will new slums be created in the process of redevelopment? What will happen to the thousands of illegal, polluting, but flourishing businesses?

There are also more fundamental questions: Why do so many people have to live like this in cities? Why do the rest of us accept that fact?

Source: Sharma (2000), Bunsha (2004) and Jamwal (2004a)

What can we learn from the Dharavi case?

Managing the urban population is becoming a bigger and more complex problem every day. In particular, how do we treat the poor, who arrive in the cities by the thousand each day? Can we provide jobs, food, shelter, water and sanitation for all of them? Such questions are likely to become more pressing in the coming decades, especially in developing countries.

This chapter looks at issues of population and urbanization with reference to environmental problems.

What is meant by population explosion?

As we saw in Chapter 1, the world population is growing exponentially. It is doubling every 45 years or so. Thus, we talk about the explosive growth in population.

About 10,000 years ago, there were approximately five million people on Earth, all hunter-gatherers. Then came agriculture and settlements, and the population started growing. By the fourteenth century the figure had reached 500 million. Thereafter, it remained steady for about 500 years, until the Scientific and Industrial Revolutions took place in Europe.

Between 1850 and 1950, the world population doubled to a figure of two billion. By that time, exponential growth had set in and the five billion mark was reached in 1987, increasing to six billion in 1999. World population is projected to reach about 9.3 billion by the middle of this century. According to the Malthusian Theory, it is beyond the Earth's capacity to support such growth.

What is the Malthusian Theory of Population?

In 1798, the British economist and demographer Thomas Robert Malthus published his 'Essay on Population', which was later to become very famous. Malthus noted that human population was growing exponentially, while our ability to grow food was only increasing arithmetically (like the compound interest and simple interest comparison we saw in Chapter 1). Ultimately, concluded Malthus, human population would outgrow the capacity of land to produce food. Famines, plagues, natural disasters, and wars would then bring the population under control.

The Malthusian theory is often quoted in argument against allowing population to grow. Malthus, of course, did not foresee new agricultural techniques, family planning methods, and the migration of people to other lands. These factors and high death rates have prevented what he predicted from coming true as early as he stated. We cannot, however, predict the future in this regard.

How is the population distributed in the world?

Land area covers only 30 per cent of the Earth's surface and of that area, 80 per cent is not conducive to human settlement. This includes deserts, the Polar Regions, the tropical rainforests, the tundra, and the like.

Most of the population lives in coastal areas, river basins, and cities. As we saw in Chapter 4, at least 40 per cent of the world's population lives within 100 km of the coast. By 2025, this figure is likely to double. Again, nearly half the population lives in the cities, many of which are on the coast.

How is the world population growth distributed? The most glaring and disturbing fact is the stark difference in the population growth pattern between developing and industrialized nations. Nearly 99 per cent of all population increase takes place in developing countries, while population size is static or declining in industrialized nations. Among the major industrialized nations, only the US has significant population growth, mainly because of immigration.

By 2050 industrialized countries are expected to increase their population by about 4 per cent. In contrast, the population of developing countries, is likely to go up by about 55 per cent. For example, even as Western European populations decrease, West Asian countries are expected to have about 186 million more people by 2050.

Table 15.1 lists the 10 most populous countries in the world and shows how the list is likely to change between now and 2050. Of the three industrialized countries, which are at present in the list, only the US is expected to remain there 2050. Russia and Japan are likely to be replaced by the Democratic Republic of Congo and Ethiopia.

Table 15.1 The ten most populous countries: 2004 status and 2050 estimates

2004 Status			2050 Estimates		
Rank	Country	Population Millions	Rank	Country	Population Millions
1	China	1300	1	India	1628
2	India	1087	2	China	1437
3	USA	294	3	USA	420
4	Indonesia	219	4	Indonesia	308
5	Brazil	179	5	Nigeria	307
6	Pakistan	159	6	Pakistan	295
7	Russia	144	7	Bangladesh	280
8	Bangladesh	141	8	Brazil	221
9	Nigeria	137	9	D.R. Congo	181
10	Japan	128	10	Ethiopia	173

Source: Population Reference Bureau

Note that, by 2050, India is expected to be well ahead of China in terms of the size of its population. Obviously, China is doing a better job of controlling its population (read Box 15.2). Even so, there are questions about China's ability to grow enough food for its people, given the environmental problems it faces. Thailand has also been successful in its efforts at controlling its population (read Box 15.3).

Box 15.2 Success: Carrot and stick for fewer births

China has 21 per cent of the world's population, but only 7 per cent of global freshwater and cropland, 3 per cent of forests, and 2 per cent of the soil. Desertification and soil erosion are serious problems (refer to Chapter 11, Box 11.4).

China introduced birth control measures in 1971, persuading young people to marry late, space out children, and have not more than two children. From 1979, the government introduced a more aggressive plan offering incentives for late marriages and having just one child. While the child was assured of free medical care, schooling, and preferential employment, the family would get cash bonuses, preferential housing, and retirement funds. If a second child were to arrive, however, all the incentives would be withdrawn and penalties would be levied.

This policy was controversial and unpopular, but it produced impressive and rapid results. Between 1972 and 2000, the crude birth rate fell by 50 per cent and the average number of births per woman came down from 5.8 to 1.8. It is said that 81 per cent of women use contraceptive devices.

One problem that China is facing is that disproportionately more male babies are being born. It is not clear whether female babies are being aborted or whether female infanticide is practised (as in India). By the middle of the twenty first century, marriageable males will outnumber marriageable females by a million.

In 1984, the one-child policy was relaxed in rural China. Now family planning is promoted more through education and publicity and there are fewer penalties.

In spite of these successes, China's population was 1.3 billion in mid-2004 and is expected to reach 1.4 billion by 2050. China should however derive some consolation from the fact that India, which had a population of one billion in 2000, will overtake them by 2050!

Source: Miller (2004) and Raven and Berg (2004)

Box 15.3 Inspiration: Asian success!

The population growth rate of the country was 3.2 per cent in 1971 and the average number of children per family was 6.4. By 2000, the figures were down to 1.6 per cent and 1.9 children! The current population of the country is about 64 million. This figure would have been reached 15 years ago, had the 1971 growth rate continued!

Thailand is a rare case of effective population control in Asia. How was this miracle achieved? A major part of the credit should go to Mechai Viravaidya, who founded the NGO, Population and Communication Development Association (PCDA) in 1974 to make family planning a national goal.

Viravaidya introduced some innovative ideas in family planning. Participants in the programme were offered loans to build toilets and water resources. The loan funds allotted to a village would increase with higher use of contraceptives. He also organized the distribution of contraceptives at festivals, movies, and even at traffic jams!

Between 1971 and 2000, the percentage of women using birth control went up from 15 to 70 per cent. A figure of 50 per cent is considered good for developing countries.

Thailand's success is attributed to several factors:

- The creativity of Mechai's programmes
- A high literacy rate (90 per cent) among women
- Good healthcare for women and children
- The people's openness to new ideas
- Support from religious leaders
- Government's readiness to allot funds and to work with NGOs like PCDA

Meanwhile, Viravaidya is not quiet. He was elected to the Senate in 2000 and since then, the 60-year-old activist has been campaigning to curb corruption in government. Will he succeed?

Source: Miller (2004) and the websites *www.pda.or.th* and *www.aegis.com*

The populations of Pakistan and Bangladesh are likely to double by 2050. Thus, China and the Indian subcontinent could together be home to about 40 per cent of the world's population. Nigeria is the most populous country in Africa and it has a huge reproductive potential since 44 per cent of the population is less than 15 years old. That is why it is likely to move to the fifth place in the list of the 10 most populous countries by 2050.

How is population growth measured?

The common measures of population growth are:

- Crude birth rate: The number of live births per 1000 people in a population in a given year.
- Crude death rate: The number of deaths per 1000 people in a population in a given year.
- Annual population growth rate =

$$\frac{\text{Crude birth rate} - \text{Crude death rate}}{1000} \times 100$$

Table 15.2 gives the birth and death rates in different areas of the world as of 2002.

What has been the growth pattern of India's population?

Table 15.3 shows how India's population has grown since 1901.

We can see four phases in the growth of the population in India:

- 1901–21: Stagnant population
- 1921–51: Steady growth
- 1951–81: Rapid high growth
- 1981–2001: High growth (with some signs of slowing down)

Some features of India's population (according to the *Census of India, 2001*) are:

- Almost 40 per cent of Indians are younger than 15 years of age.
- About 70 per cent of the people live in more than 550,000 villages.
- The remaining 30 per cent live in more than 200 towns and cities.
- According to the Census, nine states showed a definite decline in population growth rate, while eight states (mostly in North India) showed a high population growth rate.
- Between 1991 and 2001, the overall literacy rate improved from 51.6 to 65.4 per cent (an increase of 13.75 per cent). The male literacy rate went up from 64.1 to 75.8 per cent, while the female literacy rate improved from 39.3 to 54.2 per cent in 2001.

We have about 16 per cent of the world's population, but only 2.3 per cent of the world's land area and 1.7 per cent of the forests. As we have discussed in previous chapters, Indian agriculture

Table 15.2 Average crude birth and death rates in 2002

Area	Crude birth rate	Crude death rate
World	21	9
All industrialized countries	11	10
All developing countries	24	8
Developing countries without China	27	9
Africa	38	14
Latin America	23	6
Asia	20	7
USA	15	9
North America	14	9
Europe	10	11

Source: Miller 2004

Table 15.3 Growth of India's population

Year	Population millions	Annual growth percentage	Density per sq km
1901	238	—	77
1911	252	0.56	82
1921	251	—	81
1931	279	1.04	90
1941	319	1.33	103
1951	361	1.25	117
1961	439	1.96	142
1971	548	2.20	177
1981	683	2.22	216
1991	844	2.11	267
2001	1027	1.93	324

Source: Raghunathan 1999, *Census of India, 2001*

faces problems like soil degradation, water scarcity, and decreasing biodiversity. We have enough food, but millions go hungry every day because they are too poor to buy any food.

What has been India's response to population growth?

India was the first country in the world to start a family planning programme. It was launched in 1952, when our population was about 400 million. Fifty years and many programmes later, we have more than a billion people. By 2050, we will be the most populous country in the world.

Successive Five Year Plans have reflected a broadening of the family planning programme to provide comprehensive maternal, child, and reproductive health care. In 1976, when forced sterilizations were attempted, there was resistance.

The next government shifted the emphasis to voluntary family planning and its integration with overall maternal and child health programmes. The Eighth Five Year Plan (1992–97) launched the Child Survival and Safe Motherhood Programme. Efforts were made to provide ante-natal, intra-natal, and post-natal care to women. It is clear, however, that the goal of stabilizing our population is still far away.

Poverty, low literacy and education levels among women, lack of consistent support from the government, poor planning, and bureaucratic inefficiency are some of the reasons why the family planning programme has not been a big success. Poverty drives many people to have more children so that there will be more working members in the family. The desire to have male children is also a factor behind increasing family size. Though 90 per cent of all couples know of at least one birth control method, less than 50 per cent actually use one.

What is the scale of urbanization in the world?

The UN *Report on World Urbanization Prospects (2003 Revision)* provides the following key findings:

- The growth of the world's urban population continues to be higher than that of the growth of total population. As a consequence, about three billion people or 48 per cent of humankind is now living in urban settlements. The majority of urban dwellers live in smaller urban settlements, while less than 5 per cent of the world's population lives in mega-cities.
- The world's urban population is expected to rise to five billion by 2030. The rural population is anticipated to decline slightly from 3.3 billion in 2003 to 3.2 billion in 2030.
- The 50 per cent mark is likely to be crossed in 2007; then, for the first time in history, the world will have more urban dwellers than rural ones. The proportion of the population that is urban is expected to rise to 61 per cent by 2030.
- Over the next 30 years, the world's urban population is expected to grow at an average annual rate of 1.8 per cent, nearly double the rate expected for the total population of the world (which is about 1 per cent per year). At this rate of growth, the world's urban population will double in 38 years.

Almost all population growth expected for the world in the next 30 years will be concentrated in urban areas. Smaller urban settlements (with fewer than 500,000 inhabitants) in the less developed regions will absorb most of this growth.

Mega-cities, such as Tokyo, Mumbai, Mexico City, and New York will continue to dominate the urban landscape in some countries, but the majority of urban dwellers will reside in smaller cities.

What is driving urbanization and what are its implications?

For the poor rural inhabitant facing problems of drought, discrimination, and deprivation, the city offers what the person desperately needs—employment. Most of the world's economic activities take place in cities and even a person without any skills can almost always find some mans of earning a livelihood in the city. The city also offers the hope of a better life, with comforts not available in most rural areas.

Increasing urbanization, however, places enormous pressure on the local resources of a city. As the city grows, its ecological footprint (see Chapter 1) grows even faster. Environmental problems increase: water scarcity intensifies, more and more waste piles up, the air quality deteriorates, public transport systems get overloaded, traffic jams increase, and so on.

Governments and civic bodies are finding it increasingly difficult to provide and maintain adequate water supply, sanitation, sewage systems, housing, roads, transportation, power supply, and other infrastructure to citizens. Things get much worse when disasters like monsoon floods and epidemics strike cities. While there are many horror stories of urban situations, there are also hopeful ones. Examples are Curitiba, the ecocity of Brazil (read Box 15.4) and the post-plague transformation of Surat in Gujarat (read Box 15.5).

Box 15.4 Solution: Paradise on Earth?

Would you like to live in a city set amidst greenery, where the streets are clean, the air is breathable, the public transportation system excellent, where there are no traffic jams, where the citizens are happy and are fully involved in running the city, and so on? You cannot believe that such place can exist in this world? Well, take the next flight to Brazil and head for Curitiba, a very liveable city in Latin America!

Curitiba, with a population of 2.5 million, was transformed into an ecocity through the efforts of Jaime Lerner, architect and former college teacher. He has been elected Mayor of Curitiba three times since the 1970s. In 1994, he became the Governor of Parana.

Lerner set out to establish a municipal government that would find simple, innovative, and inexpensive solutions to the city's problems. He wanted an accountable, transparent, and honest government, ready to take risks and also to correct itself. Above all he wanted Curitiba to be an ecofriendly city.

Lerner introduced innovations in many aspects of the city's working:

Greenery: One and a half million trees were planted. No tree can be cut down without permission and if one tree is cut, two must be planted. There is a network of parks, in which municipal shepherds ensure the grass is cut by rotating the grazing areas of flocks of sheep. Flood-prone properties along the rivers in the city have been converted into a series of interconnected parks crisscrossed with bicycle paths. There is now less damage due to floods and more green cover.

Transportation: The city is not designed with personal cars in mind. It has perhaps the world's best public transport system, with clean and efficient buses carrying 1.9 million passengers (75 per cent of all commuters) every day at low cost. The buses run on high-speed, dedicated lanes.

Five major roadways penetrate into the heart of the city, each with two bus lanes. The bus stops are connected to bicycle paths, which extend 160 km into the city. The population has doubled since 1974, but the traffic has reduced by 33 per cent.

City planning: High-density development is largely restricted to areas along the bus routes, where low-income groups are housed. Every high-rise block includes two floors of shops, reducing the need to travel far for shopping. The city centre is a large pedestrian zone connected to the bus stations, parks, and bicycle paths. Telephone and internet-based systems are in place to respond to enquiries and complaints. A public Geographic Information System contains data about all the land in the city.

Welfare: Slums and shantytowns do exist, but there is a sense of solidarity and hope. Each poor family can build its own home. A family gets a plot of land, building materials, two trees, and a one-hour consultation with an architect. The poor receive free medical aid and childcare and infant mortality has reduced by 60 per cent since 1977. There are 40 feeding centres for street children.

Waste management: There is a labour-intensive garbage purchase programme—700,000 poor people collect and deliver garbage-filled bags in exchange for bus tokens, surplus food, or school notebooks. This encourages the proper collection of garbage from the shantytowns around the city. The city recycles 70 per cent of its paper and 60 per cent of metal, glass, and plastic. The recovered material is sold to the city's 500 units in an industrial park. The previous garbage dump has become a botanical garden.

Industries: The city developed an industrial park with services, housing, schools, etc., so that the workers can cycle or walk to work. There are strict pollution control laws.

Education: All the school children study ecology, while adults have special environmental courses. The older children are given training and apprenticeship in environment-related areas such as forestry, water pollution, and ecorestoration.

All this has happened even as the population grew from 300,000 in 1950 to 2.2 million in 1999, as rural people flocked to the city. The city expects to grow by another million by 2020.

Surely a dream city! Can one city in India follow this example? It is possible, as the Surat experience (Box 15.5) shows.

Source: Chiras (2001), Miller (2004) and Raven and Berg (2004).

Box 15.5 Transformation: Rising like a phoenix

It is a city famous for its diamond merchants. Like any other Indian city, it had its clean colonies and stinking slums, dirty drains and garbage galore. In 1994, however, it ran into a major crisis when pneumonic plague descended on the city. It was Surat in the state of Gujarat, with a population of two million.

During the monsoon that year, it had rained continuously for two months. This led to flooding and large-scale waterlogging in low-lying areas, due to a faulty drainage system. Hundreds of cattle and other animals died in the floods. In September, the whole system gave way, inviting the plague.

The four causes of the plague were: the huge immigrant influx into Surat; the inability of the city's infrastructure to bear this huge human load; the total apathy and disinterest of the bureaucracy; and the flooding of the city by the River Tapti leaving behind death and disease.

The northern part of the city, which was affected the most by the plague and from where the largest number of deaths were reported, did not have access to any type of sewage system. Despite being one of the richest civic bodies in the country, the Surat Municipal Corporation (SMC) had failed to provide basic sanitation and clean drinking water to a majority of the city's population.

It was a disaster waiting to happen. Historically, Surat had been known for its filthiness. The influx of immigrants had led to an explosive growth of slums without any facilities. The city was a fertile ground for public health epidemics, such as malaria, gastroenteritis, infectious hepatitis, cholera, and dengue.

The plague itself caused perhaps 50 deaths and affected 150 others. The media attention, however, blew it out of proportion. In a matter of days, Surat became a shunned city and 60 per cent of its citizens fled. The economy was devastated with a loss of several million rupees a day. Some of the international flights into India were temporarily suspended, and export of food grains from Surat was banned. Gradually, however, the plague scare subsided and normalcy began to return.

Until December 1994 there was a modicum of restraint among the citizens and garbage was not thrown on the streets. However, by January 1995, the century-old habits asserted themselves with vengeance and the city was as filthy as ever. The strange fact is that the occurrence of the plague was not the trigger for transformation. A miracle was to happen, however.

On a fateful day in May 1995, Suryadevra Ramachandra Rao, the new Municipal Commissioner of Surat, walked into the city dressed in simple blue denim trousers and canvas shoes. He took out some tobacco from his front pocket, rolled it into a cigarette and said in the best tradition of an American Western, 'I am S.R. Rao.'

The irony of Rao's coming to Surat was that he had not wanted to be there in the first place. However, faced with a city traumatized by the plague and a state government cringing over the adverse publicity, Rao set to work. When he left Surat two years later, it was ranked as the second cleanest city in India, after Chandigarh.

This is what Rao and the civic body did:

- Forming a Team that would work together on policy and implementation and delegating financial and administrative powers down the line.
- Requiring the senior officers to spend five hours a day in the field.
- Building community-shared water and sanitation facilities in many of the slums and improving the living conditions of the sanitation workers.
- Sorting out all the problems of the sanitation workers.
- Demolition of illegal structures beginning with those of the most powerful people.
- Enforcing hygiene and sanitation standards by raiding eating houses, sweet shops, food storage facilities, cold storage facilities, etc., again starting with the biggest and richest ones.
- Setting up a unique night cleansing system. Every street was scrubbed at night and garbage bins cleared so that the people awoke to a clean city each morning.
- Establishing 275 surveillance centres for monitoring public health.
- Launching a Surat First programme to involve citizens, companies, and institutions in working for the city's welfare.

Rao himself went to the filthiest of slums and the dirtiest of the eateries. He took on the politically influential builders lobby, with its huge disposable incomes. The initial scepticism of the citizens gave place to a spirit of cooperation, when they saw that the Corporation meant business.

The level of sanitation improved from 35 to 95 per cent, while solid waste removal increased from 40 to 97 per cent. The disease rate went down by 70 per cent. In fact, rag pickers lost their business.

The most important thing Rao did was to put a system in place. He made the civic staff realize how proper use of their powers would earn them respect. Later, it became a matter of prestige for the civic staff to continue the work that was started after the plague.

Surat continues to be clean and has arguably become the most liveable city in Gujarat. Swanky flyovers, broad clean roads, washed streets, and spotless footpaths are now a part of this industrial city.

Can the success of Surat be replicated? Do we need more Raos?

Source: Websites such as *http://yashadacgg.mah.nic.in/topdoc.asp*

How are the world's large cities growing?

The number of cities with five million or more inhabitants is projected to rise from 46 in 2003 to 61 in 2015. Among these, the number of mega-cities (with 10 million inhabitants or more) will increase from 20 in 2003 to 22 in 2015.

Most of the large cities are in developing countries. In 2003, 33 of the 46 cities with five million or more inhabitants were in less developed countries. By 2015, there will be 61 such large cities with 45 of them in the less developed regions of the world.

Table 15.4 lists the 10 largest cities in 1950 and in 2000. It also gives the expected list in 2015.

Table 15.4 The 10 largest cities of the world

Rank	City	1950 Population (in millions)	City	2000 Population (in millions)	City	2015 Population Estimate (in millions)
1	New York	12.3	Tokyo	26.4	Tokyo	36.2
2	London	8.7	Mexico City	18.1	Mumbai	22.6
3	Tokyo	6.9	Sao Paulo	18.0	Delhi	20.9
4	Paris	5.4	New York	16.7	Mexico City	20.6
5	Moscow	5.4	Mumbai	16.1	Sao Paulo	19.9
6	Shanghai	5.3	Los Angeles	13.2	New York	19.7
7	Essen	5.3	Kolkata	13.1	Dhaka	17.9
8	Buenos Aires	5.0	Shanghai	12.9	Jakarta	17.5
9	Chicago	4.9	Dhaka	12.5	Lagos	17.0
10	Kolkata	4.4	Delhi	12.4	Kolkata	16.8

Source: United Nations, *World Urbanization Prospects: 2003 Revision*

We can see the rapid growth of the cities of the developing world between 1950 and 2000. By 2015, only Tokyo and New York will represent industrialized countries in the list of the 10 largest cities in the world. Note that the three largest metros in India are on the 2015 list.

What are the environmental implications of population growth and urbanization in India? Population growth and urbanization will place greater pressure on natural resources, but there are ecofriendly alternatives that could mitigate the problems faced:

- Land: There will be increasing takeover of land for human settlements. Expanding cities will encroach on surrounding areas, converting any kind of land, including fertile fields. Land will also be needed for infrastructure, such as roads, highways, industries, tourist facilities, and educational complexes. Better land-use planning, creation of satellite towns, and similar measures are necessary.
- Food: Even as the number of mouths to feed keeps increasing, our ability to produce enough food is being tested. Unless there is a second Green Revolution, which does not demand high inputs of fertilizers, pesticides, etc., as the first one did, it will be difficult to feed the exploding population.
- Forests: We can expect increasing encroachment on forest areas and exploitation of forest resources. It will become increasingly difficult to protect reserved forests and national parks. The only way to do so is to make the local people partners in conservation.
- Water supply and sanitation: Water scarcity will become more severe and sanitation targets will not be met unless water conservation measures and ecological sanitation are aggressively implemented.
- Energy resources: Fuelwood and conventional energy resources will become scarce and their prices will rise—unless wood plantations are encouraged and renewable energy sources are promoted.

- Housing: The shortage of housing, which is already a problem, will become far more severe in the future. There is a need to implement mass housing projects using the principles of ecological architecture (read Boxes 15.6 and 15.7).

Box 15.6 Solution: Green houses

Brick and concrete houses (that are hot in summer and cold in winter), with Malaysian wood, Italian marble, and German bathroom fittings—that is the architecture in our cities and even smaller towns. The ecological footprints of such buildings are enormous and this approach is clearly not sustainable in the long run. What is the way out? Ecological architecture!

Ecological architecture, also called 'green architecture' or 'sustainable architecture', seeks to minimize the ecological footprint of the house, building, or complex that is being designed and constructed.

The primary objective of green architecture is not to reduce cost, but to minimize the burden that the unit imposes on nature. There is likely to be a saving in cost, but that should be viewed as a bonus. Often, green architecture involves more labour than mainstream architecture does, but that is desirable in a country like India.

In practice, ecological architecture follows the principles stated below:

- The ecological footprint of the house, that is, the area from which the unit draws all its building requirements, should be as small as possible. Use material available nearby, in the same town, state, or at worst in the country; do not in any event use imported material, even if it is less expensive.
- When there is a choice, use material that can be reused when the building is demolished.
- Minimize the use of material such as burnt bricks, cement, and steel that consume energy during manufacture. Replace them with natural materials such as mud. Earth blocks can be made on site using the soil from the foundation, if appropriate for block-making.
- Use of wood may be unavoidable, but choose the wood of a tree that you can plant to repay your debt to nature.
- Minimize finishing work that costs time and money without adding real value—plastering, whitewashing, sandpapering, polishing, painting, etc. Use natural herbal mixtures for protection against termites, etc.
- Wherever possible, use or dispose off waste material arising from the construction. For example, fix broken tiles on bathroom walls, bury plastic bags from the neighbourhood under the foundation, use discarded styrofoam sheets or even computer keyboards as fillers in concrete slabs.
- Use options that would save energy, examples are skylights to allow more natural light into the building, compact fluorescent lamps, solar lamps, solar water heaters, etc.
- Recycle wastewater and use ecological sanitation.

There are now many architects in India who practice green architecture, the best known being Laurie Baker (read Box 15.7).

Box 15.7 Inspiration: The man who builds for the people

Born in England, he worked during the World War II in an ambulance unit in China, Japan, and Burma. Some time in the early 1940s he had a chance encounter with Mahatma Gandhi, which changed his life. In 1945, he came to India to design buildings for leprosy missions. He is still here.

Laurie Baker, a British architect, has undertaken a wide range of projects in India from fishermen's villages to institutional complexes and very low-cost mud-housing schemes to low-cost cathedrals. Most of his work has been in Kerala and he now lives in Thiruvananthapuram.

He has designed more than 1000 houses, over 40 churches, school buildings, institutions (like the Centre for Development Studies), hospitals, and low-cost housing schemes. He also helped in designing and building new houses in quake-ravaged Latur. He was the principal architect of 'Dakshinachitra', a centre for the preservation of arts, crafts, and architecture of the southern states, located on the East Coast Road near Chennai.

When he first came to work in India, Baker and his Indian wife Elizabeth, lived among the poor in Pithoragarh for 16 years. During this period, the simple, efficient, and inexpensive methods used by the poor people to construct their homes had a profound influence on him. He realized the importance of using local materials, taking into account local climate patterns, and accommodating the local social pattern of living. He worked with, and learnt from, mountain tribes and village masons using indigenous materials for building. He built schools, hospitals, and community buildings and developed the unique Baker style of architecture.

In 1963, Baker moved to Kerala, where he evolved methods of using mud, of employing discarded Mangalore tiles to reduce the amount of concrete in roofs, of avoiding the plastering of walls, etc.

As one writer put it, 'Nature is transcendent in his work. He uses it to create stunning effects. A pool of water, a patch of shade, sunlight marooned in shadows, a clump of bushes, a wild tree–it is all grist for his mill.'

Baker has written simple books on design—handwritten with his own illustrations. He was awarded the Padma Shri in 1990.

He describes his approach in this way:

> I learn my architecture by watching what ordinary people do; in any case it is always the cheapest and simplest because ordinary people do it. They don't even employ builders, the families do it themselves. The job works, you can see it in the old buildings—the way wood lattice work with a lot of little holes filters the light and glare. I'm absolutely certain that concrete frames filled with glass panels is not the answer.

Source: Bhatia (1991)

What are the international initiatives on population-related issues?

The UN-sponsored, International Conference on Population and Development (ICPD) has been held in 1974, 1984, and 1994. While the first two conferences focused on controlling population growth in developing countries, mainly through family planning, the 1994 Cairo Conference took a broader perspective.

It was agreed in Cairo that:

- Population policies should address social development beyond family planning.
- Family planning should be provided as part of a broader package of reproductive health care.
- If we improve the status and health of women and protect their rights, lower fertility and slower population growth will follow.

The Cairo Conference set the following goals for the period 1995–2015:

- Provide universal access to a full range of safe and reliable family planning methods and related reproductive health services.
- Reduce infant mortality rates to below 35 infant deaths per 1000 live births and under-five mortality rates to below 45 deaths of children under the age of five per 1000 live births.
- Close the gap in maternal mortality between developing and developed countries. Aim to achieve a maternal mortality rate below 60 deaths per 100,000 live births.

- Increase life expectancy at birth to more than 75 years. In countries with the highest mortality rates, aim to increase life expectancy at birth to more than 70 years.
- Achieve universal access to, and completion of, primary education.
- Ensure the widest and earliest possible access by girls and women to secondary and higher levels of education.

Over the past decade, there has been uneven progress in the implementation of these goals. Many countries could not marshal the resources and organizational capacity necessary to address such a wide range of health and social concerns.

Sufficient funding has not been forthcoming from donor agencies to support these changes. The UN estimated in 2000 that donor funds were less than half the required amount. The US has stopped funding family planning programmes.

What lies ahead?

India and many other developing countries will face the prospect of unmanageable numbers unless urgent steps are taken to reduce the birth rate. Even if the birth rates are reduced now, it will take decades before the population growth starts slowing down.

We began the chapter with the case of Dharavi. There are hundreds of slums in Indian cities, most of them in a much worse condition than Dharavi. According to estimates, 20–25 per cent of India's urban families live in slums, squatter settlements, or refugee colonies. How do we handle the unceasing migration of the rural poor into cities? How can we improve the conditions in villages so that people do not migrate at all? How do we find space to live and the basic facilities for those who do come to the cities? Such questions will become increasingly pressing in the coming years.

A quick recap

- The unceasing migration from villages into cities and the lack of space and housing for the urban poor leads to slums like Dharavi.
- The slum development schemes of governments have in general failed.
- Innovative and transparent approaches like the ones used in Surat and Curitiba can mitigate urban problems.
- The world's population has been rapidly increasing, though there are now signs of this growth slowing down.
- Most of the population growth now occurs in developing countries.
- India is very likely to overtake China in population size.
- India's efforts in family planning have not produced the expected results. China and Thailand have done better in this regard.
- By 2007, half the world's population is expected to live in cities.
- By 2015, India's three largest metros will be among the 10 largest cities in the world.
- Ecological architecture is one way of providing sustainable housing.
- The Cairo International Conference on Population and Development held in 1994 set many progressive goals, but their implementation has been poor.

What important terms did I learn?

Crude birth rate
Crude death rate
Ecological architecture
Malthusian Theory of Population
Population growth rate

What have I learned?

Essay questions

Write an essay on the environmental and social problems facing any large city in India with a population of more than a million people. Do many migrants come into the city and end up in slums? What are the problems faced by the poor residents of the city? What is the situation with regard to water supply, sanitation, power supply, public transport, and other infrastructure? Are the residents, especially the poor, involved in city planning? How are citizens' difficulties redressed?

Short-answer questions

1. What are the problems that a resident of an urban slum faces?
2. What do we mean by the term population explosion?
3. Briefly describe the Malthusian Theory of Population? Did the predictions of Malthus come true?
4. What is the most disturbing fact about population growth in the world?
5. Compare the family planning programmes of India, China, and Thailand.
6. List the important features of India's population figures.
7. Trace the process of urbanization in the world. How are the large cities growing?
8. Why are people ceaselessly migrating to the cities?
9. What is ecological architecture?
10. What were the main goals set by the Cairo Conference on Population and Development held in 1994?

How can I learn by doing?

Volunteer your services to any non-governmental organization (NGO) working in a slum in your city. Spend about three weeks working with them and write a report on your experience. What are the social and environmental problems that the slum dwellers face? Is the city government trying to solve these problems? What is the NGO's contribution? What are your recommendations?

What can I do to make a difference?

1. Observe July 11 as World Population Day: World Population Day was inaugurated in 1988 by the UN Population Fund (UNFPA) to mark July 11, 1987, the day the world's population hit the five billion mark. Through the observance of World Population Day it is sought to focus attention on the urgency and importance of population issues, particularly in the context of overall development plans and programmes, and the need to find solutions to these issues. For more details, access the website *www.unfpa.org/*.
2. Observe World Habitat Day: Every year, since 1985, when it was designated as such by the UN General Assembly, World Habitat Day has been celebrated on the first Monday in October. This day has been set aside by the UN for the world to reflect on the state of human settlements and the basic right to adequate shelter and to remind the world of its collective responsibility for the future of the human habitat. Every year a theme is announced. The theme for 2004 was, 'Cities—Engines of Rural Development'. For more details, access the website *www.unhabitat.org/*.

Where are the related topics?

- Module 4: Renewable and Non-renewable Natural Resources, which deals with water, energy, land, and food.
- Module 5: Environmental Pollution, which covers air, noise, water, and soil pollution as well as solid waste management.

What is the deeper issue that goes beyond the book?

Should the Indian Government introduce strict family planning measures (as was done in China) to control population growth? Or should we stick to creating awareness and improving the status of women?

Where can I find more information?

- Both Dharavi and Surat provide fascinating stories. Dharavi is an extraordinary story of people's grit and resourcefulness in the face of daily difficulties and administrative persecution. Sharma (2000) gives a fascinating account of the people of Dharavi.
- There are many accounts of the Surat case on various websites, but the personal account of S.R. Rao makes fascinating reading (access the website *http://yashadacgg.mah.nic.in/topdoc.asp*). There is also a highly-acclaimed documentary film on Surat entitled, *Blessed by the Plague,* by Setu Films.
- Complete information on India's population can be accessed from the Census of India's official site, *www.censusindia.net*.
- The website of the Population Reference Bureau in the US, *www.prb.org*, has a wealth of information on population and related issues.
- The UN had put together much useful material when the population reached the six billion mark. You can access it at *www.un.org/esa/population/publications/*.

What does the Muse say?

The following is a poem on Dharavi by Imtiaz Dharker, a poet, artist, and filmmaker. She was born in Pakistan but now lives in India.

Blessing

The skin cracks like a pod.
There never is enough water.

Imagine the drip of it,
the small splash, echo
in a tin mug,
the voice of a kindly god.

Sometimes, the sudden rush
of fortune. The municipal pipe bursts,
silver crashes to the ground
and the flow has found
a roar of tongues. From the huts,
a congregation: every man woman
child for streets around
butts in, with pots,
brass, copper, aluminium,
plastic buckets,
frantic hands,

and naked children
screaming in the liquid sun,
their highlights polished to perfection,
flashing light,
as the blessing sings
over their small bones.

CHAPTER 16

Environment and Human Health

If you want to learn about the health of a population,
look at the air they breath,
the water they drink,
and the places where they live.

Hippocrates

The poor shall inherit the Earth…
and all the toxic waste thereof.

Greenpeace slogan

What will I learn?

After studying this chapter, the reader should be able to:

- define environmental health and explain its scope;
- understand how human activities have an adverse impact on human health;
- list the categories and examples of hazardous chemicals;
- describe how pesticides have positive and negative impacts on society, the use of DDT and Endosulfan being examples;
- explain the adverse effects of PCBs, lead, mercury, and asbestos;
- appreciate the long-term effects of chemical weapons like Agent Orange and dumps of toxic wastes like Love Canal;
- list the major industrial disasters (other than mining and nuclear accidents) that have occurred in the world;
- appreciate the dangers inherent in the manufacture and use of deadly chemical pesticides in the light of Seveso and Bhopal;
- understand how difficult it is to hold chemical companies responsible and liable for the industrial disasters caused by their negligence or their operations; and governments for their lack of appropriate and timely action; and
- recognize that poor people are disproportionately affected by industrial accidents.

Box 16.1 Disaster: The lucky ones died that night

It was five past midnight, December 3, 1984. The congested city was asleep and the winter air was heavy. Many had gone to sleep late, after watching the Sunday movie on television. The city was, however, to change forever. A deadly cloud had risen from a factory and was rapidly blanketing the city.

It was the world's worst industrial disaster. Twenty-seven tons of lethal gases including methyl isocyanate (MIC) had leaked from Union Carbide's pesticide factory in Bhopal. Grossly under-designed safety systems were either malfunctioning, under repair, or had been switched off as part of a cost-cutting exercise. The warning siren at the factory had been turned off.

A white cloud of gases flooded the city. Here is a survivor's account of what happened that night:

> The poison cloud was so dense and searing that people were reduced to near blindness. As they gasped for breath its effects grew ever more suffocating. The gases burned the tissues of their eyes and lungs and attacked their nervous systems. People lost control of their bodies. Urine and faeces ran down their legs. Women lost their unborn children as they ran, their wombs spontaneously opening in bloody abortion. Those who escaped with their lives are the unlucky ones; the lucky ones are those who died on that night.

Seventy thousand people were evacuated from the area after the accident and 200,000 more fled in panic. The gas affected an area of 100 sq km. More than 8000 died in the few days following the disaster. The gases poisoned more than 500,000 people, leaving them with a lifetime of ill health and mental trauma. Most of the victims were poor people living in the slums that surrounded the factory. Many of them worked as daily-wage labourers and were, after the gas leak, left too weak to do any strenuous work.

Today, at least 150,000 people, including children born to gas-exposed parents, continue to suffer debilitating, exposure-related health effects, such as headaches, breathlessness, giddiness, numb limbs, body aches, fevers, nausea, anxiety attacks, neurological damage, cancers, menstrual chaos, depression, and mental illness. The death toll has risen to more than 20,000, and at least 30 persons continue to die each month from exposure-related illnesses. More than 50,000 are too ill to work.

The effects of the gas have lingered on. Of babies born to pregnant women living in the area, 25 per cent have died soon after birth and many others were underweight. Over the years families have accumulated crippling debts merely trying to cover weekly medical costs.

How did the accident occur? Most probably water entered the storage tank and caused a runaway chemical reaction that led to an increase in temperature, which converted the liquid MIC into gas. Investigations revealed that there had been six earlier accidents at the plant and that workers had complained of exposure to dangerous substances. Yet, proper safety mechanisms were not put in place.

Union Carbide accepted only moral responsibility for the disaster and not any liability. The Government of India filed a case against the company for US$ 3 billion compensation, but strangely accepted US$ 470 million as settlement in 1989. Nearly 95 per cent of the survivors have received just Rs 25,000 for lifelong injury and loss of livelihood. That works out to less than 9 US cents a day for 20 years of unimaginable suffering.

The disaster occurred because Union Carbide had installed a poorly designed factory, and had cut costs by compromising on safety and maintenance systems and slashing staff and training levels. For this, a criminal case was filed in the Chief Judicial Magistrate's court in Bhopal, against Union Carbide and its then Chairman Warren Anderson. They failed to attend court and Anderson has been proclaimed a fugitive from justice by the court.

The story did not end with the disaster. When Union Carbide finally left Bhopal in 1998, around 5000 tons of their wastes, including deadly chemicals, stayed behind. The toxins have oozed and leached into the soil and water in and around the factory. About 20,000 local people, 70 per cent of them gas-affected, are facing soil and water contamination.

In 2001, the American corporation Dow Chemical purchased Union Carbide, thereby acquiring its assets and liabilities. However Dow Chemical has steadfastly refused to clean up the site, provide safe drinking water, compensate the victims, or disclose the composition of the gas leak, information that doctors could use to properly treat the victims. Incidentally, Dow Chemical was one of the manufacturers of Agent Orange (read Box 16.3).

Even the small amount given by Union Carbide to the Indian Government was not fully disbursed. In 2004, the Supreme Court of India ordered the government to distribute the balance of Rs 15 billion in the compensation fund among 572,000 victims.

For over 20 years, survivors' organizations in Bhopal have refused to give up their fight for justice, proper compensation, economic rehabilitation and adequate health care. They are helped by the International Campaign for Justice in Bhopal (ICJB), a coalition of all the public interest organizations and individuals that have joined forces to campaign for justice for the survivors of the Bhopal disaster. Every year, ICJB organizes the observance of December 3, the anniversary of the disaster, as the Global Day of Action Against Corporate Crime (read Box 16.8).

Source: Pope et al. (1991) and the website *www.bhopal.net*

What are the lessons from the Bhopal Tragedy?

The tragedy showed that poor communities are disproportionately affected by toxic materials discharged into the air, land, and water. When a crisis occurs or an accident happens, these poor people cannot easily get justice from the polluters or from government. The poor often do not know how dangerous their workplaces or neighbourhoods are. Perhaps voluntary organizations should have a Factory Watch in every area, to maintain a vigil over the activities and operations of production units, educate the workers and the people about potential hazards, and also have contingency plans ready for any crisis.

As we will see in Chapter 20, the Bhopal Gas Tragedy triggered the passing of several environmental laws in India. These laws should be strictly implemented.

This chapter takes up the question of environmental health, especially in the context of hazardous chemicals.

What is the connection between the environment and human health?

Humans have always been affected by the natural environment. Changes in climate and any extreme weather conditions affect us. A change in season increases pollen in the air, causing respiratory problems in susceptible people. This happens to many people in Bangalore, for example.

Sometimes, harmful substances and organisms like viruses and bacteria get into the body and cause discomfort or disease. They enter the body through the air, food, or water. Normally, the body's immune system handles the invading organisms. When the person's immune system is weak, or the invading organism is strong, the body succumbs to the disease.

Since the beginning of the Scientific and Industrial Revolutions, however, human activities have been affecting human health in a big way. We have seen a number of examples in the previous chapters.

The World Health Organization (WHO) defines environmental health as those aspects of human health that are determined by physical, chemical, biological, social, and psychosocial factors in the environment. The term also refers to the theory and practice of assessing, correcting, controlling, and preventing those factors in the environment that can potentially affect the health of present and future generations.

The areas covered by environmental health include:

- Water and soil pollution (Chapter 12).
- Indoor and outdoor air pollution (Chapter 11).
- Noise pollution (Chapter 11).
- Radioactive contamination (Chapter 14).
- Electromagnetic fields (EMFs): In recent years, there has been concern about the impact of EMFs. Television receivers, computer monitors, overhead electric lines, and mobile phones are examples of EMF sources. Such fields of all frequencies represent one of the most common and fastest growing environmental influences. Humans are now exposed to varying degrees of EMFs and the levels will continue to increase as technology advances. There is still inadequate knowledge of the effects of an EMFs on health.
- Contamination caused by hazardous chemicals: This topic was partially covered in Chapter 14 and is continued below.

What are the common hazardous chemicals?

In modern societies, synthetic chemicals dominate people's lives. Textiles, plastic goods, detergents, cleaners, toiletries, batteries, packaging material, vehicles, daily newspapers, medicines, computers—almost every item that you use has either gone through chemical processes or contains some amount of chemicals.

Most of the chemicals used by humankind are hazardous to some part of, or to the entire biosphere. Some chemicals affect all species, while others affect only a few. Some cause only minor problems, some can kill instantly. Some are dangerous only when exposure to them is prolonged, while others are toxic only in high concentrations. We do not have complete information about the effects of many of these chemicals on humans and other organisms.

There are broadly two categories of hazardous chemicals that can significantly affect humans and the environment:

- Pesticides and herbicides that are deliberately introduced into the environment: Beginning with DDT, we have experimented with a whole range of chemicals to kill pests and unwanted weeds.
- Industrial chemicals that are disposed off as waste or discharged accidentally into the environment: These include organic solvents, waste oil, polychlorinated biphenyls (PCBs), paints, glues, preservatives, and metal residues.

The use and impact of some of these chemicals and substances, such as asbestos, lead, and mercury are discussed below.

What are the chemicals used as pesticides and what is their impact?

Pesticide is a general term for any chemical used to kill unwanted organisms. The term covers insecticides, herbicides, fungicides, and other chemicals that kill pests (spiders, mites, worms, rodents, etc). Modern agriculture and sanitation depend heavily on pesticides. Pesticides have saved millions of lives by killing disease-carrying insects. They have also increased our crop yields by eliminating pests that attack plants.

Until the middle of the last century, most pesticides in common use were simple, naturally-occurring substances such as arsenic and nicotine. Then came new chemical compounds like DDT that were much more effective against crop pests and disease-carrying insects. The new chemical pesticides, however, remain in the environment for much longer than the natural ones. Many are also toxic to humans.

As far back as 1995, the WHO estimated that every year there were three million cases of pesticide poisoning worldwide, including 220,000 deaths. According to another WHO report in 1991, 25 million farm workers in developing countries were suffering from pesticide poisoning. An example is the Endosulfan case in Kerala, India (read Box 16.2).

Box 16.2 Controversy: Poison from the sky?

Seven-year-old Shruti cannot go to school. She was born with three deformed limbs and she hops around on one leg. Her mother died of cancer and her father is an agricultural labourer. In fact, every family in that area of four sq km has people suffering from diseases that were never seen before locally.

Mohana Kumar, a doctor practising for a decade in that area, the Padre village in Kasaragod District of Kerala, was a puzzled man. Among the patients coming to him from just two wards of the panchayat, he found a very high incidence of cancer, psychiatric problems, mental retardation, epilepsy, congenital anomalies, to name a few. Surprisingly, almost all the ailments were restricted to people under the age of 25 and all were difficult to cure.

Several people in Padre believed that the local guardian spirit was angry with them and had let loose the diseases on them. Dr Kumar, however, had to find a scientific explanation. He began keeping detailed records and felt that the effects could be due to the pesticide Endosulfan. He wrote to some leading doctors and also to the *Kerala Medical Journal* asking for help, but there was no response.

Endosulfan is effective against a variety of pests that attack crops of cereals, coffee, potato, tea, and vegetables. It is easily absorbed by the stomach and lungs and through the skin. Being anywhere near the area of use is enough to get contaminated. It is highly toxic to humans, birds, and animals. The US, Canada, 12 European countries, and some countries in the Asia-Pacific region have banned it, while others have restricted its use.

Since 1976, the government-owned Plantation Corporation of Kerala (PCK) had been carrying out aerial spraying of Endosulfan over cashew plantations in an area of nearly 4700 acres in Kasaragod, including the hills around Padre. The pesticide residues settled on the soil and got washed away into the drinking water streams below. Though it is mandatory to cover all water sources, such as wells, tanks, and other water bodies during the spraying of such a toxic pesticide, this was not done.

After cows in the area started giving birth to deformed calves, there were sporadic protests against aerial spraying. In December 2000, the villagers tried to prevent spraying, but the police thwarted them.

Meanwhile, Kumar and a farmer-journalist Shree Padre found enough published evidence to connect Endosulfan with the people's ailments. Samples of blood, fruits, and animal tissues from Padre were tested at the Indian Institute of Technology (IIT), Kanpur, and found to contain extremely high levels of the pesticide.

Kumar and Padre began holding public meetings to explain their findings to the public. A media war ensued with PCK and pesticide manufacturers denying the role of Endosulfan in causing the ailments. The Kerala Government banned the aerial spraying of Endosulfan.

The issue became a national one with the intervention of the National Human Rights Commission and a study by the National Institute of Occupational Health. The latter study showed that Endosulfan was the causative factor in the incidence of illnesses in Padre. There were counter studies too, absolving Endosulfan. The pesticide industry was quick to declare that Endosulfan had nothing to do with the health problems.

In August 2001, the Kerala government suspended the use of Endosulfan, but lifted the ban in March 2002. The ban on aerial spraying was continued. Some months later, the Kerala High Court banned its use, pending a final decision by the Central Insecticides Board. The Dubey Committee set up by this Board gave Endosulfan a clean chit in March 2003.

In November 2004, the committee set up by the Union Agriculture Ministry reviewed various reports on the crisis, including those of the Dubey Committee and the National Institute of Occupational Health. The members agreed that Endosulfan was a carcinogen. In December 2004, on the basis of a Kerala High Court Order, the State Pollution Control Board banned its use.

The last word on this controversy has not been said yet. However, what about the people of Padre? Who is going to help them or compensate them?

Source: Joshi (2001), Yadav and Jeevan (2002), Yadav (2003), and several reports on the website *www.indiatogether.org/*.

Chemical pesticides also kill non-target species, including the predators of the very organisms they are supposed to eliminate. Moreover, many pests are able to develop a resistance to these chemicals. This leads to the use of even greater amounts of the chemicals or more powerful chemicals, resulting in more problems. The only way out of this vicious cycle is to shift to organic pesticides and biological control.

There is great risk in the manufacture and storage of large amounts of pesticides. Several disasters have already occurred, the Bhopal Gas Tragedy (Box 16.1) being the worst on record.

Many of the new pesticides have been banned in industrialized countries. However, the same countries continue to manufacture and sell banned pesticides to poorer countries, where regulations and enforcement are often lax. The US is the world's largest exporter of pesticides, followed by Germany and the UK. Often, whole pesticide operations are shifted or sold to poorer countries to avoid environmental regulations.

It is not that the populations of developing countries alone suffer from pesticide poisoning. With freer and increasing global trade, the pesticides come back to the industrialized countries in imported food products and in other ways. (This is known as the 'boomerang effect'.)

Herbicides have also been used in wars with disastrous effects. An example is the use of Agent Orange by the US in Vietnam (read Box 16.3).

Box 16.3 Disaster: The deadly agent

The assignment given to Agent Orange was to clear the countryside of all vegetation, thus denying cover to guerrilla forces. The agent did the job only too well. Not only was the vegetation destroyed, so was the health of thousands and thousands of people for years to come. In fact, it was a case of a 'double agent': as the soldiers on the agent's side were also affected.

Agent Orange was not the codename of a spy. It was a compound herbicide, a defoliant used extensively by the US Army in the Vietnam War. As admitted by the US, 72 million litres of Agent

Orange were sprayed over parts of Vietnam (between 1962 and 1970) to defoliate the jungle and flush out the Vietcong guerrillas. The name Agent Orange was derived from the orange markings on the drums in which the chemical was shipped.

The herbicide, however, contained one of the most virulent poisons known to man, a strain of dioxin called TCDD. A small 80-gram tin of TCDD can destroy New York City. The US dropped 170 kilograms of it on parts of Vietnam!

After stripping the jungle bare, the dioxin gradually spread into the food chain. This has had serious effects on the health of the Vietnamese people as well as that of the US soldiers: defective births, Down's syndrome, skin diseases, liver cancer, mental disorders, and so on.

In a village in the heavily sprayed Cu Chi district, it is a perpetual struggle for 21-year-old Tran Anh Kiet. His feet, hands, and limbs are twisted and deformed. He can only make plaintive and pitiful grunts. He is an adult stuck inside the stunted body of a 15-year-old, with a mental age of around six. He is what the local villagers refer to as an Agent Orange baby.

There are 150,000 other children like him, whose birth defects can be traced back to their parents' exposure to Agent Orange during the war, or the consumption of dioxin-contaminated food and water since 1975. About three million Vietnamese were exposed to the chemical during the war, and at least one million suffer serious health problems today. Some are war veterans, who were exposed to the chemical clouds. Many are farmers who lived off land that was sprayed. Others are second- and third-generation victims, affected by their parents' exposure.

In the early 1980s, 2.5 million Vietnam War veterans in the US filed compensation claims against the US government and the manufacturers of Agent Orange. The companies paid out US$ 180 million in an out of court settlement, but never admitted any liability. American victims of Agent Orange get up to US$ 1500 a month. However, most Vietnamese families affected receive around 80,000 Dong a month (just over US$ 5) in government support for each disabled child.

Since the end of the Vietnam War, Washington has denied any moral or legal responsibility for the toxic legacy said to have been caused by Agent Orange in Vietnam. In January 2004, three Vietnamese filed cases against Monsanto Corporation, Dow Chemicals, and eight other companies that manufactured Agent Orange and other defoliants used in Vietnam.

The efficient Agent Orange refuses to stop working. Some of the victims live near former US military bases such as Bien Hoa, where Agent Orange was stored in large quantities. Dr Arnold Schecter, a leading expert in dioxin contamination in the US, sampled the soil there in 2003, and found it contained TCDD levels that were 180 million times above the safe level set by the US Environmental Protection Agency.

How long will the Vietnamese continue to suffer? How many US war veterans have been victims of their own Agent Orange? To what purpose?

Source: Pope (1991) and several news items on the website of the BBC, *http://news.bbc.co.uk/*

What is special about DDT? Dichloro-diphenol-trichloroethane (DDT) is an insecticide, which was first introduced in the US in the 1940s and later in many parts of the world to protect crops and human beings from insects. It is cheap, easy to make, and chemically stable.

The spraying of DDT was a very important part of the Green Revolution package (see Chapter 12). It is effective against a wide spectrum of insects in agriculture. It also kills the typhus-carrying body lice, and mosquitoes, which transmit malaria and yellow fever.

DDT is initially very effective against pests, but over a period of time the insects develop a resistance to the chemical. It is now recognized as a dangerous pollutant and is also a suspected carcinogen (cancer-causing substance). It is highly toxic to vertebrate mammals and some species of fish and birds and is passed along the food chain. It affects the breeding patterns of some predatory birds.

DDT was banned in the US in 1973, thanks primarily to the book *Silent Spring* (Read Module 3: Prologue) by Rachel Carson. Many other industrialized countries have followed suit. In most developing countries, however, DDT continues to be used against malaria.

What are the alternatives to using pesticides from the chemical industry? In the long run, chemical pesticides will only ruin the soils and the health of nations. The only alternative is to shift to organic farming and the use of traditional organic pesticides.

In California in the US, where environmental awareness is greater than in other parts of the country, more than 40 farmers are using giant vacuum machines to suck insects off the plants. These machines move slowly over the rows of crops and remove certain types of insects. They remove some beneficial insects too, but fewer than the numbers killed by chemical pesticides. Each vacuuming eliminates the need for an application of pesticide.

What are Polychlorinated Biphenyls (PCBs)?

PCBs are chemical compounds that are very stable, have good insulation qualities, are fire resistant, and have low electrical conductivity. They are widely used in electrical capacitors, transformers, hydraulic systems, fluorescent lamps, etc.

PCBs persist for long in the environment and are extremely toxic. They can cause cancers in animals and liver and nervous disorders in humans. When burnt, they leave toxic ash and when buried, they leach into the groundwater. Rainfall deposits PCBs in the atmosphere into rivers and seas.

PCBs are found in high concentrations in many marine mammals. They have a serious effect on the reproductive capacity of these animals and could lead to the extinction of many species.

Humans who mainly consume fish are at risk from PCBs. These chemicals are said to have caused defects in children in the Great Lakes area of Canada. Surprisingly, PCBs have even been found in the breast milk of Arctic women.

PCBs have been banned in industrialized countries. Yet, thousands of tons of PCBs are still present in storage, landfills, or abandoned electrical equipment. A substantial portion is in developing countries, with a high risk of contamination.

How dangerous is contamination by lead or mercury?

Lead is an extremely poisonous metal that accumulates in organisms. Large doses of lead can cause paralysis, blindness, and even death in humans. If a pregnant woman carries even a small amount of lead in her body, the mental development of the baby could be affected. Lead in the atmosphere contaminates leafy vegetables and fruits.

The lead that we breathe in comes mostly from the exhaust of vehicles. For long, lead has been added to petrol to prevent the 'knocking' of the engine. Lead-free petrol is now available in India and is mandatory for most new cars. Lead is also found in batteries, paints, bullets, and certain alloys. There is a global effort to phase out lead from products and processes.

Mercury is a liquid metal that is poisonous for humans and animals. While small doses cause headaches, large ones could lead to death. If inhaled, swallowed, or absorbed through the skin, mercury can damage the human nervous system. It accumulates easily in the body and is non-excretable.

The chemical and plastics industries use large amounts of mercury and release it along with other effluents into rivers, lakes, and seas or they dump it on the land as waste. This has led to the poisoning of fish and people (read Box 16.4 on the Minamata case).

Even small amounts of mercury used in thermometers and scientific and medical equipment can cause poisoning, if leaked under warm conditions (read about the Kodaikanal mercury case in Chapter 20, Box 20.5 of this book).

Box 16.4 Disaster: The dancing cats

It started out in the 1950s quite simply, but somewhat strangely: Cats were found 'dancing' in the streets and somtimes collapsing and dying. Friends or family members occasionally shouted uncontrollably, slurred their speech, or dropped their chopsticks at dinner. When such scattered, apparently unconnected, and mildly mysterious events began to haunt the fishing town of Minamata, Japan, the people did not realize that they were the first signs of one of the most dramatic and emotionally moving cases of industrial pollution in history.

A chemical plant of the Chisso Corporation had been dumping mercury into Minamata Bay. Between 1953 and 1983, 300 inhabitants of Minamata died by eating fish contaminated with methyl mercury. Hundreds of others suffered from blindness, convulsions, and brain damage. Children of victims were born with the disease.

Minamata is an extraordinary tragedy with environmental, social, and political aspects:

- Chisso persistently denied any connection between the mercury discharge and the disease and the government went on supporting the company. By 1959 Chisso had knowledge of the connection, but withheld the information.
- The dumping of mercury went on for a long time, even after the strange disease started spreading among the fisherfolk. Neither the company, nor the local and central governments acted to stop the discharge of mercury.
- Chisso was very much a part of the town, providing employment and bringing prosperity. The loyal workers refused to blame the company for causing the disease. In fact, the company doctor who discovered the connection revealed it only on his deathbed.
- Since it was a strange disease, the sufferers were stigmatized by their neighbours, until of course the latter too caught the disease.
- Denying any responsibility, the company paid a token compensation only to those victims who signed away their right to further legal action. Many victims took the company to court, but they had to wait a long time for justice. The first damages were awarded only in 1973. In fact, the Japanese government officially accepted mercury discharge as the cause of the disease only in 1983!
- About 3000 people were recognized as victims of mercury poisoning, making them eligible for a variety of health benefits, but over 16,000 were refused recognition.
- A persistent group of sufferers kept up pressure on Chisso through continued petitioning, recruiting of grassroots support across Japan, months of sit-ins at Chisso headquarters, etc. They ultimately did get some justice, but it required an enormous amount of patience and effort.
- Nearly 50 years passed before the victims scored a moral victory: In October 2004, Japan's Supreme Court held the government responsible for the spread of the disease. Giving its judgement

in the last pending case, the Court awarded a compensation of US$ 650,000 to 37 victims.

One fish in Minamata, one night on a street in Bhopal, or a breath of air near Chernobyl was enough to cause for a lifetime of suffering for generations. How can we still continue to support the use of toxic pesticides or nuclear power?

Source: Pope et al. (1991), a news item in *Down To Earth,* November 15, 2004, and Douglas Allchin, 'The Poisoning of Minamata', on the website *www1.umn.edu/ships/ethics/minamata.htm*

What is the problem with asbestos?

Asbestos is a fibrous silicate mineral. The fibres are woven into a cloth, a binder like cement is added, and the resulting rigid material can be shaped into many forms. It is a non-corrosive, non-flammable, and non-conducting material and is inexpensive.

Asbestos is widely used in construction, most commonly as corrugated roofing and sometimes also for doors and partitions. Asbestos wool is used as insulation.

Asbestos is very dangerous for the health if its fibres are inhaled. They can be as short as 0.0000025 cm in length and they get lodged in the lungs and bronchial tubes. They cause a disease called asbestosis, which affects the respiratory tissues. Chronic shortage of breath and sometimes premature death is the result. The fibres can also cause lung and intestinal cancer. Diseases caused by asbestos exposure take a long time to develop. Most cases of lung cancer or asbestosis in asbestos workers occur 15 or more years after initial exposure to asbestos.

In 2003, asbestos was included in the Rotterdam Convention list. This means that all forms of asbestos will now be subject to trade controls. The exporting countries must provide importers with information on its potential health and environmental effects.

There is now a worldwide movement against the use of asbestos. In 2003, conferences held in Ottawa, London, and Dresden called for a ban on asbestos, and for providing assistance and compensation to those suffering from asbestos-related diseases.

There is still an active and large asbestos industry in India. Out of 125,000 tons of asbestos used in India, 100,000 tons are imported, mainly from Russia and Canada. Efforts to ban at least some forms of asbestos in India have so far not been successful.

What are dioxins?

Dioxins are a group of chemical compounds that occur accidentally as contaminants in a number of industrial processes and products. They are carcinogens and are highly toxic to humans and animals.

Dioxins are formed due to the incomplete incineration of waste and the burning of plastics, coal, or cigarettes. When fuel is partially burnt in a vehicle, dioxins are released. Dioxins are deposited on plants and soil and in water thereby entering the food chain.

Dioxins are also produced during the manufacture of paper. The pearly white colour of paper is the result of bleaching, which uses highly toxic chemicals. Bleaching uses chlorine, which produces toxins including dioxins. Ultimately the toxic effluents from the paper factory end up in water bodies everywhere.

What were the major industrial disasters?

Table 16.1 gives a chronology of the major industrial disasters that have occurred since the mid-twentieth century. It does not include mining and nuclear accidents.

In many of these disasters, there have been inspiring stories of individuals who went to great lengths to help the victims. Some individuals have also led the struggles for justice, compensation, and government action to prevent such events in the future. Examples are Lois Gibbs in Love Canal (read Box 16.7) and Rashida Bee and Champa Devi Shukla in Bhopal (read Box 17.6, Chapter 17 of this book).

Table 16.1 Chronology of Industrial Disasters

Year	Location	Description
1947	Texas, the US	On April 16, the SS Grandcamp, carrying ammonium nitrate fertilizer, exploded in the Texas City harbour, followed the next morning by the explosion of the SS High Flyer. The disaster killed 576 and injured several thousand. The explosion was felt 75 miles away in Port Arthur, and created a 15-foot tidal wave.
1950s	Minamata, Japan	Minamata Case (read Box 16.4).
1976	Seveso, Italy	Seveso Disaster (read Box 16.5).
1977	New York, the US	Love Canal Case (read Box 16.6).
1980	Basel, Switzerland	Fire at the Sandoz chemical plant near Basel—1200 tons of pesticides were burnt and 30 tons of highly toxic chemicals were washed away into the Rhine.
1982	Missouri, the US	Dioxin found in soil and 2242 residents evacuated from Times Beach. (In 1996–97, 265,000 tons of dioxin-contaminated material from Times Beach and 26 other sites in eastern Missouri was incinerated.)
1984	Bhopal, India	Bhopal Gas Tragedy (read Box 16.1).
1993	Thailand	Fire in toy factory killed 188 women and injured over 400.
1998	France	Explosion at a fertilizer factory near Toulouse.
1998	Nigeria	Pipeline at Jesse explodes, instantly killing more than 500 people and severely burning hundreds more. Up to 2000 people had been lining up with buckets and bottles to scoop up oil. The fire spread and engulfed the nearby villages, killing farmers and villagers sleeping in their homes. (Several foreign oil companies like Shell, Chevron and Mobil split their oil revenues with the Nigerian National Petroleum Company.)
2001	France	Explosion at an agricultural chemicals factory near Toulouse. Thirty one people were killed and at least 650 people were hospitalized.
2004	Scotland	ICL Plastics plant in Glasgow explodes, killing nine and injuring more than 40.

Source: The website *www.endgame.org*

Box 16.5 Disaster: A cloud of poison

On July 10, 1976, a chemicals factory near the Italian village of Seveso released a cloud of toxic TCDD gas—the same dioxin that was in Agent Orange (see Box 16.3)—into the atmosphere. The cloud visibly settled over the homes and gardens of Seveso. For two weeks, the residents did not know that it was a deadly cloud of poison. The company, a unit of the Swiss chemicals giant Hoffman La Roche, kept quiet.

Only after the mayor was informed of the deadly nature of the leak, were 700 people were evacuated. It was too late. By that time, 500 residents had symptoms of poisoning including headaches, swellings, and skin diseases. Eventually, 70,000 animals and 80,000 domestic fowl died or were slaughtered. Thousands of birds and countless small creatures died.

The Seveso plant had released between two kg and 130 kg of TCDD. The cause of the leak was overheating combined with a valve failure. A Parliamentary Commission laid the blame on the company and the local administration for their failure to anticipate the risks of herbicide manufacture. In 1983, the company paid US$ 100 million to the village of Seveso as compensation and closed its operations. It had already paid US$66 million to the government and 250,000 individuals.

The Seveso disaster was seen at the time as an indictment of the chemical industry. Did the industry learn its lesson? Just four years later came the Sandoz accident (Table 16.1), followed by the biggest industrial disaster of all in Bhopal in 1984 (Box 16.1).

Source: Pope et al. (1991)

Box 16.6 Disaster: Dump and disappear!

This is how the story began, in the words of Lois Gibbs, housewife and mother of two children:

> In 1978, my neighbours and I discovered that our neighbourhood in Love Canal, New York, had been built next to 21,800 tons of buried toxic chemicals. When we bought our homes, none of us knew that Hooker Chemical Corporation had dumped 200 tons of a toxic, dioxin-laden chemical called trichlorophenol and 21,600 tons of various other chemicals into Love Canal. We just knew we were getting sick. We knew there were too many miscarriages, too many birth defects, too many central nervous system problems, too many urinary tract disorders, and too much asthma and other respiratory problems among us.

In the 1880s, William Love began digging a canal from the Niagara River to divert water for a power plant. The canal was never completed, but a small part remained unused. Between 1942 and 1952, Hooker bought the canal and used it as a dump for toxic waste packed in barrels. The company covered the dump with a clay cap and soil.

In 1952, when Niagara Falls County wanted land for building homes and a school, Hooker sold it to the city for one dollar and left—without disclosing the real danger under the ground! About 950 homes, a school, and playfields were built on the site.

In the late 1950s, the barrels started leaking, children playing in the school ground suffered burns. Gradually, things got worse. Trees lost bark, gardens died, and smelly pools of toxins welled up. In the 1970s, home basements were flooded with thick black sludge. Many health problems surfaced and finally the residents realised they had a serious problem on their hands.

Tests of the water, soil, and air showed 82 different contaminants, most of them carcinogens. There was a public outcry, the canal was fenced off, and several hundred families were evacuated. Ultimately, all but 72 of the owners moved out of the area. President Carter declared it a disaster area.

The clean-up began in 1987 and by 1995 the state and federal governments had spent US$ 272 million on clean-up operations, relocation of families, compensation, etc. After 15 years of litigation, OxyChem, which had bought Hooker, agreed to a settlement with the state and took responsibility for the future treatment of the wastes.

The site was covered with a new clay cap and surrounded by a drainage system. In 1991, the families started coming back, but the long-term effects remain unknown.

It turned out that Love Canal was just the tip of an enormous iceberg. In 1989, the US Environmental Protection Agency estimated the number of such sites at 32,000. Other government agencies put the number as high as 100,000 to 400,000, not counting 17,000 military hotspots.

Meanwhile, Lois Gibbs and others had formed an Association to fight against the dumping of toxic wastes. Lois raised the status of the the hazardous wastes issue to a national debate. She campaigned ceaselessly and wrote books on the issue. Her work spurred the US government to action.

The US Congress established a Superfund in 1980 to locate, investigate, and clean up the worst sites nationwide. The programme, however, has run into many problems and over 20 years, only a small number of sites have been cleaned up. Further, the 'Polluter Pays' principle is also being given up in collecting funds for the programme (read Box 15.7).

There are of course thousands of such dumps in western and eastern Europe. How many sites are there in India?

Source: Chiras (2001), Enger and Smith (2004), Miller (2004), and the website *www.chej.org*

Box 16.7 Inspiration: Mother of Superfund

In 1977, she was a young housewife with two children living near Love Canal (Box 16.6). She had no experience in environmental or social work—until she became concerned when her children began experiencing unexplained illnesses. She began investigating the cause, became an activist and ultimately a national figure.

When Lois Marie Gibbs discovered that toxic chemicals from the Love Canal dumpsite were causing the problems, she organized her neighbourhood into the Love Canal Homeowner's Association. Apart from fighting for their own rights and safety, Lois raised the general question of handling hazardous wastes.

Later, Lois moved to Washington, D.C., and set up the Citizens' Clearinghouse for Hazardous Wastes, later renamed as Center for Health, Environment, and Justice (CHEJ). This organization has helped over 7000 communities protect themselves from hazardous wastes. Lois brought this issue to the national level and her work led to the establishment of the federal Superfund for cleaning up hazardous waste sites.

Lois has become a symbol of what can happen when citizens, provoked by injustice and emboldened by outrage, stand up for themselves and their families. Her story has become a legend because her relentless demand for the truth opened the eyes of an entire nation. Her actions, and the actions of her neighbours who formed the Love Canal Homeowner's Association, demonstrate how one committed person or one committed community can change the course of history.

Lois continues to work for a safe environment and there are new battles to be fought. The US Superfund toxic waste clean-up programme has run out of polluter-contributed funds, leaving taxpayers to shoulder the financial burden. The 'Polluter Pays' taxes expired in 1995, and President Bush has refused to support this landmark principle and thus shifted the burden back to the taxpayer. Center for Health, Environment, and Justice (CHEJ) and other organizations like Sierra Club and the National Environmental Trust are together campaigning for reinstating Superfund's Polluter Pays' taxes to fulfill the original intent of a fair and equitable approach to cleaning up toxic waste sites.

With 1240 toxic waste sites in the US still in need of clean-up, Lois and others have more than enough to do!

Source: Miller (2004) and the websites *www.chej.org* and *http://arts.envirolink.org/arts_and_activism/LoisGibbs.html*

Industrial accidents continue to occur in India. Since the Bhopal Gas Tragedy, over 40 accidents have occurred, killing more than 250 people, not counting the deaths in mining accidents. Several of the accidents took place in chemical or gas factories.

What is being done about toxic chemicals?

Europe has been at the forefront of environmental conservation in the industrialized world. By 2003, the European Union (EU) had withdrawn from the market nearly 500 substances used in agriculture, including many pesticides, fungicides, and herbicides. The manufacturers did not contest the decision because they were spending large sums of money on safety tests required by the EU. This is significant since the EU accounts for 25 per cent of global pesticide consumption.

By 2008, the EU will have reviewed all the 850 chemicals used in Europe. The EU plans to promote integrated pest management, organic farming, and biological control of pests.

Box 16.8 Make a Difference: Observe the Global Day of Action Against Corporate Crime

The International Campaign for Justice in Bhopal (ICJB) declared December 3, 2003, the anniversary of the 1984 Union Carbide disaster in Bhopal as the Global Day of Action Against Corporate Crime. Since then, this day has been observed as such every year. Many non-governmental organizations (NGOs) like Greenpeace and the Sierra Club have joined the worldwide observance of this Day.

Besides coordinating protests against Dow-Union Carbide facilities worldwide, the ICJB invites all groups fighting corporate crime to take action on December 3 against the human, environmental, consumer, and labour rights violations by private or public corporations.

The Day is observed through protests at Dow Chemical facilities and offices worldwide: teach-ins, vigils, phone-ins, petition drives, celebrations, and media events.

You can use the Day to get students thinking about polluting and dangerous industries, finding out about industries in their locality, what they produce, raw materials used, and disposal of waste, etc.

Source: The website *www.bhopal.net/gda/*.

A quick recap

- While pesticides and industrial chemicals have brought benefits, they are also capable of inflicting damage on humans and other organisms, immediately and over a long period.
- The effect of chemicals on humans can extend to succeeding generations.
- Given the nature of the chemicals we use, serious accidents will continue to occur.
- It is difficult to hold chemical companies responsible and liable for the industrial disasters caused by their negligence or their operations; and governments for their lack of appropriate and timely action.
- Poor people are disproportionately affected by industrial accidents and often do not receive proper compensation and justice.

What important terms did I learn?

Agent Orange
DDT
Dioxin
Electromagnetic fields
Environmental health
PCBs

What have I learned?

Essay question

Discuss the issue of responsibility and liability in the case of industrial accidents, pesticide poisoning, and chemical warfare. Using the case studies described in this chapter, explain why it is difficult for the victims, especially the poor, to get timely and adequate compensation and long-term support. How have the governments responded in these cases?

Short-answer questions

1. What is meant by environmental health?
2. Explain through examples how pesticides and industrial chemicals have positive and negative impacts on society.
3. Discuss health issues in connection with the use of, or contamination by, any two of the following substances: DDT, Endosulfan, PCBs, lead, mercury, asbestos.

How can I learn by doing?

Identify a case of an industrial accident, pesticide or chemical poisoning, dumping of toxic waste, or water/soil contamination reported in your area. Study the incident and write a report on the causes, impact, relief and compensation to victims, concerned rules and regulations, steps taken for the future, etc.

What can I do to make a difference?

1. Examine every item you use at home and try to find out if it contains hazardous chemicals or if it has been processed using such chemicals. If the answer is yes, try to replace the item with a more ecofriendly and safer product.
2. Avoid the use of chemical mosquito repellents in your home and try to convince your friends to stop using them too. Chemicals like allethrin used in mosquito coils, mats, and liquids are toxic. Covering windows with nets and using a mosquito net are far safer methods of protecting oneself against mosquitoes. If you must use a repellent, choose natural substances like lemongrass (citronella) oil available in the market.
3. Observe December 3 as the Global Day of Action Against Corporate Crime (read Box 16.8).

Where are the related topics?

- Chapter 12: Water, Soil, and Marine Pollution
- Chapter 13: Solid Waste Management
- Chapter14: Disaster Management

What is the deeper issue that goes beyond the book?

Is it possible to get rid of all factory-made pesticides and go back to organic and natural pesticides? How do we balance the need to produce more food with the desire to avoid chemical inputs in agriculture?

Where can I find more information?

- There is an enormous literature and many websites on the Bhopal Gas Tragedy. Some of the websites are: *http://homepages.gs.net/~aaswell/bhopal/*, which contains a list of web links on Bhopal, *www.bhopal.net*, *www.scorecard.org*, and *www.studentsforbhopal.org*.

- For a recent analysis of the various aspects of the disaster and its aftermath, refer to *Seminar,* no. 544, December 2004. This issue of the monthly is devoted to Bhopal under the title *Elusive Justice* and includes a number of articles on the topic. Lapierre and Moro (2001) is an engrossing recent book on the subject.
- For more information about Lois Marie Gibbs and her campaign against the dumping of hazardous wastes, read Gibbs (1982, 1995). There is also a documentary entitled, *Lois Gibbs: The Love Canal,* made by the US television channel CBS.

What does the Muse say?

Here are two poems, both by persons directly affected by toxic substances. Read what a Bhopal survivor has to say:

Torture Me

torture me.
aim a blowtorch at my eyes
pour acid down my throat
strip the tissue from my lungs.
drown me in my own blood.
choke my baby to death in front of me.
make me watch her struggles as she dies.
cripple my children.
let pain be their daily and their only playmate.
spare me nothing. wreck my health
so I can no longer feed my family.
watch us starve. say it's nothing to do with you.
don't ever say sorry.
poison our water. cause monsters
to be born among us. make us curse God.
stunt our living children's growth.
for seventeen years ignore our cries.
teach me that my rage is as useless as my tears.
prove to me beyond all doubt
that there is no justice in the world.
you are a wealthy american corporation
and I am a gas victim of Bhopal.

This is how a Minamata fisherman suffering from mercury poisoning expressed his love for the sea:

When I thought I was dying
and my hands were numb
and wouldn't work—
and my father was dying too—when
the villagers turned against us—
it was to the sea
I would go to cry.
No one can understand
why I love the sea so much.
The sea
has never abandoned me.
The sea
is the blood of my veins.

CHAPTER 17

Women and Child Welfare

You can tell the condition of a nation
by looking at the status of its women.

Jawaharlal Nehru

Women are a tremendous force
if you empower them.
They're as difficult to organize as doctors,
but once you get them going they're unstoppable.
Women are much more open with their feelings and the truth,
and thcy're one of the golden keys to the salvation of this planet.

Helen Caldicott

What will I learn?

After studying this chapter, the reader should be able to:

- understand how environmental degradation has a greater impact on women than on men;
- appreciate how, as compared to men, women shoulder a greater part of the household work and toil for longer hours;
- comprehend how rural women are exposed to dangerous levels of Indoor pullution from burning biomass fuels in traditional cookstoves and how the problem can be handled;
- understand how women have special problems with regard to water supply and sanitation;
- describe the vulnerability of children and the large extent of child labour (including bonded labour) in hazardous occupations in developing countries;
- recall that India has the largest number of child workers and that the relevant laws to curb child labour are not enforced;
- explain what is HIV and AIDS and how the infection is passed on;
- describe the current world AIDS scene;
- recognize that the face of HIV/AIDS is increasingly that of young women and children;
- appreciate the special problems that the HIV/AIDS epidemic brings to women and children; and
- appreciate the role played by women environmentalists and how ordinary women have done extraordinary things for environmental conservation and justice.

Box 17.1 Suffering: Billion-dollar business, millions of poor women

She is up at 3 a.m., cooks for her family of five, and feeds her cow. By 5 a.m., she is in the forest collecting fuelwood. She looks for dry branches and cuts them almost to equal sizes before making a neat bundle. She continues this search and collection for two hours, walking about a kilometre inside the forest.

She pays Rs 5 as bribe to the forest guard and walks a long distance to the market carrying the 30–35 kg load. She has to sell the bundles the same day, at whatever price she can get. At the end of the day, after buying some food, she is left with Rs 15.

Phulmai Phulo, a 35-year-old woman of Chauda Village of Jharkhand has been following this routine for 15 years. Her lifeline is a patch of the Khellari Forest, where 30–40 other women do the same thing. These are among the millions of headloaders of India—all women.

How did Phulmai end up as a headloader? It took just three successive droughts and a lack of irrigation to make her one-acre plot barren. Headloading was the only option. Her husband migrated to Gorakhpur in search of a job and she hardly hears from him. In fact, there has been a mass migration of men from the area, leaving the women to fend for themselves.

During the monsoon, the headloaders do less business and many starve. In years of drought, more women join the activity, and everyone earns less. With more people collecting from the same area, there are also conflicts.

What is the popular perception of these women? They are engaged in an illegal activity. They destroy the forest. They are the environmental villains.

Their story is different. They are driven to this work only because agriculture was destroyed. They do not cut whole trees, only dry branches. They have to bribe the forest guards, panchayat officials, bus and train conductors, and finally they sell the bundles to middlemen at throwaway prices.

'Every day we sell about 100 bundles from this patch of forest. But we never cut a tree. That needs money, time, and a huge bribe,' Phulmai says. She herself never uses fuelwood or kerosene, only twigs and leaves.

India is the world's largest consumer of fuelwood. It is, in fact, a multi-billion-rupee business. The headloader is the foundation of this thriving economy. Why then do Phulmai and millions of other women walk long distances every day, work hard, and yet earn so little? Why are they considered to be the destroyers of the forest, engaged in an illegal activity? Can anything be done to change the rules of the game and enable these women to get a fair wage and a decent life?

Source: Mahapatra (2002b)

What is the significance of the Phulmai case?

We have discussed fuelwood collection in Chapter 8 with reference to energy sources. A return to the topic here was necessary, since we cannot emphasize enough the impact of environmental degradation on women, and their role in environmental conservation.

Women feel the adverse effects of environmental and occupational problems far more than men do. This is especially true in relation to inadequate fuel, water supply, and sanitation. The main reason is their role in society as determined by social, economic, and political structures. At the same time, women often do a much better job than men when it comes to environmental conservation or fighting for environmental justice. In this chapter, we will see examples of both these aspects.

Children too are very vulnerable to the consequences of environmental degradation. Toxicity in the mother's body can result in birth defects among children. Children are also easily affected by pollution. Recall, for example, the case of juvenile asthma due to air pollution (Chapter 11, Box 11.1).

Women and children are also the prime sufferers in wars and conflicts and from the environmental consequences of these wars and conflicts. For example, according to a study made by the Johns Hopkins University in the US, most of the 100,000 civilians who died in the Iraq War in 2003–04 were women and children.

The topic of women/children and environment is a vast one with many ramifications. We will confine ourselves to the following aspects:

- collection and burning of household fuel by women;
- women's special problems with regard to water supply and sanitation, especially in urban areas;
- women and children in hazardous occupations;
- the HIV/AIDS situation with regard to women and children; and
- women in the movement for environmental conservation and justice.

What are the women's issues related to collection and burning of fuel?

Millions of women like Basumati Tirkey (see Chapter 8, Box 8.1) and Phulmai Phulo (see Box 17.1 above) in developing countries make a living from collecting fuelwood. Millions of other women spend substantial amounts of time and effort every day collecting fuelwood for their own use. We rarely see men doing this work.

The women's problem does not stop with the collection of fuelwood. When they burn the wood or any biomass in their cookstoves, they inhale poison. It is a shocking fact (Mishra et al. 2002, Narain 2003) that most rural women in India inhale everyday carcinogens equivalent to smoking about 100 cigarettes!

Urban, outdoor air pollution is all too visible and affects the health of all who breathe the air. Yet, in the rural areas of developing countries, indoor air pollution can be much worse than the pollution levels in big cities. It accounts for much ill-health and well over a million deaths annually. In India alone, 400,000 to 550,000 women and children die prematurely every year because of indoor air pollution.

The World Health Organization (WHO) has ranked various risk factors in terms of the percentage of ill-health they account for in developing countries. In this list, indoor air pollution ranks fifth, behind malnutrition, AIDS, tobacco, and poor water/sanitation. The estimate is that there are more than 1.6 million premature deaths every year due to cookstove pollution. Despite such data, it does not attract much public attention. It is a quiet killer, affecting mostly poor women and young children.

Why is the problem of cookstove pollution so widespread and severe? Nearly half of the world's households use unprocessed biomass fuels like wood, animal dung, crop residues, and grass for cooking and heating. In the developing countries of South Asia and sub-Saharan Africa, 80 per cent of all homes cook using biomass fuels.

Pollution from cookstoves is serious because of the following reasons:

- Burning biomass fuels in simple indoor cookstoves release large amounts of pollutants like fine particles of matter (that can get into the lungs), carbon monoxide, nitrogen oxides, formaldehyde, and dozens of toxic hydrocarbons. A study in Gujarat showed that women were exposed to total suspended particulate matter (SPM) of 7000 micrograms per cu. m in each cooking period. The safe limit for outdoor air is 140 microgrammes per cu. m!
- The problem is aggravated when cooking areas are inadequately ventilated and the dwelling lacks a separate kitchen.
- Even when the cookstoves are used outdoors, they can increase pollution in the surrounding neighborhood to unhealthy levels.
- Fire from biomass fuels requires more or less continual feeding, resulting in extended exposure to the smoke generated. Many people are exposed for three to seven hours daily, and even longer in winter months when houses must also be heated. Every winter, deaths occur in North India when people close all openings in their houses and let the smoke circulate within the dwelling.

Inhaling the deadly smoke leads to a host of respiratory diseases, including acute respiratory infections, chronic bronchitis, asthma, and tuberculosis. It has also been linked to lung cancer, adverse pregnancy outcomes, cataract, and blindness.

What can be done to solve the problem of cookstove pollution?

The following are suggested as an obvious strategy to reduce the inhalation of smoke.

- Creating awareness about the risks of exposure to cookstove smoke.
- Providing more efficient and better ventilated cookstoves. For over 50 years, universities and research establishments in India have designed many 'smokeless chulhas', but the programmes for promoting the use of such stoves have never fully succeeded. The most successful cookstove programme has been in China, where some 200 million improved stoves have been introduced in recent decades.
- Promoting the use of cleaner fuels. When urban India gets subsidized LPG (Liquefied Petroleum Gas), there is no reason why villages should not get the same in small cylinders. There are other ecofriendly options too, like biogas.

What are the issues concerning water supply and sanitation?

In Chapter 7 we covered the issue of water scarcity and in Chapter 12 we noted the severe shortfall in providing sanitation facilities to the people of developing countries.

In these countries, it is typically women who collect and manage water. In rural areas, they have to walk long distances to the water sources and bring back heavy loads of water. In urban areas, they have to queue up for long periods at water points. In the process, they have to either get up very early or go to bed late.

It is the women again who have to manage the inadequate water supply, rationing it carefully for washing, cleaning, and cooking. Often they have to take great trouble to collect drinking water separately.

Lack of sanitation facilities affects women far more than it does men. There is a severe shortage of toilets in the slums and poorer areas of cities. In the rural areas, men use the open fields, but the women have to wait for darkness to provide cover. The suppression of natural urges for long hours causes serious health problems.

The daily struggle for water and toilet facilities in crowded urban slums is perhaps the worst burden that women have to bear. If you have never had to rely on public taps or public toilets, you cannot appreciate how tiring, stressful, and inconvenient it can be. The first-person accounts of their experience can be very moving (read Box 17.2).

Box 17.2 Struggle: Water! Water!

In 2003, the UN Human Settlements Programme (UN-HABITAT) published a report entitled, *Water and Sanitation in the World's Cities.* The Report carries first-person accounts on water and sanitation from the women of Dharavi, the slum in Mumbai that we described in Chapter 15 (Box 15.1). These are the voices of some of the women:

Paliniamma:

I have been here for 15–20 years. We have a lot of problems. There is no outlet for the drains. We dig holes near our houses and collect our washing water in it. The building says no water should come out on the path, so we collect the dirty water in drums and we take the drums and throw the water in the drain along the road. Children ask us when we will get a house. The other day I filled some containers with water and it was stinking. I could not drink any water at night after my food.

Kalyani:

I have been here for the past 39 years, since I got married. I came here before the highway was built. There was no toilet, no drains. There was no water. We had to go and beg for water. We would not bathe because there was no water. If we had to go to the toilet there were just two toilets, one for men and one for women. Once we went in the other people in the queue would shout and we had to come out in two minutes. At night we would go across the road.

Bhagwati:

I have been here for the past 18 years. Eighteen years ago we had to go to the Ganesh temple for water. We went at 4 a.m. and stood in line until 6 a.m. and got two containers of water. We had to leave the children at home. My child once fell into the drain and I thought he had died, but the neighbours picked him up and bathed him and he was okay. Five years back we put in a tap, but when they put in the borewell they broke the pipe. Now the water is dirty and we can only wash clothes in it.

The stories speak for themselves. No comments are needed.

Source: UN-HABITAT (2003)

What is the status of women's employment in the world?

According to the International Labour Organization's (ILO) *Report on Global Employment Trends for Women, 2004,* more women work today than ever before: in 2003 out of the 2.8

billion people that had work, 1.1 billion were women. However, for women there is no real socio-economic empowerment, including an equitable distribution of household responsibilities, equal pay for work of equal value, and gender balance across all occupations.

Women are more likely than men to find employment in the informal economy, outside legal and regulatory frameworks, without social security benefits. They, therefore, experience a high degree of vulnerability. This also means that the units that employ women are not likely to follow any environmental regulations and the work itself may be hazardous.

Women have a higher share than men in agricultural employment in Asia, sub-Saharan Africa, the Middle East, North Africa, in some Latin American economies, and the Caribbean. In all developing countries, women's share in industry is lower than men's.

How are poor women employed in India? Eighty-four per cent of the economically-active women in India currently derive their livelihood from agriculture. Women's work input into agriculture is more than that of men. In general, women work twice as many hours as men do. In spite of this, women's work is rarely recognized.

In Indian cities, women from the poorer groups are employed in labour-intensive work like garment manufacturing, nursing, the retail sector, and in small industries, and also in hazardous work like construction. A large number of poor urban women also work as domestics and as sellers of vegetables, flowers, and other small items. Hundreds of women are engaged in the particularly hazardous work of scarp collection (Box 17.3).

Box 17.3 Inequity: Toiling for an invisible economy

Janakibai Salve's life is similar to that of Phulmai Phulo, whom we met at the beginning of this chapter (Box 17.1). Janakibai also wakes up at 4 a.m., completes all the household chores, and goes out for her daily work—collecting garbage from the dustbins of upmarket neighbourhoods of Mumbai.

By noon, Janakibai is ready to sort the garbage and salvage the scrap. She carries the scrap (15 kg or more) to the trader's shop 4 km away. She makes about Rs 50 on which she, her husband, and their three children manage to live.

Chandrakala Adagale, Janakibai's neighbour, is also a scrap collector. Every day she goes to one of the municipal garbage dumping grounds. Chandrakala and hundreds of other women and children wait for the garbage trucks to arrive from the city. For six hours, she walks around in the dump, braving the glass and sharp metals, breathing in toxic fumes, just to collect waste paper. When she delivers the load to the trader, she too gets Rs 50.

From the trader upwards, dealing in scrap is a lucrative business. Very few scrap traders pay sales tax or income tax. It is an invisible economy, whose only visible face is that of the woman collector standing knee-deep in garbage!

There are about 50,000 scrap collectors in Mumbai, and a large number of them are women and children. They are mostly dalit migrants from the drought-prone areas of Maharashtra, Karnataka, and Tamil Nadu.

Scrap collection is an unhealthy and hazardous occupation. Women develop respiratory problems or meet with accidents. Most collectors do not have any footwear or any protective gear. They have no social security like pensions, even if they deal with the same trader for generations. If a scrap collector falls sick, her children take over the job. They are ignored or even shunned by society and there are no labour laws to protect them. On top of all this, they are often harassed by the police.

The ecosystem services provided by the scrap collectors are tremendous. They are in fact providing a free waste separation and removal service to the municipal waste management system. Further, the urban recycling industry for paper, board, glass, plastic, metal, etc., is heavily dependent on them. A significant part of the recycling process is over by the time the collector hands over the material to the trader. A large number of small industries use the scrap to reduce their cost of production.

When garbage disposal is privatized (some municipalities have done so), the collectors lose their occupation. Non-governmental organizations (NGOs) working with scrap collectors suggest ways of solving the garbage problem while improving the livelihood of the collectors. For example, one NGO in Mumbai has arranged for the municipality to dump waste from the vegetable market at a specific dumping ground. A group of women scrap collectors converts it into organic manure that they sell in the market.

Ideally, the scrap collectors should be formally involved in designing a waste management system for the city. They should get a better livelihood, even while the burden on the municipality is reduced. It can be a win-win situation, but will it ever happen?

Source: Katakam (2001)

How widespread is child labour in hazardous occupations?

The ILO has estimated that 250 million children between the ages of five and 14 work for wages in developing countries. At least 120 million of them are employed on a full time basis. Over 60 per cent of the child workers are in Asia, 32 per cent in Africa, and seven per cent in Latin America. Most working children in rural areas are found in agriculture; many children work as domestic help. Urban children work in trade and services, with fewer in manufacturing and construction. In general, they work for too many hours and too many days, for little or sometimes no pay, subject often to physical abuse, exposed to dangerous chemicals and polluted environments, and made to work with dangerous tools.

Most child workers are denied an education and a normal childhood. Some are confined and beaten or reduced to slavery. Others are denied the right to leave the workplace and go home to their families. Some are abducted and forced to work.

Child labour has consequences for adults too. Poverty is one of the causes of child labour, but it is also one of the consequences. Because it is so cheap, it causes adult unemployment and wage suppression.

The worst form of child labour occurs when a family hands a child over to an employer in return for a small, but urgently needed, loan. In most cases the child cannot work off the debt, nor can the family raise enough money to buy the child back. The workplace is often structured so that 'expenses' and/or 'interest' are deducted from a child's earnings in such amounts that it is almost impossible for a child to repay the debt. In some cases, the labour is generational, that is, each generation provides the employer with a new worker, often with no pay at all. Often, the families who become bonded in this way are dalits.

India has the largest number of working children in the world, with estimates ranging between 60 and 115 million. Most of them work in agriculture, while the rest pick rags, make bricks, polish gemstones, roll beedis, pack firecrackers, work as domestics, and weave silk saris and carpets. More than half of them remain illiterate. Many of them have been working since the age of four or five, and by the time they reach adulthood they are likely to be irrevocably sick or deformed. They will be old at the age of forty, and most likely dead at fifty.

Some of the worst cases of child labour in India can be found in the silk industry in Karnataka and Tamil Nadu, matches and firecracker units in Sivakasi in Tamil Nadu, and carpet weaving in Uttar Pradesh. Tamil Nadu has identified more bonded labourers than any other state.

There are many Indian laws against bonded and child labour, but enforcement is poor. In December 1996 the Supreme Court of India issued a groundbreaking decision outlining a detailed framework for punishing employers of children in hazardous occupations and for rehabilitating the children. In 1997 the Court ordered India's National Human Rights Commission (NHRC) to supervise states' implementation of the law against bonded labour.

The NHRC then began appointing special officers who applied pressure in certain regions and industries. State governments were obliged to conduct surveys on bonded labour and child labour, although the numbers reported seemed to be gross underestimates. A few of the employers were prosecuted, but almost no employer actually went to prison.

There has been international public awareness of the high incidence of child servitude in the carpet industry of South Asia. The Western public now associate child labour with the image of small children chained to carpet looms, slaving away to make expensive carpets for the wealthy. In particular, the case of bonded children in the Pakistani carpet industry caught the attention of the media (read Box 17.4).

Box 17.4 Inspiration: One million knots, one carpet, one child

On April 16, 1995, Iqbal Masih, a Pakistani boy, was shot to death while visiting his relatives. He was only 13, but he had already become famous the world over.

Iqbal was one of the 500,000 or more children between the ages of four and fourteen who work full-time as carpet weavers. As per UNICEF estimates, children make up 90 per cent of the workers in Pakistan's carpet industry. Boys aged seven to 10 are preferred for their dexterity and endurance. They earn one-quarter to one-third the salary of adult weavers, and they are obedient. They are from Pakistan's poorest families, and they have been sold by their parents.

The children work for 14 hours a day, six days a week, often chained to their looms. They make beautifully intricate carpets by tying thousands of knots with fingers gnarled and callused from years of work. A single carpet may have a million knots.

What is the work environment and how does it affect these children? They work in a polluted atmosphere and often have difficulty breathing due to cotton dust—many contract tuberculosis. They are thin, undernourished, and small for their age. Their backs are curved from lack of exercise and from bending over the looms. Their hands are scarred from the repetitive work. The monotony of tying thousands of knots is torture, like a death sentence, which it is for many of them. Most suffer from the 'captive-child syndrome', which kills half of Pakistan's working children by the age of 12.

Though the Government of Pakistan passed a law in 1992 prohibiting the bonded labour system, it does not enforce the law. Moved by the children's' plight, Ehsan Ulla Khan set up the Bonded Labour Liberation Front (BLLF). This NGO has liberated 30,000 adults and children from brick kilns, farms, tanneries, and carpet factories. In addition, the BLLF has established its own primary schools and is educating more than 11,000 children in them.

Iqbal Masih was one of those freed from slavery by the BLLF. He proved to be a special child. He became a BLLF worker and freed many children as he himself had been freed. Under Ehsan Khan's guidance, Iqbal became a spokesperson for the bonded children of Pakistan, and travelled to the US and Europe to convince customers not to buy Pakistani carpets. As a result, Pakistani carpet sales started falling.

Iqbal won the Reebok Human Rights Youth in Action Award, 1994. The very next year, however, Iqbal was allegedly killed by the 'carpet mafia', who were keen on continuing the use of bonded child labour in their factories. Iqbal's killers were never brought to justice. In fact, BLLF was raided and Ehsan Khan was hounded, until he fled to Europe.

What is the price of a carpet?

Source: The websites *www.thirdworldtraveler.com/Life_Death_ThirdWorld/Carpets.html* and *www.digitalrag.com/iqbal/*

Child labour is prevalent in many countries. In El Salvador, for example, child labour is pervasive on sugar plantations. Children as young as eight use machetes to cut cane, working up to nine hours each day in the hot sun. Injuries are common, but medical care is often not available. Children frequently do not attend school during the harvest season, which runs through the first few months of the academic year. The sugar that comes out of their labour ends up in soft drinks.

The children and women of the world, however, now face a much bigger threat—the HIV/AIDS epidemic. Here again, women face greater challenges than men.

What are AIDS and HIV?

The Acquired Immune Deficiency Syndrome (AIDS), is a condition in which the body's immune system is weakened and is therefore less able to fight certain infections and diseases. It is caused by infection with the Human Immunodeficiency Virus (HIV).

When the immune system breaks down the body is attacked by 'opportunistic infections'. When a person dies of AIDS, it is in fact the opportunistic infection, which causes the fatality. In itself, AIDS does not kill its victims. Of the people with HIV only some will develop AIDS as a result of their infection.

It is still not known where HIV originated, but the earliest known case was probably in 1959 in the Democratic Republic of Congo. Genetic analysis of the person's blood sample suggested that HIV might have stemmed from a single virus in the late 1940s or early 1950s. We know that the virus has existed in the US since at least the mid-1970s. The term AIDS has been in use since 1982.

The virus is passed from one person to another through blood-to-blood and sexual contact. In addition, infected pregnant women can pass HIV to their babies during pregnancy or delivery, as well as through breastfeeding.

HIV is not transmitted by day-to-day contact in any social setting. It is not passed by shaking hands, hugging, etc. One cannot become infected from a toilet seat, dishes, drinking glasses, food, or pets. The virus is not airborne or food-borne and it does not live long outside the body.

Table 17.1 World AIDS status 2004

Category	People living with AIDS in 2004 (in millions)	People newly infected in 2004 (in millions)	AIDS deaths in 2004 (in millions)
Men	19.6		
Women	17.6		
Total adults	**37.2**	**4.26**	**2.60**
Children under 15 years	2.2	0.64	0.50
Total	**39.4**	**4.90**	**3.10**

Source: www.unaids.org and estimates from UNAIDS, *Epidemic Update of December 2004*.

What is the current status of the epidemic?

UNAIDS brings out an annual update on the occasion of World AIDS Day on December 1 every year. Table 17.1 presents the key figures from the UNAIDS *Epidemic Update of December 2004*.

The highlights of the UNAIDS *Epidemic Update of December 2004* are as follows:

- The number of people living with HIV has been rising in every region of the world, as compared with two years ago, with the steepest increases occurring in East Asia, Eastern Europe, and Central Asia.
- The number of people living with HIV in East Asia rose by almost 50 per cent between 2002 and 2004, an increase that is attributable largely to China's swiftly growing epidemic.
- In Eastern Europe and Central Asia, there were 40 per cent more people living with HIV in 2004 than in 2002.
- Sub-Saharan Africa remains by far the worst-affected region, with 25.4 million people living with HIV at the end of 2004, as compared with 24.4 million in 2002. Just under two-thirds (64 per cent) of all people living with HIV are in sub-Saharan Africa, as are more than three quarters (76 per cent) of all women living with HIV.

How is AIDS spreading among women? The most disturbing aspect of the AIDS scene is that the epidemic is affecting women and girls in increasing numbers.

- Nearly 50 per cent of the 37.2 million adults living with HIV in 2004 were women—up from 35 per cent in 1985.
- Women and girls make up almost 57 per cent of all people infected with HIV in sub-Saharan Africa, where a striking 76 per cent of young people (aged 15–24 years) living with HIV are female.
- In most other regions, women and girls represent an increasing proportion of people living with HIV, as compared to five years ago.
- AIDS is the leading cause of death for African-American women between the ages of 25 and 34 years.
- Adolescent girls face HIV-infection rates five to six times higher than those faced by adolescent boys.

Many women are dangerously unaware of the risks of HIV infection and of the ways in which they can protect themselves. A recent UNICEF survey found that up to 50 per cent of young women in high-prevalence countries did not know the basic facts about AIDS. They also lack adequate access to prevention services and methods.

Young women are especially vulnerable to HIV for both biological and social reasons. They are physically more susceptible to infection than men are and they often lack the self-confidence to resist sexual advances.

The vulnerability of women and girls to HIV infection, however, arises more from gender inequality than from just ignorance. Most women around the world become HIV-infected through their partners' high-risk behaviour, over which they have almost no control.

Further, in many countries, gender discrimination means that the treatment needs of males often come first. Families are also hesitant to send women to clinics for fear of disrupting the 'care economy' that these women provide through their household duties. Often, these duties include taking care of other family members with AIDS.

Just as in family planning, better education and greater empowerment of women are the key necessities in AIDS prevention. We have to safeguard women's legal rights and provide them with equal access to health care.

How many children are affected with AIDS?

The UNICEF *State of the World's Children 2005* report states that more than half the world's children are suffering extreme deprivations from poverty, war, and HIV/AIDS.

The number of AIDS orphans has grown from 11.5 million in 2001 to 15 million in 2003. Children, however, suffer the adverse effects of HIV/AIDS long before they are orphaned. Millions live in households with sick and dying family members.

Many children whose families are affected by AIDS, especially girls, are forced to drop out of school in order to work or care for their families. They face an increased risk of engaging in hazardous labour and of being otherwise exploited.

Families that have taken in orphans become poorer because the household income will have to sustain more dependents. Grandparents are increasingly shouldering the burden of care for orphans.

Children or adolescents are often forced to assume the burden of caring for sick parents or for their younger siblings. Even where children have been taken in by grandparents or other relatives, they may be required to work to help sustain the family.

AIDS is also destroying the protective network of adults in children's lives. Many teachers, health workers, and other adults on whom children rely are also dying. Because of the time lag between HIV infection and death from AIDS, the crisis will worsen for at least the next decade.

What is the AIDS situation in India? The estimated number of HIV-infected persons in India has increased from 200,000 in 1990 to 5.1 million in 2003. In terms of cases reported to the National Aids Control Organization (NACO), Tamil Nadu had the maximum number of cases (over 40,000), followed by Maharashtra and Andhra Pradesh. The other states now considered vulnerable to the spread of AIDS are Assam, Bihar, Delhi, Himachal Pradesh, Kerala, Madhya Pradesh, Punjab, Rajasthan, West Bengal, and Orissa.

Where can hope spring from for the women and children of the world?

We can perhaps draw some hope from the fact that a number of women activists and environmentalists have joined the struggle to conserve the environment and prevent human-induced disasters and accidents. They are also trying to hold the polluters responsible for the consequences.

In the previous chapters, we reviewed the contribution of some of the better-known women environmentalists:

- Elisabeth Mann Borgese, who worked for the sustainable and peaceful use of the ocean (Chapter 4, Box 4.5).
- Wangari Maathai, who founded the Green Belt Movement in Kenya (Chapter 9, Box 9.6).
- Julia Butterfly Hill, who accomplished a tree sit-in of two years, to save the California redwoods (Chapter 9, Box 9.9).
- Lois Marie Gibbs, who raised the issue of toxic dumps to the national level in the US (Chapter 16, Box 16.7).

We have also seen how ordinary and poor women made remarkable contributions, like the Chipko activists in Tehri-Garhwal (Chapter 9, Box 9.5) and Thimmakka, the tree planter (Chapter 9, Box 9.8). This chapter carries the stories of Marina da Silva, who is trying to save the Amazonian forests (Box 17.5) and the twosome, Rashida Bee and Champa Devi Shukla, who are fighting for justice in the Bhopal Case (Box 17.6). There are of course many other women environmentalists in India and abroad.

Box 17.5 Hope: From rubber tapper to Environment Minister

She was born one of 11 children in a rubber tapper's family. At the age of 16, she was still illiterate. In 2003, she became Brazil's Environment Minister.

Marina da Silva went to school at 16 and in three years completed the examinations required for entering university. She graduated in history, studying at night and working during the day as a maid. Marina joined the Workers' Party founded by Chico Mendes (Chapter 9, Box 9.7). His assassination in 1988 was a personal tragedy for her.

She became a local councillor in the state of Acre and then in 1994 became a Senator. When President Lula appointed her Minister for Environment, she promised 'to introduce all aspects of sustainable development into the heart of the government.'

'I want to see the best of modern times, but also the best of tradition. It is the marriage between tradition and modernity, between city and forest, sky and earth that will make Brazil into the nation we seek,' she said.

Committed to the protection of the Amazonian rainforests, Marina set up an environmental police academy in November 2004. Camouflage-clad federal police agents began learning how to raid illegal mining and squatter camps, and to stop the destruction and theft of plants, animals, and natural medicines that costs Brazil billions of dollars every year.

Like Lutzenberger (Chapter 13, Box 13.4), Marina is experiencing difficulties in shifting from activism to government. We can only hope that she has greater success than Lutzenberger.

Source: The websites *www.ipsnews.net/fsm2003/28.01.2003/nota12.shtml*, *www.tierramerica.org*, and *www.naturalist.com*

Box 17.6 Inspiration: Two women take on big business

The two women first met as employees at a stationery factory in 1986 where they founded an independent union to fight for better labour conditions and wages (traditionally male-dominated unions would not accept them). In 1989, they led a 750 km march to New Delhi, presented a petition to the Prime Minister, and won a wage raise and other important concessions. Rashida Bee and Champa Devi Shukla, however, were destined to fight a much bigger battle—to seek justice for the victims of the Bhopal Gas Tragedy (read Chapter 16, Box 16.1).

Bee and Shukla are themselves survivors of the tragedy. Since 1984 Bee has lost six family members to cancer. Shukla, who has one grandchild born with congenital deformities, lost her husband and her health.

Buoyed by their success with the labour union, the two women decided to join the Bhopal struggle. The long-suffering victims have found new hope in Bee and Shukla. They have ignited the international campaign to seek justice for the survivors.

The two have drawn poor and illiterate women like themselves from the margins of society to the centre of an unequal struggle. The aim is to hold Union Carbide and Dow Chemicals accountable for the disaster, bring their executives to trial, get them to provide long-term health care and economic support to survivors and their children, and to clean up the site.

In 2002, Bee and Shukla organized a 19-day hunger strike in New Delhi to press for their demands. In 2003, they confronted Dow officials at their offices in Mumbai and the Netherlands with hand-delivered samples of toxic waste. They also toured more than 10 cities across the US ending with a passionate protest at Dow's shareholder meeting in Michigan and a 12-day hunger strike on

New York's Wall Street. They are also party to the case against Union Carbide filed in the US by the survivors.

Their leadership has energized the International Campaign for Justice in Bhopal and catapulted the issue onto the global stage once more. In their journey from disaster victims to grassroots activists, Bee and Shukla have had to overcome the enormous stigma of their poverty, their status as women in a male-dominated society, and, in Bee's case, illiteracy. They have also had to struggle with chronic health problems. During their 2003 hunger strike in the US, both women had to be rushed to the emergency room.

The two women have worked together in a complementary way. Bee's oratory ability and Shukla's quiet diligence and strength have together made a powerful combination. When asked how they were able to sustain the struggle, Bee said, 'A woman's life involves discarding relationships that she has known from infancy and adopting strangers as her own. If she can face the world outside at such a fundamental level, then why should any other struggle for empowerment scare her?'

For their selfless contribution to the cause of the Bhopal survivors, they were awarded the Goldman Environmental Prize in 2004.

Whether Bee and Shukla ultimately succeed or not, they have made the struggle for justice for the Bhopal survivors a powerful validation of women's role on the frontline of India's civil society. More Bees and Shuklas will surely rise up in defence of justice.

Source: The website *www.goldmanprize.org/recipients/byyear.html*

Box 17.7 Make a difference: Observe World Aids Day

Since 1988, December 1 is being observed as World AIDS Day. The idea emerged at the London World Summit of Ministers of Health held in January 1988 on AIDS Prevention. World AIDS Day has received the support of the World Health Assembly, the UN, and governments, communities and individuals around the world. It is the only international day of coordinated action against AIDS.

The Day celebrates progress made in the battle against the epidemic and brings into focus the remaining challenges. In the process, it creates greater awareness of AIDS among the people.

Each year, a theme is chosen to highlight the ways in which HIV/AIDS affects different groups of people. The first theme was 'Join the Worldwide Effort' and some of the subsequent themes were 'Women and AIDS' (1990), 'Children Living in a World with AIDS' (1997), and 'Men Make a Difference' (2000).

World AIDS Day 2004 focused on 'Women, Girls, and AIDS', which was also the theme for the World AIDS Campaign 2004. The campaign explored the gender inequality that fuels the AIDS epidemic and the vulnerability of women and girls to the disease.

Information on the year's theme, posters, etc., can be seen on the website cited below.

Source: The website *www.worldaidsday.org/*.

A quick recap

- Environmental degradation has a greater impact on women than on men.
- Compared to men, women shoulder the greater burden of the household work, and toil for longer hours.
- Rural women are exposed to dangerous levels of indoor pollution from burning biomass fuels in traditional cookstoves and this problem has not been adequately addressed.
- Millions of rural women collect and sell fuelwood.

They get very little, though the trade is worth billions of rupees.

- Scarcity of water and lack of sanitation facilities place women under particular stress, especially in urban slums.
- Though more and more women are joining the workforce, real socio-economic empowerment eludes them.
- There are 250 million child workers in the world, with the largest number found in India. Many work as bonded labourers.
- HIV/AIDS is affecting women and children in increasing numbers.
- Many women activists have been making great contributions to environmental conservation and justice.

What important terms did I learn?

AIDS
HIV
Opportunistic infection

What have I learned?

Essay question

Write an essay on the challenges faced by poor (rural and urban) women in India. What is their role in the household? How do environmental problems affect them? Where do they find employment and what are the hazards they face in their work? How does the HIV/AIDS epidemic affect them?

Short-answer questions

1. How does the fuelwood business work in India and what is the role of women in it?
2. What is the cause of the high levels of indoor pollution in rural households and what is its effect on women?
3. Describe the problems faced by a child working as a bonded labourer in one of the hazardous industries.
4. What are the problems faced by children who are either AIDS victims or live in AIDS-affected families?
5. Describe the current status of the AIDS epidemic, with particular reference to women and children.
6. Describe the work of any two women activists working for environmental conservation and justice.

How can I learn by doing?

1. Offer your services as a volunteer in an organization working either to eradicate child labour or to help AIDS patients. Study the problems faced by the target group, the attempts to help the group or to solve the problems, any preventive actions that are being taken, and so on. Write a report on your work.
2. Visit an organization that has been promoting 'smokeless chulahs'. Study the different designs that have been tried. What were the difficulties faced in promoting the use of efficient cookstoves? What is the feedback from the women who use them? What can be done to popularize such stoves in a big way? Write a report on your study.

What can I do to make a difference?

1. Find out more about child labour in India in occupations such as carpet weaving, the silk industry, and matchbox and firecracker units. Beginning with yourself, campaign for a boycott of products that come out of child labour. Join groups that campaign against firecrackers from the environmental and child labour angles.

2. Observe the following days:
   (a) International Women's Day on March 8
   (b) National Children's Day on November 14
   (c) World AIDS Day on December (Box 17.7)

Where are the related topics?

- Chapter 19, which deals with the problems of displacement of people due to development projects, their resettlement, and their rehabilitation.

What is the deeper issue that goes beyond the book?

What is it that drives business owners to mistreat and abuse children (for example, the way Iqbal Masih and others were treated in the carpet units)? How can we even begin to change the situation?

Where can I find more information?

- Mishra et al. (2002) give a detailed account of the indoor air pollution created by cookstoves. The Indian situation is described by Narain (2003).
- For information on child labour, access the website of Human Rights Watch *www.hrw.org/children/labor/*.
- For information on HIV/AIDS, access the website *www.unaids.org*.

What does the Muse say?

Here is a poem by Stephanie Ray, who was born in 1986 with AIDS, not just infected with HIV. Her mother, who contracted the disease from a transfusion, passed it on to her and died in 1992. Doctors expected Stephanie to survive only till the age of two or three. Stephanie not only survived, but became a young advocate for people with AIDS. This poem was taken from the website *http://earthrenewal.org/children.htm*.

Listen to my heart speak....
Please look at me and see
I am just a child
Trying to live with aids/hiv

It was given to me by my
Mother passed on to me
At birth and now I'm trying
To live my life giving it my
Best while here on earth

Listen to my heart speak
To those who are afraid
Showing kindness touching
And hugging me will not
Give you hiv or aids

I know what it feels like to live
With pain when people I love
Are sick and go away
Smiling and laughing
Sometimes can be a strain

Listen to my heart speak
Please, please hear
What I say let us love one
Another for I, like you
May be here just for today.

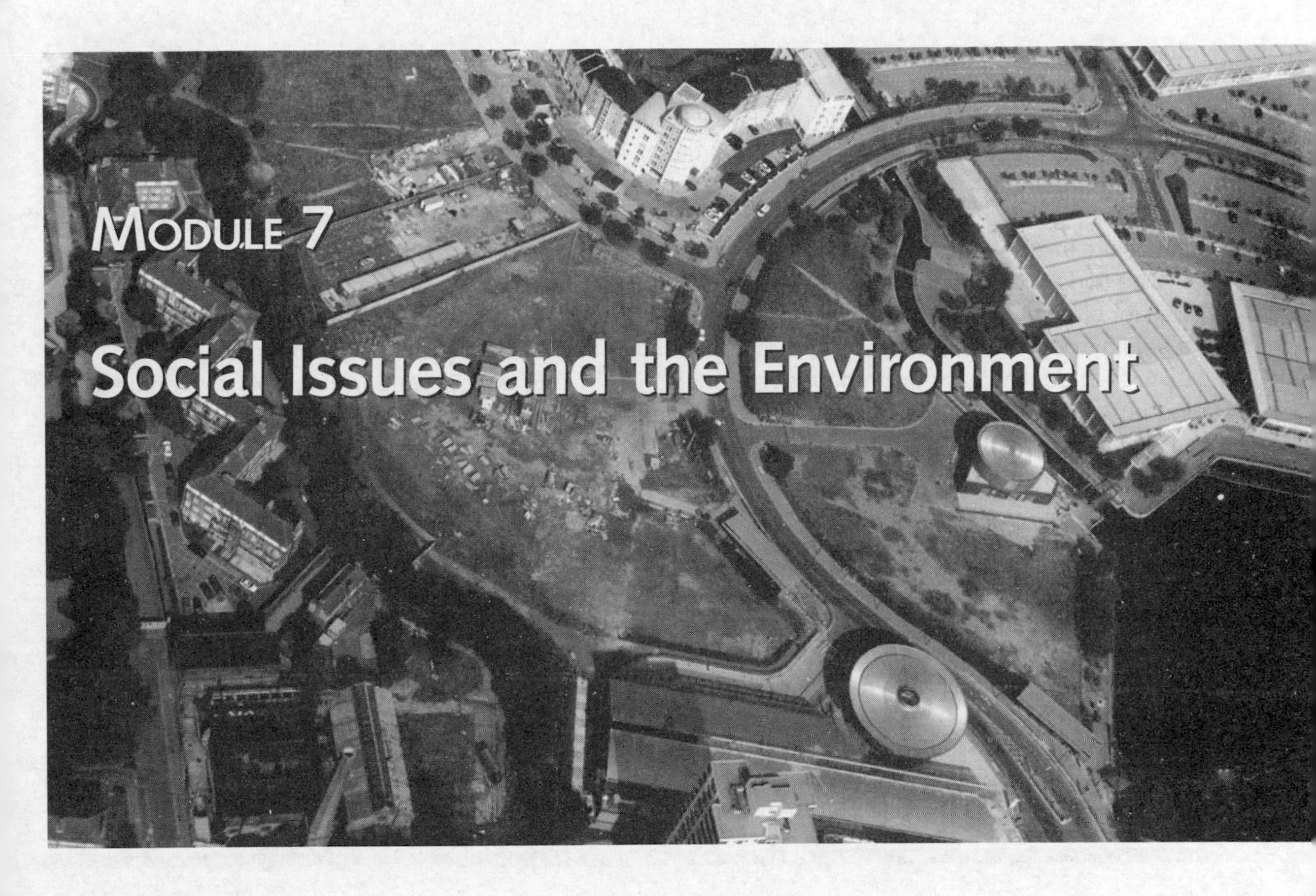

MODULE 7

Social Issues and the Environment

Prologue

The lesson of the hummingbird

EKNATH EASWARAN

Often, as I eat my breakfast, I see a flash of iridescent orange zip by the kitchen window and hover in midair at the lip of a flower. A hummingbird threads its long, delicate bill into the centre of the flower, not even touching the petals, and sips its breakfast. A moment later it is gone, having drunk only what was necessary and leaving the flower pollinated. Precise, efficient, agile, respectful: I think humanity can find no better teacher in the art of living.

To me, the hummingbird holds out a promise: this is how we all can live, gradually outgrowing a way of life in which we gulp down all the nectar, spoil the flower by pulling off the petals, and finally uproot the plant. 'Such a way of life,' writes E.F. Schumacher, referring to our overuse of fossil fuels, 'could have no permanence and could therefore be justified only as a purely temporary expedient. As the world's resources of non-renewable fuels—coal, oil, and natural gas—are exceedingly unevenly distributed over the globe, and undoubtedly limited in quantity, it is clear that their exploitation at an ever-increasing rate is an act of violence against nature, which must almost inevitably lead to violence between men.' The same could be said about any of our precious resources, from bauxite to rain forests.

To put it in economic terms, we are frittering away our capital when we should be living wisely on the interest, leaving the capital to bear rich dividends for future generations. This is what Gandhi meant by commerce without morality, a way of life in which all our nobler goals and aspirations are subsumed in the desire to produce and consume more and more.

One thing is certain: nothing will happen if we all wait for others to do it first. The first step in creating a healthy, peaceful post industrial era is for a few of us to start peeling back the iron mask of conditioning ourselves, basing our lives on a higher image of who we are, and a deeper understanding of what we need for a satisfying life. In the midst of a quickly changing world, such 'evolutionaries' can provide an inspiring example of what Schumacher calls 'a viable future visible in the present': a life that is not only ecologically sound but vastly more fulfilling than modern industrial life. There may be only a handful of such people to start with, but that should not deter us.

Our present way of life is characterized by a lack of sensitivity and inventiveness, by a lack of freedom, by hypnotization, and by the profit motive. We need men and women who can think and invent with a mind filled with compassion, charged with the kind of creativity that finds a place for the smallest songbird and the largest elephant. We need people with artistry to live in simplicity as the hummingbird does, enjoying the nectar without bruising the flower. We need men and women who delight in working together for a common goal.

This is how we can heal the environment. We have the answer to the environmental crisis already present inside us; it does not have to be invented. But neither do we have to do all the work: biologists tell us there are powerful natural forces that can reverse the damage we have done to the ecosystem. Just as the human body has healing capacities, nature is filled with restorative processes that can heal all the wounds we have inflicted, if we will only give her a chance. Where attempts have been made to reverse environmental illnesses, nature has been quick to respond. It is only when she is overwhelmed by pollution and abuse that she begins to fail.

These are extracts from Eknath Easwaran, *The Compassionate Universe: The Power of the Individual to the Heal the Environment.*

Each chapter of the book focuses on one of the 'seven social sins' made famous by Mahatma Gandhi:

- Knowledge without character
- Science without humanity
- Wealth without work
- Commerce without morality
- Politics without principles
- Pleasure without conscience
- Worship without self-sacrifice

What is ahead in Module 7?

This module focuses on the complex relationships between people, environmental conservation, governments, and environmental laws. Chapter 18 explores the issues of environment, development, and employment. It also presents examples of 'green livelihoods' and community-based conservation. Chapter 19 covers phenomena like acid rain, ozone layer depletion, and global warming and discusses their impact. It also addresses the question of displacement of people due to environmental and other reasons. Chapter 20 presents the status of environmental legislation in India and also covers Public Interest Litigation (PIL).

CHAPTER 18 People and Environment

The solution cannot be that which bans the development of those who need it the most; the fact is that everything that contributes to underdevelopment and poverty is an open violation of ecology.

Fidel Castro

If I were a Brazilian without land or money or the means to feed my children, I would be burning the rainforest too.

Sting

What will I learn?

After studying this chapter, the reader should be able to:

- appreciate the complexity of interactions between the government, conservationists, the judiciary, and the people;
- understand the connections between development, environment, and employment;
- recognize the need for alternate models of economy to enable society to move from unsustainable to sustainable development;
- list the common objectives of alternate models of economy;
- explain why unemployment is increasing in India despite economic growth;
- suggest examples of green livelihoods that would provide more employment and at the same time conserve the environment;
- describe the traditional approach to water harvesting in India; and
- give examples of community involvement in rural water harvesting in India.

Box 18.1 Crisis: Men versus mangroves

November 12, 2002 began as a normal day in the middle of the fishing season in the Sunderbans area of the West Bengal coast. Weather reports had predicted a clear day. Yet, suddenly the sea turned rough and the fishermen knew they were caught in a cyclone.

Fishermen Shishu Ranjan Das and Lakhikanta Das recount:

> To escape the cyclone, some of the boats headed for Jambudwip, the nearest island. When the boats neared the creek, however, armed policemen and guards, who pointed guns at us, turned us away. The

government had erected pillars on the creek and iron chains were drawn from one pillar to another. The storm was raging and the sea was a monster. Two boats went down that day and 12 of our people died.

Why had Jambudwip become out of bounds? Who are these fishermen? The 10,000-strong community had migrated from Chittagong to the Indian side after Partition. For the last 50 years, they had been using Jambudwip for drying fish.

The 1950-hectare island contains mangrove forests. It is also ideal for drying fish, because it has a natural harbour that allows the boats to come in twice a day. The harbour was also a good shelter during cyclones. Between October and February, the fishermen would bring all the catch to the island in mechanized boats. Along the creek, a non-forest area spread over 100 hectares served as the fish drying beds.

The fishermen had a unique system of division of labour. There were two groups in this transient fishing community—those who spent the entire six months on the seas and the others who stayed on the island in temporary shelters to dry and process the fish.

Trouble started in 2002, when the Ministry of Environment and Forests (MoEF) asked the states to implement a Supreme Court order to clear all forest encroachments. In August 2002, the state Forest Department swooped on the fishermen, burnt their boats, fishing gear, and shelters and evicted them from the island. They also put up concrete pillars at the mouth of the creek. In one swoop, the livelihoods of 10,000 fishermen and their 100,000 dependents became endangered.

The Forest Department says that the fishermen were cutting the mangroves to clear space for drying fish and had created the artificial creek. In fact, the Chief Conservator claims that the so-called transient community of fishermen did not exist. There were only a handful of fish merchants operating on the island and no one lived there except for watchmen.

Do the fishermen and their activities pose a grave threat to the mangrove forests? The fishermen say that they are aware of the relationship between mangroves and fish catch, since their fishing ground is dependent for its productivity on the debris of the mangroves flushed out by the monsoon waters.

After the eviction came the cyclone incident. The fishermen, with the support of the National Fish Workers Forum (NFF), staged a massive protest near the creek. The West Bengal Chief Secretary conceded the right of the community to use the island, but the Forest Department continued to oppose their entry. The Department suggested another island, Haribhanga, for drying fish, which turned out to be a sanctuary and a part of the Sunderbans Biosphere Reserve!

In 2003, the NFF took the matter to the Supreme Court. The state Forest Department and the Fisheries Department filed contradictory affidavits! Later, the government recalled both and filed a third one supporting the NFF. The Court's response was to ban all fishing in the Jambudwip area.

Meanwhile, the fishermen had returned to the practice of fishing in the area and drying the fish on the island. In October 2003, however, three fishermen were arrested and the fishing came to an end.

As of October 2004, the fishermen were still out of work. The Supreme Court had not given its final verdict. The differences of opinion within the government continued.

Meanwhile, there is another twist to the story. A leading business group seems to have been given permission to develop Haribhanga and other places in the Sunderbans as tourist spots!

Will the Supreme Court ultimately allow the fishermen to go back to Jambudwip?

Source: Shashikumar (2004) and Mathew (2004)

What does the Jambudwip story tell us?

This is a classic case of conflict between environmental conservation and the livelihood issues of poor people. This story has been repeated many times in every part of the world, including India.

The case of Jambudwip concerned fishermen. More common is the conflict between forest tribals and the forest departments in India. The tribals are generally branded as being primitive, anti-conservation, and destroyers of the forest. They should be evicted from the forest, relocated, civilized, and thus saved!

The other side of the picture is that the forest dwellers generally respect nature and primarily use the minor produce of the forest. They do know how important trees are for their own well-being. The Jambudwip fishermen are also aware of the importance of mangroves for fishing. While there are always exceptions to this ideal picture of the indigenous people, the official view is clearly lopsided.

In this chapter, we will expand on this theme—the interaction of poor and indigenous people with environmental and conservation issues. The focus will be on the following topics:

- The connections between development, environment, and employment.
- Ways of conserving the environment along with employment creation and poverty reduction.
- Involving communities in conservation with examples of watershed management and rainwater harvesting.

What is the environment versus development dilemma?

As we saw in Chapter 1, the Idea of Progress has taken control of our minds and we believe that continuous economic growth is the only way to solve our problems, including poverty and unemployment. In brief, the theory says that if there is sufficient growth, prosperity will trickle down to the poorest groups and everybody will be happy.

When we try to conserve the environment, we will have to introduce laws for the sustainable use of natural resources, to control pollution levels, to rehabilitate people displaced due to development projects or environmental disasters, and so on. In other words, we have to pay the environmental and social costs of development.

The argument is that the conservation approach would push up costs, reduce growth, and increase poverty, making everybody unhappy. Hence, environmental concerns are posited as being against growth and prosperity.

From all that we have seen in this book, it should be clear that the dominant growth-oriented model of development is unsustainable, since it leads to the use of natural resources much faster than they can be regenerated.

Can we then have a sustainable model for the economy? The challenge is to conserve the environment and at the same time increase employment and income levels. In other words, we need to move from unsustainable to sustainable development.

Individuals and groups in several countries have been working on alternative economic models. These models go by names such as New Economics, Green Economics, and Eco-economy. They raise, and try to answer, the following questions:

- In the face of the environmental devastation being caused by current economic systems, how can we move to more sustainable economic models?
- What do the concepts of ecological sustainability and sustainable development really mean and how can they be achieved?
- Can the value of the environment be measured?
- What tools for change exist within the current system?
- How do we make the transition from a global to a local economy, from efficiency to sufficiency?

The following objectives seem to be common to many of these alternative models:

- Achieve full employment by providing an ongoing means of livelihood for all job seekers.
- Protect and restore the environment for present and future generations.
- Ensure healthy communities based on the sustainable use of local resources for local consumption and regional trade, the availability of meaningful work, and the opportunity for creative, diverse cultural expression.
- Measure the success of the economy by the improvement in the well-being of all people and the development of healthy, vibrant, diverse communities, rather than by the total monetary value of economic activity.

It is difficult to say whether any of the alternative models can be implemented in the near future. This movement, however, provides us with some hope and also with ideas that can be implemented even now. (See the end of the chapter for web links on the new economics.)

What is the employment situation in India?

We have to accept the fact that the 'trickle-down' theory has not worked in India. Even after several decades of planning, there is no sign of prosperity reaching the lower levels. Poverty and unemployment levels are high.

Economic growth has not led to a higher rate of employment growth. In fact, even as the economic growth rate increased between 1994 and 2000, the rate of growth in jobs declined. Between 1983 and 1993, the average annual employment growth rate was 2.7 per cent per year, but the figure fell to 1.07 per cent between 1994 and 2000. This happened even as the growth rate of Gross Domestic Product (GDP) increased from 5.3 to 6.7 per cent per year.

Lack of employment is the main cause of poverty in India. Successive governments have introduced many schemes to generate more employment, but no scheme has ever met the planned target. According to the National Sample Survey Organisation (NSSO), the number of unemployed increased from 20.13 million in 1993–94 to 26.58 million in 1999–2000.

The main reason for increasing unemployment is that agriculture, the major employment generator, has stopped absorbing people. Further, more and more small farmers are giving up cultivation because the cost of inputs like fertilizers and pesticides is spiralling, even as yields go down due to soil degradation. They begin working as casual labourers for agriculturists with bigger landholdings or migrate to cities looking for work. In either case, they move from full self-employment to partial or no employment.

Promoting green livelihoods: Earthern cups and leaf plates being used on Indian Railways.

What should we do in India to tackle poverty and environmental degradation?

A study (Barik et al. 2004) by the Centre for Science and Environment (CSE), New Delhi, suggests that the key to generating employment lies in promoting productive and sustainable livelihoods based on natural resources. We should renew and revive the key livelihood areas of agriculture, forestry, animal husbandry, fishing, and horticulture, by promoting sustainable occupations in these sectors. The existing impediments to such development should be removed.

The following are examples of such green livelihoods that would generate employment and income and at the same time conserve the environment:

- Bamboo: It is an exceptionally versatile plant with about 1500 uses, including use as a building material and for making paper, handicrafts, and agricultural appliances. It generates 10 million jobs now and can provide 8.6 million jobs more. The main hurdle is that it remains a monopoly of the forest department. It should be declared an agricultural item, thereby, permitting people to cultivate, harvest, process, and sell it in various forms.
- Fuelwood: Wood plantations on degraded land can be promoted and managed by the headloaders (Chapter 17, Box 17.1), or the headholder community and the fuelwood trade can be legitimized. Apart from generating huge employment, we will also be promoting renewable energy in a big way. Such plantations will also provide timber for homes and village industries and help regenerate the soil.
- Silk: From rearing silkworms to making the fabric, this sector employs about eight million people in 26 states. At least three million more can be employed in this activity if the technology is improved, quality silkworms are made available, and the use of non-mulberry sources in forests is allowed.
- Sal leaf trade: The sal leaf is collected from the forests of Orissa and certain places. It is processed into disposable cups and plates and sold all over the country. It is an unorganized and yet a profitable multi-million rupee business, employing 1.2 million people in Orissa alone. If the forest laws are rationalized and the business is organized, it can employ five million people for 100 days a year in Orissa alone.
- Tree plantations for pulp for the paper industry: An area of about one million hectares can produce the five million tons of raw material that the paper industry needs every year. This could generate employment for 550,000 families.

For such green livelihoods to succeed, changes in laws and policies are needed. There should be a shift from conservation to sustainable utilization of natural resources. Laws that prohibit the growing, transporting, and marketing of trees like bamboo should be repealed or modified.

How can we involve communities in environmental conservation?

In recent times, there has been a greater appreciation of the need to involve local communities in environmental conservation. It is increasingly clear that, instead of viewing these communities as the enemies of environment, we should make them partners in conservation and devise win-win activities that benefit them while promoting conservation.

The Joint Forest Management (JFM) initiatives (Chapter 9) and the India Eco-development Project (Box 18.2) are examples of community involvement in conservation. Many of the government programmes are not implemented properly, but their basic objectives are commendable. The green livelihoods listed above can also be practised as community activities. There are also cases of communities that have a tradition of conservation. The Bishnois of Rajasthan (Chapter 3, Box 3.3) and the unique story of Kokkere Bellur (Box 18.3) are examples.

Box 18.2 Development: One lamp, one pot, and one blanket

Thimma, the young and angry leader in Gadde was rough on the journalist:

> Are you from the press? International, national? Documentary maker? Consultant? Are you from NGO? We have dealt with them all. We are still living in this hell, seven years after the project began. They are all gone. What do you want? None of us has anything to say. Go!

Gadde, home to 56 Jen Kuruba tribal families, is in the Nagarhole National Park in Karnataka. This park was one of the seven sites chosen for the India Eco-development Project (IEDP) that ran from 1996 to 2003 and was supported by the Global Environment Facility (GEF) and the World Bank.

This project (IEDP) was meant to enhance the country's biodiversity and provide people living inside and around protected areas, such as national parks and reserves, with sustainable livelihoods. The Rs 2.9 billion project, implemented in seven parks and tiger reserves, was expected to promote village eco-development and increase people's participation in the management of protected areas. The Forest Department would provide alternative livelihoods and the people would commit to helping the department protect the forest. The project plans were to be drawn up by Eco-development Committees (EDCs), comprising villagers and Forest Department officials.

What did the residents of Gadde get from the project?

> A solar lamp and a brass pot for each family. That is all. And yes, one blanket per family. The lantern stopped working a few months after it was given to us, no one came to repair it. The pot is iron rimmed and not really copper. One cannot heat anything in it.

About 1500 families living in hamlets like Gadde in the forest have not seen any improvement, but only the disturbance of their livelihoods. They are, however, better off than the families relocated outside the forest. Though the relocation was supposed to be 'voluntary', there were protests by the people. Each family was given a house and a two-acre field to cultivate on lease. The land was not fertile enough and the people could not make a living out of cultivation. After a while, most of the tribals sub-let their fields to outsiders and worked as daily wage labourers.

The project did achieve success in some sites, the Periyar Tiger Reserve in Kerala being one example. Here, the EDCs have been active and they converted threats into opportunities. For example, the EDCs manage stalls and supply shops when five million people make the annual pilgrimage to the Sabarimala shrine in the area. The state government supported them in setting up the Periyar Foundation to sustain eco-development activities.

Among other things, the Foundation provides salaries to poachers turned protectors of the forest.

Unfortunately, IEDP has more failures than successes to report. The project illustrates what has repeatedly happened in India. The government comes with money, projects, and grandiose plans that do not work. The development NGOs want the tribals left inside the forest, quietly collecting honey and other forest produce to manage a living. The conservationists want them to be quietly shunted out of the forest to save the trees and the wildlife. No one, however, asks the tribals what they want!

Source: Sethi and Parashar (2004)

Box 18.3 People power: The pelicans are coming!

For six months in the year, it is a quiet village like any other in rural Karnataka. In December, however, a spectacular transformation occurs here with the arrival of hundreds of spotbilled pelicans and painted storks. Migrating from the lakes of south Karnataka, these birds settle down in the trees of the village for the next six months. While the village below goes about its business, there is another world up above, where the birds court, mate, and rear their young.

It is a mystery why the pelicans and the storks come every year to breed in Kokkare Bellur, located 80 km from Bangalore in Mandya District. Both species are exclusively fish eaters and there is no large water body near the village. When they stay in Kokkare Bellur, they have to feed in irrigation tanks within a 100 km radius. Yet, the birds have been coming here for at least 150 years, if not more.

What is remarkable is that the people of Kokkare Bellur provide complete protection to the birds. They believe that the birds bring them good fortune in the form of good rains and harvests. The birds are noisy with the young ones clamouring for food. There is also the fishy stench of bird droppings. The villagers tolerate all these inconveniences and continue to protect the birds. Children are discouraged from teasing the birds or stealing their eggs. Any outsider, who attempts to steal the eggs or the chicks, is punished by the village panchayat.

The pelicanry of Kokkare Bellur was discovered by the outside world only in 1976. In 1982, when the Bannerghatta National Park was being set up in Bangalore, the Forest Department wanted to carry away some of the pelican chicks to the Park. The villagers stopped their vehicle and took them to the panchayat, which reprimanded the foresters for not consulting them in advance. Since the chicks could not be put back in their respective nests, they were given away reluctantly.

Now, however, the paradise of pelicans is under threat. The growing population has led to greater demand for the tree resources and the villagers have become less hospitable to the birds. As tree felling increased, there has been a gradual decline in the number of birds nesting at Kokkare Bellur.

In the early 1980s, the Forest Department issued an order protecting the trees, but this could not be implemented. Eventually, a scheme of annual compensation was introduced. The villagers, however, are not satisfied with the quantum of compensation as well as with the operation of the scheme.

Since 1994, the NGO Mysore Amateur Naturalists (MAN) has been working in the village trying to involve the people actively in the conservation process. The programmes include formation of a local youth group, tree planting, educational activities, and a health clinic.

Other threats are looming on the horizon. The widening of roads, the setting up a holiday resort, pesticide pollution of the feeding tanks, introduction of exotic fish species, and direct poaching of pelicans are making things difficult for the birds. How long will we continue to hear the cacophony of birds in Kokkare Bellur?

Source: Manu and Jolly (2000)

Water harvesting

There are also many voluntary initiatives in which communities have worked together to conserve natural resources, when they have realized the advantages of such cooperation. The most common example in India has been community-based water conservation.

What is community-based water conservation?

Community-based water conservation is not a new concept. Until about 150 years ago, no government supplied water and communities managed their own water sources. The rulers, of course, had lakes and ponds built, but the maintenance and control of these sources were generally with the communities.

In India, the arrival of the British marked the government takeover of water sources and the emergence of the Public Works Department (PWD). Gradually, the communities lost their role in maintaining and managing water sources and became dependent on the state for their water supply.

The second change in the water scene has been the decline in use of rainwater and the increased exploitation of rivers through dams and of groundwater through borewells. Rivers and aquifers collect only a small portion of the total rainfall and are hence unable to meet the rapidly increasing demand for water.

Why is rainwater harvesting a very effective way of water conservation? It rains, it may even pour, but all the water soon disappears into the ground or into the streams and rivers making their way to the ocean. We do not realize how much water really falls on the Earth's surface.

Rainwater being sold in bottles: Guaranteed no pesticides

Take the case of the Rajasthan desert, where one can only expect 100 mm of rain annually. If you had one hectare of land there, you could collect one million litres of water in a year. This will be enough to meet the cooking and drinking water needs of about 180 people using 15 litres a day. Even in the worst drought conditions, you can collect substantial amounts of rainwater in most areas.

India receives most of its rain in just 100 hours every year. That is a strong argument in support of 'catching' the rain when and where it falls. We have a long tradition of rainwater harvesting, with a wide variety of methods being followed to suit local conditions (Agarwal and Narain 1997).

What is the recent experience in India with rural water harvesting?

In recent years, with increasing water scarcity in many parts of the country, a number of initiatives have been taken by communities to harvest and conserve water (Agarwal et al. 2001). Here are some examples:

- The watershed development and overall village development work of Anna Hazare in Ralegaon Siddhi (Box 18.4) and the Pani Panchayat initiative of Vilasrao Salunkhe (18.5), both in Maharashtra, are well known.
- Hundreds of checkdams have been built by villagers in Alwar, Rajasthan, under the guidance and inspiration of the Tarun Bharat Sangh (Chapter 7, Box 7.4). The results have been spectacular.
- The non-governmental organization (NGO), Saurashtra Lok Manch, has inspired farmers in the Saurashtra-Kutch Region to recharge more than 300,000 wells and tubewells using the monsoon rainfall. Jal yatras were organized to ensure people's participation in the programme. Water tables have gone up and bumper crops could be harvested.
- Uthan, an Ahmedabad-based NGO, has built 21 lined ponds in the Ahmedabad and Bhavnagar districts, with the active participation of local village communities, particularly women. This has substantially increased water availability in the area.
- By adding new structures to existing, traditional water harvesting systems, the Mewar Krishak Vikas Samiti has increased water availability in Rajasamand District of Rajasthan.
- The people of Laporiya in Jaipur District of Rajasthan have built dykes around degraded pasturelands to harvest rain and have brought about a dramatic ecological change. Their unique method involves rectangular plots of land, which store rainwater in the pastures. The water first collects in the lower plots and as the level rises, water flows into the neighbouring plots and so on.
- Jhabua, a poor tribal district of Madhya Pradesh, has become a model of watershed management under the Rajiv Gandhi Mission for Watershed Development. The key success factors were the involvement of people in decision-making, an area-specific approach, and priority given to indigenous knowledge. More than 200 micro-watersheds were developed and 14 million human-days of employment were created. R. Gopalakrishnan, a bureaucrat, played a major role in the project devising ways of cutting the red tape.
- Inspired by the Vivekananda Kendra, villagers in Ramanathapuram District of Tamil Nadu contributed their labour to help renovate several traditional ponds called *ooranis*.

Increased water availability in these villages is encouraging many other villages in the area to follow suit.

- Building bunds and checkdams along with large-scale tree plantation has helped arrest soil erosion in villages around Auroville near Pondicherry.
- The efforts of women in the villages in Sabarkantha District of Gujarat in arresting run-off have yielded good results. They have worked out a system of conserving rainwater and soil using stones. The women, whose husbands work as migrant labour elsewhere, now plant a variety of crops, earn more, and feel empowered.

There are many more such stories of success (read Agarwal et al. 2001). In recent times, water scarcity is forcing civic bodies to make urban rainwater harvesting compulsory.

Communities around the world are also fighting to protect their water resources (like the Plachimada Case described in Chapter 7, Box 7.2) or to resist privatization of water supply (like the Cochabamba Case recounted in Box 18.6).

Box 18.4 Inspiration: From liquor den to dream village

He was a retired army truck driver and in 1975 he went back to his village in Ahmednagar district, Maharashtra. He found the village of about 200 families reeling under drought, poverty, debt, and unemployment. Drinking water was scarce, but liquor was plenty. He decided to mobilize the people and, with the collective support of all the villagers, he began to introduce changes. Today his village, Ralegaon Siddhi, is regarded as a role model for development.

Kisan Baburao Hazare, affectionately called Anna Hazare, began with the reconstruction of the village temple using his retirement savings. He ran out of funds and still the temple was incomplete. Then the villagers pitched in and a movement began under Anna's leadership.

From temple renovation, he moved to water conservation. With the voluntary work of the villagers, he built bunds, check dams, and percolation tanks, slowing down the run-off. Large canals with ridges on either side were dug to retain rainwater. Massive tree plantation was undertaken, and the hills were terraced to check erosion.

As a result, the water table in this area is now considerably higher and the wells and tube wells are never dry, even though the average annual rainfall is only 300–350 mm. It is now possible to raise three crops a year where only one was possible before. The irrigated area has increased from 60 to 1200 acres. Onions, vegetables, and fruits are now exported from Ralegaon.

Government programmes on soil and water conservation have been fully utilized. Nobody cuts trees in the village. All wood and water is common property. There are no stray cattle or sheep and hence all trees grow up without any tree-guards.

Farming is done through co-operatives. All small pieces of land fed by a well have been formed into a co-operative, which rations the water. Loans raised for drip irrigation systems and plantations have been returned in two years

All the streets in the village are lit with solar lights. There are four large community biogas plants and a large windmill used for pumping water. A number of households have their own biogas plants.

The village is self-sufficient in many aspects and the panchayat is run by women. All major decisions are taken in the Gram Sabha held once a month in the temple. There is no unemployment.

After creating such a dream village, Anna Hazare has moved on. He has declared a war on corruption and has taken on the powerful politicians. He was even arrested once, but he still carries on in his crusade.

Source: The websites *http://www.farmedia.org/ and http://edugreen.teri.res.in/*

Box 18.5 Inspiration: Share and survive!

'If you provide food and water security to rural people, they will never leave their villages,' he used to say. In fact, his watershed development and water conservation project at Ambegaon in Pune district succeeded in reducing the migration of rural people to urban areas by 80 per cent.

Vilasrao Salunkhe was an electrical engineer and for many years worked for the Maharashtra State Government. The severe drought of 1972 in Maharashtra was a turning point in Salunkhe's life. It was then that he started working for the drought-affected people of Purandar Taluka.

Salunkhe realized that watershed development with the full participation of the community was the only solution. He initially tried his ideas on a 16-hectare plot on a hillside in Naigaon village. He got the barren and uncultivable land from a trust on a 50-year lease and built a hut where he and his family lived and worked with the community.

He raised a series of contour bunds to trap water and check soil erosion. At the base of the hill slope, a percolation tank was constructed. A well was dug below it and water pumped from there up the hill slope for irrigating the fields. Trees were planted in the rocky areas. Slowly production from the land increased. Meanwhile, Salunkhe developed his idea of a Pani Panchayat.

The basic philosophy of the Pani Panchayat was to share the water that was available. Salunkhe believed that without a system of equitable distribution of water, watershed development would remain lopsided, benefitting only those who could afford to transport water to their fields. He evolved five basic principles of the Pani Panchayat:

- Irrigation schemes are undertaken for groups of farmers, rather than for individuals.
- Cropping is restricted to seasonal crops with low water requirement. Crops like sugarcane, bananas, and turmeric cannot be cultivated.
- Water rights are not attached to land rights. If land is sold, the water rights revert back to the farmers' collective.
- All members of the community, including the landless have the right to water.
- The beneficiaries of the Panchayat have to bear 20 per cent of the cost of the scheme. They have to plan, administer, and manage the scheme and distribute water in an equitable manner.

Salunkhe relentlessly fought for the equitable distribution of water, irrespective of landholding and advocated an 800 cu. m quota of water per capita per year. He forced the government to include the principle of equitable distribution in its water programme. However, the government and vested interests somehow managed to defeat him when it came to the actual implementation.

In later years, Salunkhe was active in the tribal belts of Maharashtra. He also took a keen interest in organic farming. At the time of his death, he was engaged in spearheading the Chikotra Valley agitation in Kolhapur to secure an equal share of water for residents of 52 villages in the valley.

Salunkhe's Naigaon pattern of watershed development on wasteland won him international recognition and the Swedish International Inventor Award in 1985 and also the Jamnalal Bajaj Award for meritorious social service in 1986.

He passed away in 2002 at the age of 65.

Source: The websites *www.goodnewsindia.com/*, *www.indiaseminar.com/2002/516/*, and *www.teri.res.in/teriin/terragreen/issue21/feature.htm*

Box 18.6 People's action: Water at a price

On April 8, 2000, a 17-year-old boy, Victor Hugo Daza, was shot dead by a Bolivian army captain who opened fire into a crowd of demonstrators in the city of Cochabamba. The trouble in Cochabamba arose out of the privatization of its water supply.

The International Monetary Fund (IMF) approved a US$ 138 million loan for Bolivia in 1998 to help the country control inflation and bolster economic growth. In return, Bolivia agreed to sell off all remaining public enterprises, including national oil refineries and Cochabamba's local water agency.

In September 1999, the Bolivian government signed a US $ 2.5 billion contract with Aguas del Tunari, a multinational consortium, including a subsidiary of the US Bechtel Corporation, awarding it 40-year concession rights to provide water and sanitation services to Cochabamba.

Aguas del Tunari took over the water supply and the prices went up. Though the company had previously informed Bolivian officials that water rates would increase only by 35 per cent, the citizens found that the hike had doubled and tripled their water bills. In January 2000, protesters in Cochabamba shut down the city for four days.

Soon the people formed the Coalition for the Defence of Water and Life (La Coordinadora), led by union organizer and anti globalization activist Oscar Olivera. La Coordinadora held an unofficial referendum in which an overwhelming majority disapproved of water privatization. They demanded the termination of the contract, but the government refused.

In April 2000, protests originating in Cochabamba's central plaza spread to La Paz and other cities and outlying rural communities. Thousands clashed with riot police, erected roadblocks, and protested not only the water-rate hikes but also the country's overall economic malaise and high unemployment.

The government finally gave in. It terminated the water contract, and granted control of Cochabamba's water to La Coordinadora. Legislation that would have charged peasants for water drawn from local wells was also removed.

The story did not end there. In February 2002, Aguas del Tunari and its parent corporation, Bechtel, registered a case against Bolivia with the World Bank's International Centre for the Settlement of Investment Disputes (ICSID), seeking US $ 25 million in damages for breach of contract. We should note that Bechtel's revenues in 2000 exceeded US $ 14 billion, while Bolivia's national budget was US $ 2.7 billion.

As a result of the public pressure people around the world put on Bechtel, the company is now ready to drop the case, for a token payment. However, the company's investment partner, Abengoa of Spain, has refused to follow Bechtel's lead in abandoning the case.

Just as the Cochabamba story is coming to an end, another water revolt is set to erupt in El Alto, near Bolivia's capital, La Paz. As in Cochabamba, the public water system of El Alto was privatized in 1997 and given to the French water giant, Suez. The company has raised water prices by 35 per cent since it took over and poor families who want to connect to the water system are expected to pay more than US$ 445—an amount in excess of the national minimum wage for six months. What is more serious, however, is that the company has allegedly left more than 200,000 people with no possibility of access to water by failing to expand its service to El Alto's growing outskirts.

Will a struggle start all over again?

Source: The websites *www.pbs.org* and *www.democracyctr.org*

What lies ahead?

Water is likely to become a major issue in the coming decades. Increasing water scarcity, privatization, and higher costs will lead to conflicts and agitations. At the same time, as many examples given in this chapter show, we can also get communities to cooperate in water conservation.

A quick recap

- There are often conflicts between environmental conservation and the livelihood issues of poor people.
- There is the wrong perception that all tribal groups are anti-conservation and tend to destroy the forests where they live. In general, the tribal culture tends to respect nature.
- The growth-oriented model of economics is unsustainable because it leads to the exploitation of natural resources much faster than they can be regenerated.
- In India, economic growth has not led to a higher rate of employment growth. We have to promote green livelihoods that would generate employment and income and at the same time conserve the environment.
- There is now a greater appreciation of the need to involve communities in environmental conservation.
- There are many examples of community-based water conservation efforts in India.
- Rainwater harvesting is vital for meeting the water crisis.

What important terms did I learn?

Green economics
Green livelihood
New economics
Water harvesting

What have I learned?

Essay question

Write an essay on the need to involve communities in forest conservation. What is the dominant view about forest dwellers? How is it possible to change the perception of them as enemies of conservation, to that of partners in conservation?

Short-answer questions

1. What are the general issues concerning people and environment that arise from the Jambudwip case?
2. Explain the conflict between development and environment.
3. What are the questions raised by those who propose alternate economic models? What are the objectives of such models?
4. Briefly describe the unemployment problem in India.
5. Give two examples of green livelihoods that would generate employment and income and at the same time conserve the environment in India.
6. Give two examples of community-based rural water harvesting in India.

How can I learn by doing?

Join any group promoting rainwater harvesting in your area as a volunteer. Get involved in the actual work of installing rainwater harvesting systems in buildings. Write a report on your experience. How much did it cost? How much water will the system collect if the rainfall were to be normal? How soon will the system pay for itself? What other advantages does it offer? What are the minus points?

What can I do to make a difference?

1. Install a rainwater harvesting system in your own house or apartment block.
2. Buy natural products like earthen cups, leaf plates, bamboo dust bins, etc.

Where are the related topics?

- Chapter 7 covers water resources.
- Chapter 9 deals with forest resources.

What is the deeper issue that goes beyond the book?

What should we do to improve the lives of tribals and indigenous people who live in the forests? For example, how can we help the ancient tribes of the Andaman and Nicobar Islands without disrupting their culture? What can we learn from them about environmental conservation?

Where can I find more information?

- For a detailed account of green livelihoods, read Barik et al. (2004).
- For information on New Economics, Green Economics, etc., access the websites: *www.neweconomics.org, www.greeneconomics.net, www.greenecon.org, and www.ilsr.org*.
- If you want try rainwater harvesting, guidance is available from the websites: *www.rainwaterclub.org/* and *www.rainwaterharvesting.org/*.

What does the Muse say?

A poem by the American educator, economist, and peace activist Kenneth Boulding:

A Conservationist's Lament

The world is finite, resources are scarce,
Things are bad and will be worse,
Coal is burned and gas exploded,
Forests cut and soils eroded.
Wells are dry and air's polluted,
Dust is blowing, trees uprooted.
Oil is going, ores depleted,
Drains receive what is excreted.
Land is sinking, seas are rising,
Man is far too enterprising.
Fire will rage with Man to fan it,
Soon we'll have a plundered planet.
People breed like fertile rabbits,
People have disgusting habits.
Moral: The evolutionary plan went astray by evolving Man.

CHAPTER 19

Social Consequences of Development and Environmental Changes

In our every deliberation
we must consider the impact of our decisions
on the next seven generations.

From the Great Law of the Iroquois Confederacy

We are drawing on the world's natural capital
far more rapidly than it is regenerating.
Rather than living on the 'interest' of the 'natural capital',
we are borrowing
from poorer communities and
from future generations.

World Commission on Forests and Sustainable Development

What will I learn?

After studying this chapter, the reader should be able to:

- explain the causes and effects of acid rain;
- describe the problem of ozone layer depletion and the formation of the ozone hole over the Antarctic;
- outline the provisions of the Vienna Convention and the Montreal Protocol;
- describe the results of implementation of the Montreal Protocol and the lessons learnt;
- explain global warming, its causes, and possible effects;
- outline the provisions of the Kyoto Protocol;
- explain the reasons for the displacement of populations and its impact;
- describe the special problems of displacement caused by dams; and
- discuss the issues concerning the resettlement and rehabilitation of displaced groups.

Box 19.1 Displacement: Flood of fears

My children will never till the hill slopes. It is unthinkable. My father tilled here. I walked by his side as a child, knowing that I would be doing the same when I grow older. I cannot give my children these assurances. I cannot give them anything because whatever I have is suddenly not mine.

Jatrabhai, a 70-year-old resident of Manibeli village in north Maharashtra.

Manibeli was the first village to be affected by the Sardar Sarovar dam project on the River Narmada. That was more than 10 years ago, when the dam was only 50 metres high. The villagers shifted to a higher level then.

As the height of the dam increased, the waters have been rising higher up the hill slope. When the water level reaches too close for comfort, the villagers dismantle their homes and move further up the hill. So far, the residents have relocated their homes and fields thrice. Now they live at the very top of the hill and have literally nowhere to go if the water level rises further. The final countdown for Manibeli began with the decision to raise the height of the dam to 110.64 m. With that height, the waters would completely submerge Manibeli.

The story of displacement and distress is common to many of the affected villages of the Narmada Valley. Every year prior to the monsoon, the people have the same questions: Will our houses go under this time? Will our village become an island when the waters slowly rise? Will we be cut off from all assistance?

During the monsoon, the resettlement process becomes active with government officials forcibly evicting people from the submergence zones. During the rest of the year, the hapless and often illiterate villagers do the rounds of government offices holding tattered pieces of documents and claiming what is due to them.

Nobody knows exactly how many families have been affected by the project. Estimates are 35,000 in Madhya Pradesh and about 5000 each in Gujarat and Maharashtra, but the real numbers are likely to be much more. Further, the numbers keep going up as the height of the dam is increased. The numbers also do not tell the story of repeated relocation of some families. In general, the resettlement and rehabilitation process has been incomplete and unsatisfactory.

However, not everyone is unhappy at the relocation. At Sarlia, for example, the people are proud to have moved, as their children can now go to school. At Sadiapani, a resident gazes fondly at the railway line and says, 'Never did I imagine that we would ever get a railway near our village.' But then, there is also Ladkibai who weeps because she cannot see her beloved Narmada Mai every day. The Sardar Sarovar dam project is also expected to bring drinking water to some dry areas and to generate power.

In 1994, Bava Mahalia of Jalsindhi, one of the dam-affected villages, wrote a letter to the Chief Minister of Madhya Pradesh. Here is an excerpt:

> You tell us to take land in Gujarat. You tell us to take compensation. For losing our lands, our fields, for the trees along our fields.... But how are you going to compensate us for our forest?.... How will you compensate us for our river—for her fish, her water, for the vegetables that grow along her banks, for the joy of living beside her? What is the price for this?.... Our gods, and the support of our kin—what price do you put on that? Our adivasi life—what price do you put on that?

It is a moving letter, very much like Chief Seattle's speech (Module 4, Prologue). It makes one ask: What price do you put on development?

Source: Menon (2003) and Bavadam (2004a, 2004b)

What are the lessons from the Narmada case?

The story highlights only the displacement aspect of the Sardar Sarovar dam project and illustrates the problems faced by poor people who are displaced by development projects. Such stories are common to many large projects in different parts of the world. There are many other aspects of dams like the high costs, loss of biodiversity, radical changes in the ecosystem, siltation, impact on fishing, etc. There are also the possible benefits like generation of power, supply of water for drinking and irrigation, and promotion of tourism.

In this chapter, we will first look at the causes and consequences of major environmental changes like acid rain, depletion of the ozone layer, and global warming. Such changes will increasingly lead to the displacement of people. Hence, we will also deal with issues of displacement caused by development and economic and environmental factors, as also issues of resettlement and rehabilitation of refugees.

What is acid rain?

When atmospheric water droplets combine with a range of man-made chemical air pollutants, acid rain is formed. Other forms of precipitation like mist and snow may also be acidic for similar reasons.

The main pollutants involved are oxides of nitrogen and sulphur. In nature, volcanoes, fires, and decomposing matter emit these substances in small amounts. However, since the advent of the Industrial Revolution, human activities have been releasing such pollutants in large quantities. Such emissions are very high in the major industrial centres and have been increasing rapidly since the mid-twentieth century.

Automobiles and coal- and oil-fired power stations are major sources of acid-forming compounds. In fact, any burning of coal, oil, and (to a lesser extent) natural gas produces these compounds.

Acid rain ultimately falls on the ground, sometimes hundreds of kilometres from the area in which it formed and generally one to four days later. The effects of such acid rain are generally quite damaging.

What is the effect of acid rain?

When soil is acidified, it leads to a loss of productivity. The acidification damages plant roots and they are not able to draw in enough nutrients to survive and grow.

When trees, particularly conifers, are exposed to acid rain for several years, they begin to lose their leaves and die. This is one of several causes for the decline of forests in Europe, North America, and Japan. Plants like orchids, lichens, and mosses are also very sensitive to acid fallout.

Acid rain falling on lakes and rivers destroys the life forms that live in these water bodies. Thousands of lakes in Sweden, Norway, and Canada, for example, have been permanently affected by acid fallout. Fish populations have died and so have species, such as otters, amphibians, and birds that depended on fish for their food.

Acid rain harms people directly when they breathe in the acidic air. Acid rain can also harm people indirectly, when they eat fish caught in affected lakes or rivers.

Automobile pollution also causes acid rain

Old buildings are also threatened by acid rain. Acid fallout has caused the famous St Paul's Cathedral in London to decay more in the last 50 years than it has in the previous two centuries. Some famous statues, such as the Lincoln Memorial and Michaelangelo's statue of Marcus Aurelius, have started deteriorating because of the effects of acid rain. The same is true of many historic buildings in Europe.

The Taj Mahal was also threatened by acid rain caused by factories in Agra. Thanks to the orders of the Supreme Court many of these industries have been shifted or closed down.

A side effect of acid rain is the leaching of aluminium out of the soil into water bodies. Aluminium is very toxic for fish and the birds that prey on them. Sometimes acidification leads to the leaching of cadmium and this can also have adverse effects on animals.

What can be done about acid rain?

In most cases, acid rain formed in one country falls on some other country. It is also difficult to prove that a particular country was responsible, or to quantify the amount of contributing pollution from a source.

Pouring powdered limestone into water bodies is a rapid, but short-lived, method of reducing acidity. A more permanent, but slow and expensive, method is the liming of surrounding soils.

Some technologies for reducing emissions are flue gas desulphurization in power stations and catalytic converters and engine modifications in automobiles.

What is ozone layer depletion?

Ozone is a poisonous gas made up of molecules consisting of three oxygen atoms. This gas is extremely rare in the atmosphere, representing just three out of every 10 million molecules. Ninety per cent of ozone exists in the upper atmosphere, or stratosphere, between 10 km and 50 km above the Earth.

The ozone layer in the atmosphere absorbs most of the harmful ultraviolet-B (UV-B) radiation from the Sun. It also completely screens out the deadly UV-C radiation. The ozone shield is thus essential to protect life.

Depleting the ozone layer allows more UV-B rays to reach the Earth. The result is an increase in skin cancers, eye cataracts, weakened immune systems, reduced plant yields, damage to ocean ecosystems and reduced fishing yields, and adverse effects on animals.

In the 1970s, scientists discovered that when CFCs (chlorofluorocarbons, used as refrigerants and aerosol propellants), finally break apart in the atmosphere and release chlorine atoms, they cause ozone depletion. Bromine atoms released by halons (used in fire extinguishers) have the same effect.

The ozone layer over the Antarctic has steadily weakened since measurements started in the early 1980s. The land area under the ozone-depleted atmosphere has increased steadily to more than 20 million sq km in the early 1990s and has varied between 20 and 29 million sq km since then. In 2000, the area of the ozone hole reached a record 29 million sq km.

While no hole has appeared elsewhere, the Arctic spring has seen the ozone layer over the North Pole thinning by up to 30 per cent. The depletion over Europe and other high latitudes varies between 5 and 30 per cent.

What are the international initiatives against the depletion of the ozone layer?

Intergovernmental negotiations for an international agreement to phase out ozone-depleting substances started in 1981 and concluded with the adoption of the Vienna Convention for the Protection of the Ozone Layer in March 1985.

The Vienna Convention encourages intergovernmental cooperation on research, systematic observation of the ozone layer, monitoring of CFC production, and the exchange of information. The Convention commits the signatories to taking general measures to protect human health and the environment against human activities that modify the ozone layer. It is a framework agreement and does not contain legally binding controls or targets.

The Convention did, however, set an important precedent. For the first time nations agreed in principle to tackle a global environmental problem before its effects were felt, or even scientifically proven.

In May 1985, British scientists published their discovery of severe ozone depletion in the Antarctic. Their findings were confirmed by American satellite observations and offered the first proof of severe ozone depletion. The discovery of the 'Ozone Hole' shocked the world. It is regarded as one of the major environmental disasters of the twentieth century.

Governments now recognize the need for stronger measures to reduce the production and consumption of a number of CFCs and several halons. As a result, the Montreal Protocol on Substances that Deplete the Ozone Layer was adopted in September 1987.

Ninety-six chemicals are presently controlled by the Montreal Protocol and are subject to phase-out schedules under it. The Protocol was designed so that these schedules could be revised on the basis of periodic scientific and technological assessments.

Governments are not legally bound by the Protocol until they ratify it as well as the Amendments to it. Unfortunately, while most governments have ratified the Protocol, ratification of the Amendments, with their stronger control measures, still lags behind.

What have been the results so far?

The Montreal Protocol is working. However, even with full compliance with the Protocol by all parties, the ozone layer will remain particularly vulnerable during the next decade or so.

In 1986 the total consumption of CFCs worldwide was about 1.1 million ODP (Ozone Depleting Potential) tons; by 2001 this had come down to about 110,000 tons. It has been calculated that without the Montreal Protocol, global consumption would have reached about three million tons in the year 2010 and eight million tons in 2060, resulting in massive ozone layer depletion.

Without the Protocol, there would have been a doubling of UV-B radiation reaching the Earth in the northern mid-latitudes and a quadrupling of the amount in the south. The amount of ozone-depleting chemicals in the atmosphere would have been five times greater. The implications of this would have been horrendous: 19 million more cases of cancer and 130 million more cases of eye cataracts.

The bulk of the 1986 total, or about 0.9 million ODP tons, was consumed in developed countries, but by 2001 these countries consumed just about 7000 tonnes. The developing countries have reduced their CFC consumption by about 15 per cent between 1986 and 2001.

Scientists predict that ozone depletion will reach its worst point during the next few years and then gradually decline until the ozone layer returns to normal in around 2050, assuming that the Montreal Protocol is fully implemented.

The success of ozone protection has been possible because science and industry have been able to develop and commercialize alternatives to ozone-depleting chemicals. Developed countries ended the use of CFCs faster and with less cost than was originally anticipated.

What have been the lessons of the Montreal Protocol?

The efforts of the world community to protect the ozone layer are a fascinating example of how humanity can act as one to face a common danger. The Protocol offers many lessons that could be applied to solving other global environmental issues:

- Adhere to the 'precautionary principle' because waiting for complete scientific proof can delay action to the point where the damage becomes irreversible.
- Send consistent and credible signals to industry (for example, by adopting legally binding phase-out schedules) so that they have an incentive to develop new and cost-effective alternative technologies.
- Ensure that improved scientific understanding can be incorporated quickly into decisions about the provisions of a treaty.
- Promote universal participation by recognizing the 'common but differentiated responsibility' of developing and developed countries and ensuring the necessary financial and technological support to developing countries.
- Base control measures on an integrated assessment of science, economics, and technology.

Is the global climate changing?

Weather is the condition of the atmosphere at a particular place and time. It is characterized by parameters such as the temperature, humidity, rain, and wind. Climate is the long-term pattern of weather conditions for a given area.

Until the middle of the twentieth century, the earth's climate was generally regarded as unchanging, but it is now known to be in a continuous and delicate state of flux. Relatively small changes in climate could have a major effect on our resources like food, energy, and water.

The factors that influence global climate are the amount of solar energy the Earth receives, the condition of the atmosphere, the shape and rotation of the Earth, and the currents and other processes of the ocean.

What is global warming and how does it occur?

Recall the way the carbon cycle works (Chapter 2). During photosynthesis, plants absorb carbon dioxide and release oxygen. Organisms breathe in this oxygen, and give out carbon dioxide, which goes back to the plants. This is the carbon cycle in brief.

Normally, carbon dioxide and other gases that surround the planet let the radiation from the Sun reach the Earth, but prevent some of the heat from being reflected back out again. Without these greenhouse gases, the Earth would be far colder and largely covered with ice. The problem comes when the amounts of these gases exceed a certain limit.

By burning large amounts of fossil fuels, we release huge quantities of carbon dioxide into the atmosphere. Concurrently, deforestation also releases carbon trapped in the tissues of the trees. At the same time, loss of trees reduces the Earth's capacity to absorb carbon dioxide through photosynthesis. Natural processes like volcanic eruptions and earthquake-induced fires also contribute to carbon dioxide emissions.

Some of the other greenhouse gases are far more effective than carbon dioxide in trapping heat. Chlorofluorocarbons (CFCs) are an example, but their role is greater in the depletion of the ozone layer. Methane, released from swamps, human and animal waste, and garbage dumps, is also a greenhouse gas and its concentration in the atmosphere is increasing. Similarly, human activities are causing a rapid increase in the amounts of 30 other greenhouse gases in the atmosphere.

The abnormal increase in the concentrations of these gases leads to higher temperatures and global warming. The average temperatures around the world have risen by about 0.5ºC since the beginning of the twentieth century. If emissions continue at the current rate, a temperature rise of 1.5ºC to 4.5ºC is likely by 2030.

From which countries do the human-induced greenhouse gas emissions come? Most of the greenhouse gas emissions come from the northern hemisphere, the US being the biggest contributor. Russia is also a major source. European countries also produce substantial amounts of greenhouse gases, but they are also trying to reduce emissions.

Developing countries are fast catching up with developed countries when it comes to greenhouse gas emissions. China and India are industrializing rapidly and their emissions are likely to double within the next two decades.

What will be the effects of global warming?

An increase of just 1.5ºC in global temperature could cause changes greater than anything experienced during the last 10,000 years. Regional and seasonal weather patterns will change, with longer warm periods and shorter cold seasons.

Extreme weather conditions like floods and droughts are likely to occur more often. Many believe that this effect is already noticeable. In fact, there have been an unprecedented number of natural disasters in the 1990s. During that decade, the weather-related damage was five times greater than in the 1980s. More such events will accelerate soil erosion and desertification, which could disrupt world agriculture and threaten food security.

The ocean will become warmer, its waters will expand, and sea levels will rise. The possible melting of polar ice caps will add to the problem. Coastal areas will be flooded in places like the Netherlands, Egypt, Bangladesh, and Indonesia, necessitating the evacuation of large populations. Small islands like those of the Maldives and many of the islands in the South Pacific could disappear (read Box 19.2).

Box 19.2 Disaster: Days are numbered

I feel sad and angry at the same time, sad that eventually we will have to move, and angry because this is not our doing, but because of doings of others who don't care, who are looking after their own needs and not at the bigger picture.

The above words are that of Paani Laupepa, the Assistant Secretary for the Environment, Tuvalu. He was referring to the soon-to-come evacuation of Tuvalu before it disappears into the ocean as the world's first casualty of global warming.

Tuvalu is a chain of nine coral islands in the South Pacific with a total area of 25 sq km and a population of 11,000. It is one of the smallest and most remote countries, halfway between Hawaii and Australia. It is a paradise of chalk-white beaches and coconut palms. Soon, however, it will be a paradise lost!

Rising sea levels are already a fact in Tuvalu. The islands are made of porous fossil coral and water has started flowing up through holes in the ground. The tides are higher and the storms are more frequent and severe.

The people, who survive on subsistence agriculture and fishing, are ready to migrate. While Australia has refused them permission, New Zealand is taking in 75 persons every year.

Prime Minister Koloa Talake has hired law firms in the US and Australia to file a case against greenhouse gas emissions at the International Court of Justice at the Hague.

Leave they must, sooner or later. When they leave, who will compensate them for the loss of their entire culture?

Source: Price (2003), Ede (2003) and Lynas (2004)

Thousands of animal and plant species will go extinct, unable to adjust quickly enough to the new conditions. Polar species may be the first to go, followed by those in the coastal zones everywhere.

On the whole, a grim future awaits the Earth if the current rate of emission of greenhouse gases continues. Even if these emissions were drastically decreased now, the effects of past emissions will continue for a long time. Yet, countries like the US are unwilling to sacrifice some comforts today in order to avoid catastrophes tomorrow.

Is there any doubt that global warming is occurring?

There is no dispute about the increased accumulation of greenhouse gases in the atmosphere. There is, however, a minority who believe that our understanding of climate is insufficient to allow us to make any predictions of global warming. Most scientists agree with the environmentalists that the unchecked increase in greenhouse gases is too great a risk to take. Meanwhile, more and more studies confirm that global warming is a reality.

In 1992, over 1600 distinguished scientists (including 104 Nobel Laureates) issued a statement called the *World Scientists' Warning to Humanity*. This was a warning to the world about rapid environmental degradation. The following are two key sentences in the statement:

- 'Human beings and the natural world are on a collision course.'
- 'We must move away from fossil fuels to cut greenhouse gas emissions.'

The Intergovernmental Panel on Climate Change (IPCC), a task force of climate scientists from nearly 100 countries, also came to a clear conclusion that global warming was indeed occurring, in its 1995 report. This report, with 78 lead authors and 400 contributors from 26 countries, came to the following serious conclusions:

- Global warming is happening.
- Human activity is causing it.
- Global warming is likely to unleash unnatural and devastating storms, floods, heat waves, droughts, etc.
- Carbon dioxide emissions must be cut, particularly in the industrialized nations.

Many other studies have come to similar conclusions. In November 2004, new evidence of global warming came from a four-year study sponsored by the United Nations Environment Programme (UNEP), which was conducted by an international team of 300 scientists.

The study, called the *Arctic Climate Impact Assessment*, predicts that greenhouse gases from human activities are likely to contribute to a global temperature increase of 3º–9ºC over the next 100 years. The Arctic region is an indicator of global climate change and provides early warning of events worldwide. It is certainly getting warmer at the Poles and glaciers are melting in many parts of the world (Boxes 19.3 and 19.4).

Box 19.3 Found: Dead bodies and ancient mummies

In Norwegian cemeteries, bodies of those killed by flu and smallpox early last century are surfacing. In the Italian Alps, passing hikers found a Neolithic Iceman, a 5300-year-old mummified body.

The reason? The permafrost is melting and revealing long-kept secrets. As its name implies, the permafrost is not supposed to melt. It is a perennially frozen layer of the soil. Any rock or soil remaining at or below 0ºC for two or more years is permafrost. Almost 20 per cent of the world's land surface is made up of permafrost and most of it is in the Arctic.

Frozen organic matter in the permafrost holds one-sevenths of the world's carbon. Its release could dramatically increase the concentrations of carbon dioxide in the atmosphere.

The ice is melting at the Poles. In Alaska, temperatures have increased by an average of 2ºC since the 1950s. All permafrost below the Yukon has warmed and roads and buildings are buckling. Greenland has been losing 50 cu. km of water per year from its vast ice sheet.

The thaw at the North Pole in 2002 exceeded all records. The sea ice was the thinnest in 40 years. When the ice is thin, less sunlight is reflected back, more radiation is absorbed by the water, and temperatures go up.

In Antarctica, the temperature is behaving oddly. It is stable in some parts, dropping slightly in others, and rising on the peninsula. In March 2002, 1250 sq m section of the Larsen B ice shelf collapsed.

The world's glaciers are melting everywhere. In Glacier Park, Montana, in the US, of the 150 glaciers it had in the nineteenth century only 35 remain. The famous snows of Mt Kilimanjaro in Africa are vanishing (Box 19.4).

The snow peaks of the Andes in South America are melting fast. Bolivia and Peru depend on them for drinking water, irrigation, and hydropower. In fact, Lima in Peru, the second largest desert city in the world after Cairo, is totally dependent on the water from the glacier. Over the past 30 years, 800 million cu. km of water has been lost from the ice fields above Lima. A huge chunk of the glacier is threatening to fall into Lake Palcacocha in Peru and if it does, it will flood Huaraz, a city of 60,000.

In Bolivia, the glacier Qori Kalis is retreating at the rate of 60 cm a day. Paul Rauber writes, 'Sit beside the mighty Qori Kalis on a warm afternoon and watch the Ice Age melt before your eyes.' Scary, isn't it?

Source: Rauber (2003) and Lynas (2004)

Box 19.4 Crisis: The vanishing snows

'Wide as all the world, great, high, and unbelievably white in the sun.' That is how Ernest Hemingway once described these ice fields. They are today vanishing before our eyes.

In January 2003, amateur adventurer Vince Keipper realized a long-time goal when he trekked to the top of the mountain. 'The sound brought our group to a stop,' Keipper recalls. 'We turned around to see the ice mass collapse with a roar. A section of the glacier crumbled in the middle, and chunks of ice as big as rooms spilled out on the crater floor.'

Keipper was in the right place at the right time to get a photo of the crumbling Furtwängler Glacier on Mt Kilimanjaro. The photo is dramatic evidence of the glacier's recession. Big blocks of ice tumbled across the trail Keipper had hiked the day before.

Mt Kilimanjaro in Tanzania is the highest mountain in Africa, almost 5900 m high. The top of the mountain is covered with snow formed more than 11,000 years ago. This snowcap makes Kilimanjaro one of the most recognized landmarks of Africa.

The mountain rises above flat land, called the savannah. The land is home to many different kinds of animals. Ancient beliefs in Africa consider the mountain a holy place.

The snow and ice on the summit of Mt Kilimanjaro is now melting so fast that some scientists believe its ice cap could be gone by the year 2015. The ice cap has shrunk from about 12 sq km in 1912 to about 2 sq km in 2003, a decrease of 80 per cent.

There is concern that the loss of Kilimanjaro's ice cap could impact both the local climate as well as the availability of fresh water for local populations who depend upon the glacial melt run-off, particularly during the dry seasons. It will also affect agriculture and hydroelectric production. As Tanzania's top tourist attraction, drawing 20,000 tourists a year, the country's foreign currency earnings will also be hit.

Most predictions of global climate change suggest that early signs of warming will be seen at high elevations where these ice caps exist. It is, however, not clear if global warming explains the melting of Kilimanjaro's ice.

Some scientists believe that forest reduction in the areas surrounding Kilimanjaro, and not global warming, might be the strongest human influence on glacial recession. Fires and the clearing of land for agriculture have greatly reduced the surrounding forests. The loss of foliage causes less moisture to be pumped into the atmosphere, leading to reduced cloud cover and precipitation and increased solar radiation and glacial evaporation.

Scientists are scrambling to learn as much as they can from a vanishing resource. Kilimanjaro's glaciers are the only source of tropical ice core records for the whole continent of Africa. The glacial ice core samples hold vital atmospheric and climatic records—information that is key to understanding tropical weather patterns over the past millennia.

Can the melting be halted? Scientists are willing to try. Some have suggested covering Kilimanjaro's ice cap with a bright white cover (inspired by those used in England to protect cricket fields from the elements), a membrane, which will seal the glaciers, preventing evaporation and reflecting solar radiation.

The Furtwängler Glacier may continue to disappear in large chunks. The rest of Kilimanjaro's ice cap will follow suit, but rather than exploding, it will steadily and stealthily evaporate into the African air.

Source: The websites *http://earthobservatory.nasa.gov/*, *http://greennature.com/*, and *http://ot2.nationalgeographic.com/*

The main force behind creating doubts about warnings regarding the occurrence of global warming was a group called the Global Climate Coalition (GCC). This group represented more than 100 major corporations of the world from the coal, oil, power, automobile, and chemical industries. The GCC was set up to protect the interests of the major companies that benefit from the large-scale use of fossil fuels. Through sustained advertisement campaigns and heavy lobbying, GCC has tried to counter scientific evidence of global warming. It has effectively undermined public support in the US for efforts to stabilize climate.

The GCC, however, has lost its major members as corporations changed their attitudes toward the environment. Companies like British Petroleum, Dupont, Royal Dutch Shell, Ford, Daimler Chrysler, Texaco, and General Motors have left the group and have embraced environmental goals of their own.

What are the international initiatives to control global warming?

It is unlikely that global warming can be avoided, but we can reduce its adverse effects. Immediate and drastic reduction of emissions, as recommended by the IPCC, is the need of the hour.

Energy conservation, promotion of renewable energy sources, cleaner and fewer automobiles, greater support for public transportation, reduced deforestation, cleaner technologies, and similar measures are urgently needed. The most important international initiative in this regard is the Kyoto Protocol.

What is the Kyoto Protocol?

The Kyoto Protocol is a legally binding international agreement to reduce greenhouse gas emissions. It was initially negotiated during a meeting held in Kyoto, Japan, in 1997. The Protocol commits industrialized countries to reducing emissions of six greenhouse gases by five per cent by 2012. Some of the reduction targets are for the US (seven per cent), the European Union, Switzerland (eight per cent), Canada, and Japan (six per cent).

The agreement specifies that all Parties to the Protocol must follow a number of steps, some of which are given below:

- Design and implementation of climate change mitigation and adaptation programmes
- Preparation of a national inventory of emission removal by carbon sinks
- Promotion of climate friendly technology transfer
- Fostering partnerships in research and observation of climate science, impacts, and response strategies

Developing countries are not legally bound to emissions reduction targets as yet, because these countries have historically been responsible for only a small portion of the global greenhouse gas emissions.

Even if the Kyoto Protocol were to be fully implemented, it would do very little to solve the problem of carbon dioxide emissions. The IPCC had stated in its report that the emissions have to be cut by 60–80 per cent if the climate is to be stabilized.

Many countries including India have ratified the Protocol. The US, however, has not accepted even the modest requirement of a seven per cent reduction. In November 2004, Russia ratified the Kyoto Protocol and, as a result, the treaty will come into force by early 2005.

Global warming and other environmental problems are likely to cause displacement of large populations. In fact, displacement and resettlement are already major issues and that is the topic of the next section.

What are reasons for displacement of populations?

Millions of people have had to leave their homes and move to other places, cities, or foreign countries for a variety of reasons:

- Conflicts like wars between countries, civil wars, ethnic conflicts, and religious persecution
- Natural disasters like cyclones, hurricanes, tsunamis, floods, earthquakes, volcano eruptions, and drought
- Environmental disasters like industrial mishaps, nuclear accidents, pollution of water bodies, toxic contamination of sites, and oil spills
- Development projects in various sectors:
  - ♦ Water supply (dams, reservoirs, irrigation)
  - ♦ Urban infrastructure (roads, flyovers, beach resorts, etc.)
  - ♦ Transportation (roads, highways, canals)
  - ♦ Energy (mining, power plants, oil exploration and extraction, pipelines)
  - ♦ Agricultural expansion
  - ♦ Forests (conversion into national parks and reserves)
- To satisfy economic needs (migration in search of a livelihood or for improving the quality of life)
- Forcible relocation by the government under population redistribution schemes as has happened in South Africa under apartheid, China during the Cultural Revolution, and the Soviet Union in the Stalinist era

The reasons for displacement are often interconnected. For example, a family displaced by the construction of a dam loses its livelihood and hence moves to another place for economic reasons.

When people leave an area, they may leave behind environmental problems like abandoned agricultural land. Their migration may also cause environmental problems in the area into which they move. There are large refugee camps lacking in basic amenities at many places in the world, which have become permanent settlements. In other cases, the refugee group upsets the balance of the area by making unsustainable demands on the natural resources.

The United Nations High Commission for Refugees (UNHCR) predicts that economic and environmental refugees will continue to make up a significant percentage of the total refugee populations around the world, and that the number will only increase as more damage is done to the environment and as more countries become globalized and offer better opportunities than their home countries.

What are the displacement problems created by development projects?

People who are forced to flee from a disaster or conflict usually receive sympathetic attention and international aid. The same cannot be said about the millions of people worldwide who have been displaced by development, even though the consequences they face may be every bit as grave as those faced by people displaced by other forces.

In the past, the dominant view was that large-scale development projects accelerated progress toward a brighter and better future. If people were displaced along the way, that was deemed a necessary evil.

In recent times, however, a new development paradigm has been articulated, one that promotes poverty reduction, environmental protection, social justice, and human rights. In this paradigm, development is seen as bringing both benefits as well as imposing costs. Among its greatest costs has been the displacement of millions of vulnerable people.

The World Bank estimates that every year since 1990, roughly 10 million people worldwide have been displaced by development projects for a variety of reasons. During the last 50 years, development projects have displaced 25 million people in India and 40 million in China (of whom, 13.6 million were displaced in the 1990s alone). Read about China's Three Gorges Dam in Box 19.5.

Box 19.5 Controversy: A dragon tamed by technology?

Until recently, Huang Zongjin was a poor farmer cultivating rice and vegetables on the banks of the Yangtze river near Wushan town in China. Then his world changed, suddenly and totally. He has now become Huang the boatman, ferrying tourists up the Yangtze.

The government had decided to build a dam on the Yangtze (known in China as 'The Dragon'), just downstream from the Three Gorges. Huang and more than 700,000 others were asked to move, leaving their homes and fields, which were to be submerged in water. Many more will have to move at the next stage.

If completed, the Three Gorges Dam on the Yangtze will be the largest hydroelectric dam in the world. It will stretch nearly 1.5 km across and tower 200 m above the world's third longest river. Its reservoir will extend over 560 km upstream and force the displacement of close to 1.9 million people. Construction began in 1994 and is scheduled to take 20 years and cost over US$ 24 billion to complete. The river is no longer seen as an unstoppable force, but as a dragon, which technology can tame.

The expected benefits of the dam include the prevention of the devastating Yangtze basin floods, which have killed millions of people, and the hydrogeneration of 18,200 MW of electricity, supplying a tenth of China's needs. The power generated would be equivalent to operating more than a dozen nuclear power plants or burning 50 million tons of coal. It is a remarkable feat of engineering.

Outside China, the project has been strongly criticized for the corruption, secrecy, financial incompetence, human rights violations, and possible environmental consequences. As a result, initially the dam failed to win financial support from the World Bank or overseas investors. The dramatic expansion of the Chinese economy in recent years, however, again altered the general picture: a headlong rush by Western business to participate in China's emerging economy has brought Western governments under increasing political pressure to signal backing of the Three Gorges Dam.

By October 2003, the second stage was over, the water had risen, 700,000 people had been relocated, and the dam was finally producing electricity. So is it the disaster everyone predicted?

Most of the people seem to be either positive or fatalistic about the dam. Huang Zongjin used his compensation money to buy his dilapidated boat. The government has built the family a new, far bigger, hillside home, but as yet it has no running water or electricity. The land is too steep to be cultivated.

Huang, however, is sanguine. 'I may float over my old home every day, but I never think about it. What's the point?' he says. 'We can't change anything. And besides, life is better now. We have a new home, more space and more money. The dam has been good for us'.

Is it good for everybody? Is it good for the environment? Will the dam surprise everybody with its secrets?

Source: The websites *www.hrw.org, www.irn.org/programs/threeg/* and *www.guardian.co.uk/china/*

They may not have crossed a border and may not be regarded as living in 'refugee-like' circumstances within their own country. Nevertheless, they have been evicted from their homes or places of habitual residence, have had their lives and livelihoods disrupted, and face the uncertainties of resettling in unfamiliar and often inhospitable locations.

What are the special problems with regard to displacement by dams?

After a multi-year study, the World Commission on Dams (WCD) listed the following social impacts of dams in its report published in 2000:

- Dams have physically displaced some 40–80 million people worldwide.
- Millions of people living downstream from dams—particularly those reliant on natural floodplain functions and fisheries—have also suffered serious harm to their livelihoods and the future productivity of their resources has been put at risk.
- Many of the displaced were not recognized (or enumerated) as such, and therefore were not resettled or compensated.
- Where compensation was provided it was often inadequate, and where the physically displaced were enumerated, many were not included in resettlement programmes.
- Those who were resettled rarely had their livelihoods restored, as resettlement programmes have focused on physical relocation rather than the economic and social development of the displaced.
- The larger the magnitude of displacement, the less likely it is that even the livelihoods of affected communities can be restored.
- Even in the 1990s, the impacts on downstream livelihoods were, in many cases, not adequately assessed or addressed in the planning and design of large dams.

The WCD report noted that the poor and other vulnerable groups and future generations were likely to bear a disproportionate share of the social and environmental costs of large dam projects without gaining a commensurate share of the economic benefits:

- Indigenous and tribal peoples and vulnerable ethnic minorities have suffered disproportionate levels of displacement and negative impacts on livelihood, culture, and spiritual existence. The outcomes have included assetlessness, unemployment, debt-bondage, hunger, and cultural disintegration.
- Affected populations living near reservoirs as well as displaced people and downstream communities have often faced adverse health and livelihood outcomes from environmental change and social disruption.
- Among affected communities, gender gaps have widened and women have frequently borne a disproportionate share of the social costs and have often been discriminated against in the sharing of benefits.

The WCD report concluded that, 'impoverishment and disempowerment have been the rule rather than the exception with respect to resettled people around the world.' The report gives specific recommendations to national governments, NGOs, affected people's organizations, the private sector, aid agencies, and development banks on all social aspects of large dam projects, including displacement and resettlement.

What are the issues concerning the resettlement and rehabilitation of displaced groups?

The majority of displacement and resettlement programmes involving large populations have occurred in developing countries. The reasons are the massive development projects undertaken in these countries, financed by foreign aid and international agencies, and their high population densities.

The following are some of the problems faced by displaced people:

- The compensation for the lost land is often not paid or the payment is delayed, and even if it is paid, the amount is usually pitifully small. The oustees are rarely able to start new lives with the compensation they receive. To add to the problem, agents and corrupt officials deprive the poor of the full compensation.
- Generally, the new land that is offered is of poor quality and the refugees are unable to make a living.
- Basic infrastructure and amenities are not provided in the new area. Very often, temporary camps become permanent settlements.
- When tribal groups are displaced, they do not get any compensation since they have no legal title to the land.
- Ethnic and caste differences make it difficult for the refugees to live peacefully with the communities already living in the area.

While the general rehabilitation scene is depressing, there are some success stories, like the one in Madhya Pradesh (read Box 19.6).

Box 19.6 Success: From forests to fisheries

Most people displaced by development projects end up losing everything—their lands, livelihood, and shelter. In most cases they do not get any compensation for their lands and even if they do, the amount is grossly inadequate. This story, from Madhya Pradesh, however, is one of hope.

A dam on the River Tawa, a tributary of the Narmada, submerged 20,000 hectares and displaced over 4000 adivasi families in 44 villages. More than half the villages were on forestland and hence the people got no compensation at all.

Normally these oustees would have migrated to other villages or towns in search of livelihoods. This group, however, campaigned for and secured fishing rights in the reservoir created by the dam. The adivasis learnt how to spread nets, use boats, and carry on fishing operations.

They formed the Tawa Matsya Sangh (TMS), a federation of primary fishermen's cooperative societies with 1200 members. In 1996, TMS was given a five-year lease by the government. It was in charge of overall management and was responsible for stocking the reservoir with fishes. It also took care of the transport and sale of the fish catch, regulatory measures for conservation, and members' welfare.

The former hunter-gatherers became good fishermen, their catch was good, and TMS made profits and shared it with its members. It even paid wages to them during the lean season, between June and August when there was an agreed ban on fishing. Moreover, it was also able to pay the agreed royalty to the government.

It was a rare case of dam oustees doing well in a new occupation, but troubles were ahead. When the lease was renewed the government steeply increased the royalty amount. In return, however, the government agency did little to improve the infrastructure. Even the promised ice factory was not built. Under the new terms, TMS was also subject to greater government control. In fact, the government appointed a review committee to evaluate the performance of TMS, but did not share the report with them.

Difficult times are ahead for a group that has achieved success against great odds!

Source: Menon (2002)

What lies ahead?

The encouraging fact is that displacement, caused by development and economic and environmental factors, is now attracting greater attention than it did in the past. In 1998, a team of international legal scholars presented to the UN a set of Guiding Principles on Internal Displacement. These are the first guidelines developed within the context of human rights and humanitarian law to address the issue of internal displacement. In 2002, the UN set up an Internally Displaced Persons Unit in the Office for the Coordination of Humanitarian Affairs.

A quick recap

- In recent times, acid rain has resulted from the emissions of pollutants from human activities. Acid rain has a number of adverse effects on the environment.
- Massive displacement of populations takes place due to development projects like dams, environmental and natural disasters, etc.
- The ozone layer in the atmosphere is getting depleted and an Ozone Hole has developed over the Antarctic.
- The depletion of the ozone layer has adverse effects on life.
- The Montreal Protocol on Substances that Deplete the Ozone Layer has been successful in reducing the production and consumption of ozone depleting substances.
- There is clear evidence that global warming is occurring due to emissions from human activities.
- Global warming will have serious effects on the environment and human society.
- The Kyoto Protocol has set modest targets for the reduction of greenhouse gases. The US has so far not signed the Protocol.
- There is increasing displacement of populations due to a variety of reasons.
- The resettlement and rehabilitation of poor and indigenous displaced people have been unsatisfactory in many projects.

What important terms did I learn?

Acid rain
Chlorofluorocarbons (CFC)
Global warming
Greenhouse gas
Ozone Depleting Potential (ODP)
Ozone layer

What have I learned?

Essay question

Write an essay on the impact of dams on people. What are the usual expected benefits? What are the negative impacts on people? What does the World Commission on Dams (WCD) have to say on the topic? What do the cases of Narmada and the Three Gorges dams illustrate?

Short-answer questions

1. Explain the causes and effects of acid rain.
2. What is the ozone layer and why is it getting depleted?
3. Outline the provisions of the Vienna Convention and the Montreal Protocol
4. What have been the results of the implementation of the Montreal Protocol?
5. What is global warming and how does it occur?
6. What are the possible effects of global warming?
7. Outline the provisions of the Kyoto Protocol.
8. Explain the reasons for the displacement of populations and its impact.
9. Describe the special problems of displacement caused by dams.
10. Discuss the issues concerning the resettlement and rehabilitation of displaced groups.

How can I learn by doing?

Visit the site of any project that has displaced people. Study the project documents and examine the plans and reports regarding resettlement and rehabilitation. Interview some of the oustees and record their experiences. Write a report outlining what was promised and what was delivered as far as resettlement and rehabilitation were concerned. What are your recommendations?

What can I do to make a difference?

Observe September 16 as the International Day for the Preservation of the Ozone Layer: The Day commemorates the date on which the Montreal Protocol on Substances that Deplete the Ozone Layer was signed in 1987. The UN invites all member states to devote this special day to promotion, at the national level, of concrete activities in accordance with the objectives and goals of the Montreal Protocol and its amendments. Every year an appropriate theme is chosen for the Day. The theme for 2004 was 'Save Our Sky: Ozone Friendly Planet, Our Target'. You can get more information on the year's theme and promotional material from the UNEP website.

Where are the related topics?

- Water resources were covered in Chapter 7.
- Forest resources were covered in Chapter 9.
- Natural disasters were discussed in Chapter 14 and industrial disasters in Chapter 16.

What are the deeper issues that go beyond the book?

An African diplomat once told the delegates of the rich countries, 'If you do not share your wealth with us, we will share our poverty with you.' Currently, the rich countries are responsible for most of the greenhouse gas emissions. How should they support the poorer countries in developing and using cleaner technologies?

Where can I find more information?

- The website of the Intergovernmental Panel on Climate Change *www.ipcc.org*.
- The UNEP website, *www.unep.org/ozone/*, gives detailed information on ozone depletion and the international response to it.
- You can access the full report of the World Commission on Dams (WCD) published in 2000 on the website *www.dams.org*.
- The website *www.displacement.net/* covers displacement issues.

What does the Muse say?

A poem from Our Earth Music taken from the website *http://homepage.mac.com/ourearthmusic/*.

Global Warming, Can't You See?

Hope for the earth and
Hope for the human race.
Lets do something for change,
Its within our choice and range.

We must make it cool to be green.
To live, air and water must be clean.
Don't let our leaders convince us to fear
All but what will always be here.

Our earth, our atmosphere
Discrediting scientists. Refusing to hear
Anything against the status quo
Glaciers melting, what else is there to show?

Don't let terrorism and money sway you
Into thinking the Earth doesn't need you.
Join with those working for life
A sustainable future will end the strife.

CHAPTER 20

Environmental Law and Regulations

I am not against anyone at any time,
as I am often perceived to be.
I am just for the environment at all times.

M.C. Mehta

What will I learn?

After studying this chapter, the reader should be able to:

- outline the constitutional provisions in India regarding the environment;
- describe the beginnings of environmental legislation in India;
- recall the main provisions of the:
  - Environment (Protection) Act of 1986
  - Air (Prevention and Control of Pollution) Act of 1981
  - Water (Prevention and Control of Pollution) Act 1974
  - Forest Conservation Act of 1980
  - Wildlife Protection Act of 1972;
- explain the requirement of an Environmental Impact Assessment (EIA) for certain types of projects;
- appreciate the role played by Public Interest Litigation (PIL) and judicial activism in enforcing environmental laws in India;
- explain environmental ethics and the three different ways of viewing environmental responsibility; and
- recall the initiatives to bring environmental ethics into industry.

Box 20.1 Crisis: The river that catches fire

It all began with a simple matchstick. Someone dropped a burning match by chance into the river and a whole 1 km stretch caught fire. In fact, the flames rose to a height of about seven metres and could not be extinguished for three hours. That was the River Ganga near Haridwar in 1984, when the effluents from two factories were so toxic that a match could start a fire.

Environmental lawyer M.C. Mehta heard this story and he filed a petition in the Supreme Court of India in 1985 against the two polluting factories. The scope of the case was later broadened to include all the industries and all the municipal towns in the river basin—from the beginning to the end of the Ganga.

Every Friday for over two years, Justice Kuldip Singh heard the Ganga Pollution Case, which became a landmark event in Indian PIL and judicial activism. Justice Singh brought to his court 'an informal style rooted in common sense'. He issued

The polluted River Ganga

directions every week in an attempt to clean up the mess that was the Ganga.

There were several hundred parties to the case:

- Hundreds of polluting industries and numerous small units such as tanneries in Uttar Pradesh, Bihar, and West Bengal
- A number of small municipalities that discharge untreated sewage into the river
- Half a dozen municipal corporations in the aforementioned three states
- The Eastern Railway
- Giant thermal plants in Bokaro and Patratu

Dismayed at the persistent violation of environmental laws by the polluters and the continued inaction of the various state Pollution Control Boards, the Court took an unusual result-oriented approach. It bypassed many formal court procedures and statutory requirements in favour of quick decisions.

With the help of Mehta, the Court identified the polluters. Each polluter was given three months to meet the effluent standards. The result of non-compliance would be closure of the unit. What is more, the order was issued without hearing the party concerned.

The respective Pollution Control Board had to serve the Court's notice on the polluter and, at the end of the period, report back to the Court on the status. Willing units could get an extension of time, but it was closure for the obstinate.

What was the result? Hundreds of factories installed effluent treatment plants, the two thermal power plants ordered equipment to reduce their pollution, Eastern Railway agreed to treat its wastewater, and even municipalities started working on finding a solution to the problem.

Where the boards had failed, the Court succeeded in putting pressure on industries and civic bodies. It also created environmental awareness among administrators, the subordinate judiciary, the police, and the municipal officials, all of whom were involved in implementing the Court's orders.

Yet, there are limits to what a court can do in the face of lack of political will, an indifferent bureaucracy, and poor budgetary allocations to the enforcing agencies and municipalities. Further, how long can the judiciary assume the massive administrative tasks that should ideally be carried out by the executive? How can the judges avoid being accused of arbitrariness, the same criticism earlier directed at the administrators?

Justice Kuldip Singh retired in 1997 and the Friday hearings on the Ganga stopped.

Meanwhile, soon after the filing of the case by Mehta, the Government of India initiated a Ganga Action Plan (GAP) to clean the river. Five billion rupees were spent on GAP without significant results.

Mehta had this to say in 1997, 13 years after he filed the first case:

> Many parts of the Ganga river are totally dead; the water is so polluted that it is unfit for drinking, washing, bathing or irrigational purposes. Thirty per cent of the Ganga pollution is being caused by industry. Seventy per cent is caused by cities and towns which discharge domestic waste straight into the Ganga. The municipal bodies say that they do not have money to treat the waste, and you can't close down a city.

Laws could not be enforced. Judicial activism could not be sustained. The grand action plan did not succeed. In 2003, *sadhus* refused to take the traditional holy dip at the Sangam because of the pollution. Holy and polluted flows the Ganga, even today! Is there a way out?

Source: Divan and Rosencranz (2001)

What does the Ganga Pollution case teach us?

When we try to clean up the pollution in rivers and other places, we really have a complex and tough problem on hand. A question arises from the Ganga case: Why do the thousands who take a dip in the dirty river, keep quiet? Why are they not demanding a clean river and why are they not doing something about it? Apart from some stray individuals and one or two non-governmental organizations (NGOs), there is no public pressure to clean up the river. This apathy is everywhere, in our towns and in our villages.

The most common response would be that we need laws and regulations that ensure conservation of the environment. In this chapter we will review the many environmental laws that we have. We will also discuss judicial activism and environmental ethics.

What are the provisions of the Indian Constitution regarding the environment?

The Constitution (Forty-Second Amendment) Act of 1976 explicitly incorporates environmental protection and improvement. Article 48A, which was added to the Directive Principles of State Policy, declares: 'The state shall endeavour to protect and improve the environment and to safeguard the forests and wildlife of the country.'

Article 51A(g) in a new chapter entitled 'Fundamental Duties', imposes a similar responsibility on every citizen 'to protect and improve the natural environment including forests, lakes, rivers and wildlife, and to have compassion for living creatures.'

In addition, Article 21 of the Constitution states: 'No person shall be deprived of his life or personal liberty except according to procedure established by law.' This Article protects the right to life as a fundamental right. The courts have interpreted this Article to mean that the enjoyment of life, including the right to live with human dignity, encompasses within its ambit the protection and preservation of the environment.

Do we have an environmental policy? In 1980, the Union Government established the Department of Environment. It became the Ministry of Environment and Forests (MoEF) in 1985. The Ministry initiates and oversees the implementation of environmental policies, plans, laws, and regulations.

MoEF prepared the first National Environmental Action Plan in December 1993, laying down India's environmental priorities. In 2004, the MoEF unveiled a new Draft Environmental Policy.

What was the beginning of environmental legislation in India?

Prime Minister Indira Gandhi gave a speech at the UN Conference on the Human Environment held at Stockholm in June 1972. The participating countries agreed to take appropriate steps to preserve the natural resources of the Earth. In consonance with this decision, India began enacting various environmental laws.

Initially, the laws were not very different from the general body of law. For example, the Water Act of 1974 was very much like other laws and created another agency-administered licensing system to control effluent discharges into water.

The Bhopal Gas Tragedy changed the situation. In the 1990s, a spate of laws were passed covering new areas, such as vehicular emissions, noise, hazardous waste, transportation of toxic chemicals, and environmental impact assessment.

Further, the old licensing regime was supplemented by regulatory techniques. The new laws included provisions such as public hearings, citizens' right to information, deadlines for technology changes (in motor vehicles, for example), workers' participation, and penalties on the higher management of companies for non-compliance.

The powers of the enforcing agencies like the various Pollution Control Boards of the states were also enhanced. For example, previously a board would have to approach a magistrate to get a factory shut down. Now, it was given the power to order closure, leaving the polluter to challenge the order in court.

Let us now briefly review the main environmental laws of the country.

What powers does the Environment (Protection) Act of 1986 give to the Central Government?

The Act is an umbrella legislation. It is an *enabling* law that provides the executive with powers to frame various rules and regulations.

The Act authorizes the Central Government to protect and improve environmental quality, to control and reduce pollution from all sources, and to prohibit or restrict the establishment and operation of any industrial facility on environmental grounds.

The Act defines terms such as environment, environmental pollutant, and hazardous substance. According to the Act, the Central Government has the power to:

- take measures to protect and improve the environment;
- give directions (for example, to close, prohibit, or regulate any industry, operation or process); and
- make rules to regulate environmental pollution (air and water quality standards, prohibiting or restricting the handling of hazardous materials, siting of industry, etc.).

The Chapter of the Act on the prevention, control, and abatement of environmental pollution includes: controlling discharge of environmental pollutants, enforcing compliance with procedural safeguards, power of entry and inspection, power to take samples, setting up of environmental laboratories, appointing environmental analysts, and prescription of penalties for contravening the Act.

What are the Environment (Protection) Rules, 1986? These rules set the standards for emission or discharge of environmental pollutants. In addition, more stringent standards may be laid down for specific industries or locations.

There are rules prohibiting and restricting the location of industries and the carrying on of processes and operations in different areas. Factors to be taken into consideration include: the topographic and climatic features of an area, environmentally compatible land use, the net adverse impact likely to be caused by an industry, proximity to areas protected under various other laws, proximity to human settlements, etc.

What is meant by Environmental Impact Assessment?

In 1994, Environmental Impact Assessment (EIA) was made mandatory for certain types of projects. The regulations require the project proponent to submit an EIA report, an environmental management plan, details of the public hearing, and a project report to MoEF.

There are 30 categories of projects that require an EIA, such as nuclear power, river valley projects, ports, harbours (except minor ports and harbours), all tourism projects between 200 and 500 m of the High Water Line and at locations with an elevation of more than 1000 m, with investment of more than Rs 50 million, thermal power plants, mining projects (major minerals with leases of more than five hectares), highway projects, and thermal power plants.

The Ministry's Impact Assessment Agency evaluates EIA reports. The assessment is to be completed within 90 days of receipt of the requisite documents and data from the Project Authorities and completion of the public hearing and the decision must be conveyed within 30 days thereafter. The clearance granted is valid for a period of five years from the commencement of the construction or operation of the project.

What are the provisions of the Air (Prevention and Control of Pollution) Act of 1981?

The objective of this Act is to provide for the prevention, control, and abatement of air pollution. The Act defines air pollution as the presence in the atmosphere of any solid, liquid, or gaseous substance (including noise) in such concentrations as may be injurious to human beings, other organisms, property, or the environment.

The provisions of the Act are to be implemented by the Central Pollution Control Board (CPCB) along with the various state boards. The Act lists a number of functions of the CPCB including: setting of air quality standards, collecting data on air pollution, organizing training and awareness programmes, establishing laboratories, etc. The CPCB can specify air pollution control areas and set standards for vehicular emissions.

The Act lays down penalties for the violation of its provisions. This applies to companies and their owners and managers as well as to government departments. Citizens can file complaints with the CPCB.

What are the water-related environmental laws in India?

The early Acts concerned with water issues were the following:

- The Easement Act of 1882, which allowed private rights to use groundwater by viewing it as an attachment to the land. It also states that all surface water belongs to the state and is state property.
- The Indian Fisheries Act of 1897, which established two sets of penal offences whereby the government can sue any person who uses dynamite or other explosive substance in any way (whether coastal or inland) with intent to catch or destroy any fish or poison fish in order to kill them.
- The Merchant Shipping Act of 1970, which deals with waste arising from ships along the coastal areas within a specified radius.

What are the objectives of the Water (Prevention and Control of Pollution) Act 1974?

The objectives of the Act are to prevent and control water pollution and the maintenance or restoration of the wholesomeness of water. The Act defines water pollution as the contamination of water, alteration of its physical, chemical or biological properties, or the discharge of any sewage or trade effluent or any other liquid, gaseous, or solid substance into water, which may render such water harmful to public health, or to domestic, commercial, industrial, agricultural or other legitimate uses, or to the life and health of organisms.

The Act establishes an institutional structure for preventing and abating water pollution. It establishes standards for water quality and effluent discharge into water. Polluting industries must seek permission to discharge waste into effluent bodies.

The Act's implementation is similar to the Air (Prevention and Control of Pollution) Act, 1981, with the CPCB being the main enforcing agency. In fact, the CPCB was constituted under this Act.

What does the Forest Conservation Act of 1980 specify?

The Forest Conservation Act of 1980 and the Forest (Conservation) Rules of 1981 provide for the protection and conservation of forests. The Act specifies the requirements that should be met before declaring an area a Protected Forest, a Wildlife Sanctuary, or a National Park.

Under the Act, a state government may regulate or prohibit, in any forest, the clearing of land for cultivation, pasturing of cattle, or the clearing of vegetation for any of the following purposes:

- Protection against storms, winds, floods, and avalanches, for the preservation of the soil on the slopes, the prevention of landslips or the formation of ravines and torrents, or the protection of land against erosion
- Maintenance of a water supply in springs, rivers, and tanks
- Protection of roads, bridges, railways, and other modes of communication
- Preservation of public health

The Act makes it mandatory for the owner of a forest to seek permission before converting it to any non-forest use, such as the cultivation of tea, coffee, spices, rubber, palms, oil-bearing plants, horticultural crops, or medicinal plants.

What are the objectives of the Wildlife Protection Act of 1972?

The Wildlife Protection Act of 1972 defines wildlife to include any animal, bees, butterflies, crustaceans, fish, and moths; and aquatic or land vegetation, which form part of any habitat. The Act along with the Wildlife Protection Rules of 1973 provides for the protection of birds and animals and for all matters that are connected to this whether it be the habitat, the waterhole, or the forest that sustains them.

What are the other Acts and Rules concerning the environment?

The following are the other important environment-related Acts and Rules:

- The Hazardous Waste (Management and Handling) Rules of 1989, which control the generation, collection, treatment, import, storage, and handling of hazardous waste

- The Manufacture, Storage and Import of Hazardous Chemical Rules of 1989, which set up an Authority to inspect, once a year, the industrial activity connected with hazardous chemicals and isolated storage facilities
- The Manufacture, Use, Import, Export and Storage of Hazardous Micro-organisms, Genetically Engineered Organisms or Cells Rules of 1989, which were introduced with a view to protecting the environment, nature, and health, in connection with the application of genetic engineering
- The Public Liability Insurance Act and Rules of 1991 and the Amendment Act of 1992, drawn up to provide for public liability insurance for the purpose of providing immediate relief to persons affected by accidents while handling any hazardous substance
- The National Environmental Tribunal Act of 1995, which was created to award compensation for damages to persons, property, and the environment arising from any activity involving hazardous substances
- The National Environment Appellate Authority Act of 1997, which was created to hear appeals with respect to the restriction of areas in which classes of industries, etc., are carried out subject to certain safeguards under the Environment (Protection) Act
- The Biomedical Waste (Management and Handling) Rules of 1998, which constitute a legal control over health care institutions, with a view to streamlining the process of proper handling (segregation, disposal, collection, and treatment) of hospital waste
- The Coastal Regulation Zone Notification of 1991, which regulates various activities, including construction, in the coastal zone. It gives some protection to the backwaters and estuaries

How are the environmental laws being enforced?

We saw that there is no dearth of laws for protecting and conserving the environment. However, the implementation of these laws continues to be poor. Government agencies have vast powers to regulate industries and others who are potential polluters. They are, however, reluctant to use these powers to discipline the polluters.

The Parliament and the State Assemblies are ready to pass environmental laws, but they do not provide funds for their implementation. Nor do they demand that the governments enforce the laws strictly.

The poor performance of the government agencies in enforcing the laws has compelled the courts to play a pro-active role in the matter of the environment. They have directly responded to the complaints of citizens or PILs and assumed the roles of policy makers, educators, administrators, and, in general, friends of thc cnvironment. The environmental lawyer M.C. Mehta has played a remarkable role in this process (Box 20.2).

Many petitions filed in various courts by individual citizens, groups, and voluntary organizations have led to major decisions given by the courts forcing the state governments to act against pollution and degradation of the environment. The Supreme Court, for example, has in numerous cases issued directions to close down or shift factories, change to less-polluting technologies, implement environmental norms, etc. Examples are the Ganga Pollution Case (Box 20.1) and the Shrimp Aquaculture Case (Box 20.3).

Box 20.2 Inspiration: The Green Avenger

He visited the Taj Mahal for the first time in early 1984 and was shocked to see that the monument's marble had turned yellow and was pitted as a result of pollutants from nearby industries. Being a lawyer, he filed an environmental case in the Supreme Court of India on the issue.

In 1993, after a decade of hearings, the Supreme Court ordered 212 small factories surrounding the Taj Mahal to close because they had not installed pollution control devices. Another 300 factories were put on notice to do the same.

Mahesh Chandra Mehta did not stop with the Taj Case. In 1985, he filed a petition in the Supreme Court against the factories that were polluting the River Ganga (Box 20.1). Many positive actions to prevent the pollution of the river resulted from this case.

Mehta has continued his crusade and at the last count, his cases numbered more than 50 in the Supreme Court alone. He has successfully fought cases against industries that generate hazardous waste and has succeeded in obtaining a court order to make lead-free gasoline available. He has also been working to ban intensive shrimp farming and other damaging activities along India's coast.

Through his work, Mehta has set the national agenda in the fields of water and air pollution, vehicular emission control, conservation of the coastal zone, and the translocation of heavy industry away from urban areas. Almost single-handedly, he has obtained more than 40 landmark judgements and numerous orders from the Supreme Court against polluters. In addition, responding to his petition, the Court has ordered that environmental studies be taught as a compulsory subject in all schools, colleges, and universities in the country.

No other environmental lawyer in the world has perhaps done this much. He has inspired many lawyers in the lower courts to take up environmental cases. Mehta won the Goldman Environmental Prize in 1996 and the Ramon Magsaysay Award for Public Service in 1997.

Source: Divan and Rosencranz (2001) and the website *www.goldmanprize.org/*

Box 20.3 Judicial activism: Mediating in prawn wars

> We live on this land. The sea is our mother. It takes care of us. This is our livelihood. If the prawn farms start operating it will ruin us, but we will destroy them. Wherever it is, we will not spare them. We will wipe them out.

Why are the fisherfolk and other villagers on the eastern coast of India so much against prawn farms?

Shrimp aquaculture production in India increased from 30,000 tons in 1990 to 102,000 tons in 1999, made possible by a rapid increase in the area under semi–intensive farms. The expansion was driven by the high profitability of shrimp farming and it attracted a wide range of investors, from individual farmers converting paddy fields to multinational companies investing in large-scale semi-intensive and intensive shrimp farming.

Coastal areas are ideal for shrimp aquaculture since it needs both freshwater and seawater. The rapid expansion of shrimp aquaculture along the coast, particularly in Tamil Nadu and Andhra Pradesh, has created many social and environmental problems:

- Polluted salty water from the prawn tanks was discharged into fields, degrading the soil and killing the crops. Wells were contaminated and the water became undrinkable. Villagers were forced to walk many miles to fetch fresh water.
- The salty water resulted in salinization leading to the collapse of the mud houses of the poor.
- The wastes from prawn farms containing pesticides, fertilizers, and antibiotics were often pumped back into the sea polluting the breeding zones of the fishes.

- The inlet and outlet pipes extending into the sea caused problems for fishing nets and boats.
- In many places, the farms blocked the fisherfolk's access to the sea.
- Mangrove areas were converted into aquaculture farms. During the 1990s, shrimp aquaculture accounted for about 80 per cent of the conversion of mangrove land.

In 1994, S. Jagannathan, a Gandhian social worker from Tamil Nadu, filed a writ petition in the Supreme Court in the public interest. The petition sought: the enforcement of the Coastal Zone Regulation Notification of 1991; the stoppage of intensive and semi-intensive prawn farming in the ecologically fragile coastal areas; prohibition of the use of wastelands/wetlands for prawn farming; and the constitution of a National Coastal Management Authority to safeguard marine life and the coastal areas.

The Supreme Court bench, headed by Justice Kuldip Singh (Box 20.1), issued many injunctions and directions. The shrimp aquaculture industry was prohibited in the coastal zone and it was directed that existing farms be demolished. The workers, rendered unemployed, were to be compensated by the owners. Agricultural lands, salt-pan lands, mangroves, wetlands, forestlands, and land for village common purposes were not to be converted into shrimp ponds. The Court also directed that the government set up an Aquaculture Authority to regulate the industry.

After Justice Kuldip Singh retired in 1997, the aquaculture lobby filed a petition seeking review of the judgement. The Court issued a new interim order on 19 August 1997, which stated that the farms that were to be demolished as per the 1996 judgement would not be demolished until further order, but that no fresh seeds could be put in these farms.

The last word on this issue has not been said.

Source: Divan and Rosencranz (2001)

As a result of this 'judicial activism', hundreds of factories have installed effluent treatment plants and there is greater environmental awareness among the bureaucracy, police, and municipal officials, all of whom are involved in implementing the Supreme Court's orders. Judicial activism has certainly helped the cause of the environment, but there is also the view that the courts should not be taking over the role of the executive.

Laws, regulations, and PILs cannot save the environment if there is lack of environmental ethics among citizens, government officials, companies, and industries.

What are environmental ethics?

Ethics are concerned with what is wrong and what is right, irrespective of the culture and society. For example, it is ethical to have reverence for all forms of life and any killing is unethical.

Morals reflect a culture's predominant feelings on ethical issues. For example, in most cultures it is morally right to kill enemies in a war, though it is unethical.

Environmental ethics try to define the moral basis of environmental responsibility. There are three possible viewpoints here:

- Anthropocentric view: Our environmental responsibility is to ensure that the Earth remains hospitable and pleasant for human beings. This is the development ethic, the basis of the Idea of Progress (Chapter 1). In this view, nature has value only when humans utilize it.
- Biocentric view: All forms of life—humans, animals, and plants—have an inherent right to exist and live without hindrance. This is the preservation ethic, which recognizes that nature has an intrinsic value apart from its use as a resource for humans.

- Ecocentric view: The environment deserves care and consideration in itself and not because it serves the interests of humans, animals, and plants. This is the conservation ethic, which extends the preservation ethic to the entire Earth and for all time.

Can there be an environmental ethic for industry?

Important decisions affecting the environment are increasingly made by large and powerful corporations. The executives of these companies are primarily driven by the profit motive and they often regard environmental laws and regulations as a hindrance to their expansion and making of increased profits.

Companies use lobbying, political pressure, legal loopholes, and time-consuming legal action to circumvent or delay compliance with regulations. When their actions actually lead to environmental problems or even accidents, they use all their power to evade responsibility and liability, delay justice, and even to shift the responsibility to the victims. Often, for various reasons, governments also collude with them. There are numerous examples of such irresponsible and unethical corporate behaviour including the Bhopal Gas Tragedy (Chapter 16, Box 16.1), Minamata (Chapter 16, Box 16.4), Jadugoda uranium mine (Chapter 14, Box 14.6), and Love Canal (Chapter 16, Box 16.6). In this chapter, we describe the Texaco case (Box 20.4) and the Kodaikanal Mercury case (Box 20.5).

Box 20.4 Trial: Living in toxic sludge

On October 29, 2003, there was a spectacular pre-dawn lightning storm in Lago Agrio, Ecuador. It marked the end of the testimony in a landmark case involving corporate accountability and environmental justice. The verdict is still awaited.

In the 1960s oil was found in the pristine rainforest area in the Andean region of Ecuador, called Oriente. The US oil giant Texaco was invited to drill for the oil and between 1971 and 1992 the company's operations covered one million hectares of undisturbed rainforest. The company even built an 800 km trans-Andean pipeline to bring the oil to the coast.

The contract expired in 1992 and Texaco left after handing over the assets to the national company Petroecuador. Texaco, however, also left behind another deadly asset: hundreds of unlined toxic waste pits with oil sludge and billions of litres of toxic wastewater on the land and in water bodies.

The drilling operations resulted in large amounts of toxic wastewater and oil spills. Instead of injecting this waste back into the deep sub-soil, the company saved money (US$ five billion in all) by dumping it into local streams and landfills or spreading it on the local dirt roads. Over 21 years, 75 billion litres of wastewater and oil were discharged into the area. There are more than one hundred swamps contaminated with oil.

There is a permanent layer of sticky crude on the roads, which are regularly topped off with sludge suctioned out of the toxic waste pits. It also comes from the area's farms, which are often polluted with raw crude oil when pipelines rupture. People are regularly smeared with grease and have no way of cleaning themselves, except with gasoline-soaked rags.

Three indigenous communities live in the area: the Cofan, the Secoya, and the Siona. These peoples have developed distinct cultures and traditions that are inextricably linked to the abundance of the rainforest where they have lived for thousands of years. Texaco's toxic discharges, however, have literally propelled these cultures to the brink of extinction. All three have seen dramatic decreases in their populations (from several thousands to a few hundreds) due to disease and forced migration.

The three tribes have always depended on two rivers, the Aguarico and the Napo, for fish, bathing, and transport. In two decades of oil development, these rivers have been rendered virtually useless as sources of nourishment. Since the people could no longer fish in the rivers, they turned to hunting. The population of game has decreased to the extent that there is not enough food for adequate nourishment.

There has been a dramatic increase in skin diseases, particularly in children, and abnormally high rates of cancer and miscarriages. The growth of livestock is stunted and vegetation is withering. Many families spend hours each day searching for drinkable water, or hunting for animals, leading many to abandon their traditional lands.

In November 1993, several dozen tribal leaders filed a billion-dollar lawsuit against Texaco in a US court. The plaintiffs wanted Texaco to clean up the mess and compensate the people for the hardships they suffered. After 10 years of hearings, a New York court ordered Texaco to submit to the laws of Ecuador and commit itself to complying with any judgement given against it.

The historic trial began in Lago Agrio in 2003. After the testimony ended in October, the judge was to inspect hundreds of the contaminated sites. That inspection is probably still going on.

Texaco has not disputed the claim that it dumped the toxic wastewater. It insists that its practices were not wrong and that the waste pits complied with Ecuadorian laws in effect at the time. It claims that its presence in Ecuador actually raised the quality of life for the persons who are now suing the company.

The environmental devastation created by Texaco was only second to that of the Chernobyl disaster. Thousands of people continue to suffer. Will the trial go the way of the people? Will Texaco come back and spend millions of dollars in a clean-up?

Source: Koenig (2004) and the website *www.texacorainforest.org./*

Box 20.5 Solution: Send it back!

Tons of mercury waste from broken thermometers were dumped on a dirt lot near the factory. Normally, the waste would have slowly contaminated the soil and caused problems for years. In this case, however, the mercury was sent back to the US for recycling—thanks to a citizens' campaign and environmental litigation.

Hindustan Lever Ltd., a subsidiary of the Anglo-Dutch multinational Unilever, shut down a mercury thermometer factory in Watertown, New York and moved it to the hill station of Kodaikanal, in Tamil Nadu in 1983. It imported mercury from the US and exported the finished thermometers back to the US. By 1997, the plant was making 125,000 thermometers every year.

When liquid mercury is spilled, it forms droplets that can accumulate in the tiniest of spaces, which then emit vapours. Health problems caused by mercury depend on how much of it has entered the body. Hence, regardless of quantity, all mercury spills should be treated seriously.

Local people in Kodaikanal discovered the waste in March 2001. The company had sold the mercury to a scrap dealer over four years. The community immediately purchased the mercury as evidence and informed the Tamil Nadu Pollution Control Board (TNPCB). The company initially denied it was their waste, but later admitted that it had been 'inadvertently sold'.

The TNPCB ordered the company to collect and dispose of the waste. Environmental lawyers found that the company's proposed clean-up plan contained numerous flaws and lacked important environmental safeguards.

The plan did not adequately control public access to the contaminated area and would have allowed persons cleaning up the waste to work without respirators. In addition, the plan would have covered the contaminated area with a tarpaulin sheet, which would trap hazardous mercury vapours that would be undetectable by the plan's instruments.

In March 2003, the TNPCB ordered the company to ship the mercury-laden waste back to the US for proper disposal. Under close supervision, the waste was packed into containers and brought to a port in Tamil Nadu. A ship, carrying some 300 tons of mercury-contaminated waste, departed May 2003 for Bethlehem Apparatus, a mercury recycling plant in Pennsylvania in the US.

The story is not over yet. At a public hearing held in September 2002, the former workers of the factory said they had never been told about the poisonous nature of mercury and demanded that their health problems be addressed. Meanwhile, studies have shown that mercury levels in the Kodaikanal atmosphere are abnormally high.

In September 2004, the Hazardous Waste Monitoring Committee appointed by the Supreme Court visited Kodaikanal. It has ordered a clean-up of the site as well as medical aid for the affected people.

Just imagine if all waste is sent back to the countries of origin! Industrialized countries would become the dirtiest places on Earth and developing countries would be a lot cleaner, until of course their own industries get into the act!

Source: The website *www.elaw.org* and Subramanium (2004).

There are of course responsible groups in the business world who are concerned about the environment and would like to see a more environment-friendly attitude among corporations. An example is the Coalition for Environmentally Responsible Economies (CERES), formed in 1989 to promote responsible corporate environmental conduct. It (CERES) is a network of over 80 organizations including environmental groups, public interest and community groups, and also investors.

Shortly after the *Exxon Valdez* disaster, CERES drew up a set of 10 principles to represent an environmental ethic for corporations (Box 20.6). CERES encourages companies to endorse these principles. By 2002, over 70 US companies, including large multinational corporations, mid-sized companies, and small firms had endorsed the CERES Principles.

Box 20.6 Hope: Greening the corporation

The Coalition for Environmentally Responsible Economies (CERES) offers the following principles as a comprehensive statement of environmental values for businesses within any sector of industry.

They are intended to help companies formalize their dedication to environmental awareness and accountability, and actively commit to an ongoing process of continuous improvement in environmental performance, dialogue, and comprehensive, systematic reporting.

The 10 CERES Principles are:

- Protection of the Biosphere: We will reduce and make continual progress toward eliminating the release of any substance that may cause environmental damage to the air, water, or the Earth or its inhabitants. We will safeguard all habitats affected by our operations and will protect open spaces and wilderness, while preserving biodiversity.
- Sustainable Use of Natural Resources: We will make sustainable use of renewable natural resources, such as water, soils, and forests. We will conserve non-renewable natural resources through efficient use and careful planning.
- Reduction and Disposal of Wastes: We will reduce and where possible eliminate waste through source reduction and recycling. All waste will be handled and disposed of through safe and responsible methods.

- Energy Conservation: We will conserve energy and improve the energy efficiency of our internal operations and of the goods and services we sell. We will make every effort to use environmentally safe and sustainable energy sources.
- Risk Reduction: We will strive to minimize the environmental, health, and safety risks to our employees and to the communities in which we operate through safe technologies, facilities and operating procedures, and by being prepared for emergencies.
- Safe Products and Services: We will reduce and where possible eliminate the use, manufacture, or sale of products and services that cause environmental damage or health or safety hazards. We will inform our customers of the environmental impacts of our products or services and try to correct unsafe use.
- Environmental Restoration: We will promptly and responsibly correct conditions we have caused that endanger health, safety, or the environment. To the extent feasible, we will redress injuries we have caused to persons or damage we have caused to the environment and we will restore the environment.
- Informing the Public: We will inform in a timely manner everyone who may be affected by conditions caused by our company that might endanger health, safety, or the environment. We will regularly seek advice and counsel through dialogue with persons in communities near our facilities. We will not take any action against employees for reporting dangerous incidents or conditions to management or to appropriate authorities.
- Management Commitment: We will implement these Principles and sustain a process that ensures that the Board of Directors and Chief Executive Officer are fully informed about pertinent environmental issues and are fully responsible for environmental policy. In selecting our Board of Directors, we will consider demonstrated environmental commitment as a factor.
- Audits and Reports: We will conduct an annual self-evaluation of our progress in implementing these Principles. We will support the timely creation of generally accepted environmental audit procedures. We will annually complete the CERES Report, which will be made available to the public.

Source: The CERES website *www.ceres.org*

In 1997, CERES and the United Nations Environment Programme (UNEP) convened the Global Reporting Initiative (GRI) to develop and disseminate globally applicable Sustainability Reporting Guidelines. These are meant for voluntary use by organizations, including companies, for reporting on the economic, environmental, and social dimensions of their activities, products, and services: 322 organizations in 31 countries are now publishing reports about themselves according to the GRI Guidelines.

Another initiative is the World Business Council for Sustainable Development (WBCSD), a coalition of 170 international companies united by a shared commitment to sustainable development via the three pillars of economic growth, ecological balance, and social progress. Its members are drawn from more than 35 countries and 20 major industrial sectors.

WBCSD also benefits from a global network of 50 national and regional business councils and partner organizations involving some 1000 business leaders globally. Its activities reflect its belief that the pursuit of sustainable development is good for business and business is good for sustainable development.

What lies ahead?

In India, the trend toward 'judicial activism' may not continue for long. There are a large number of pending cases and the judges are aware that judicial intervention alone cannot bring about systemic change in a country as large and complex as India. Though it is one of the many tools for promoting the environmental agenda—public education, lobbying, and political action by organizing citizens may be far more effective in some cases.

In the long run, there must be the political will to conserve the environment, demonstrated in the form of substantial budgetary allocations for enforcing environmental laws. Courts can only supplement, and not replace, the administrative enforcement mechanism.

A quick recap

- India has many laws to protect the environment, but implementation has been poor.
- Responding to PILs by citizens and groups, the courts have been taking a proactive role in enforcing environmental laws in India.
- We have constitutional provisions to safeguard the environment.
- The Bhopal Gas Tragedy gave impetus to the passing of several environmental laws incorporating regulatory mechanisms.
- The Environment (Protection) Act of 1986 is an enabling law, providing powers to the executive to frame various rules and regulations.
- An Environmental Impact Assessment (EIA) is mandatory for certain types of projects.
- We have specific laws concerning air, water, forests, wild life, hazardous waste, etc., but the enforcement of all these laws is poor.
- Environmental ethics try to define the moral basis of environmental responsibility.
- Some initiatives have been taken to promote environmental ethics in industry on a voluntary basis.

What important terms did I learn?

Environmental ethics
Environmental Impact Assessment
Judicial activism
Public interest litigation

What have I learned?

Essay question

Write an essay on judicial activism in India. Give examples of cases in which the courts have assumed the role of policy maker, educator, or administrator, and have given directives that the government or the enforcement agency should have issued in the first instance.

Short-answer questions

1. Outline the Indian constitutional provisions regarding the environment.
2. Describe the beginnings of environmental legislation in India.
3. Explain the requirement of an Environmental Impact Assessment for certain types of projects.
4. What are the main provisions of the following Acts?
   (a) Environment (Protection) Act of 1986
   (b) Air (Prevention and Control of Pollution) Act of 1981
   (c) Wildlife Protection Act of 1972
   (d) Water (Prevention and Control of Pollution) Act of 1974
   (e) Forest Conservation Act of 1980

5. Give two examples of Public Interest Litigation and judicial activism in enforcing environmental laws in India.
6. Explain environmental ethics and the three different ways of viewing environmental responsibility.
7. What are the international initiatives to bring environmental ethics into industry?

How can I learn by doing?

Get in touch with any individual or group of environmental lawyers in your area. Volunteer your services for a few weeks and help them with a case. Write a report on the case. What were the issues involved? Who are the parties to the case? Which of the laws are involved?

What can I do to make a difference?

Acquaint yourself with the basic environmental laws and take action when you come across any violation. File complaints with the appropriate authorities.

Where are the related topics?

Almost all the chapters are related to the issues discussed here. The cases of Plachimada (Chapter 7, Box 7.2), the Aravalli Hills (Chapter 10, Box 10.6), pollution in Delhi city (Chapter 11, Box 11.5), pesticides in packaged water (Chapter 12, Box 12.3), the *Exxon Valdez* (Chapter 14, Box 14.7), Bhopal (Chapter 16, Box 16.1), Agent Orange (Chapter 16, Box 16.3), Minamata (Chapter 16, Box 16.4), Love Canal (Chapter 16, Box 16.6), Cochabamba (Chapter 18, Box 18.6), Narmada (Chapter 19, Box 19.1) and the cases discussed in this chapter are relevant.

What are the deeper issues that go beyond the book?

Is judicial activism to be welcomed, given that the governments are not implementing the existing environmental laws properly? How long can the courts continue to play the role of administrators? Will the fear of the courts force the governments to enforce the laws strictly or will they let the court take on that task?

Where can I find more information?

- Divan and Rosencranz (2001) give an authoritative and detailed account of environmental law, cases, and policy in India.

What does the Muse say?

A poem for reflection by the Chinese philosopher Lao Tzu:

The more laws and restrictions there are,
The poorer people become,
The sharper men's weapons,
The more trouble in the land.

The more ingenious and clever men are,
The more strange things happen,
The more rules and regulations,
The more thieves and robbers.

MODULE 8

Fieldwork

CHAPTER 21

Project Work

The way to get started is to quit talking and begin doing.

Walt Disney

Why should I carry out project work?

In this book, we have covered a wide variety of environmental problems facing India and the world. You may not have imagined that the state of the environment was so serious. You may not even believe that all that has been said in this book is true.

How does one verify and validate the various descriptions and ideas presented in the book? One way of doing this would be to embark on project work. It can be either undertaken individually or as a group activity.

Project work gives you an opportunity to find out for yourself how things are in the real world. In doing the project, you will involve yourself in a micro-level situation. You will then begin to appreciate the real nature and complexity of the environmental problems facing us today. You will also have the satisfaction of moving from being a part of the problem to becoming a part of the solution.

A group project also gives you to chance to work together with others who may not always agree with you. The lessons you will learn in working in a team will be of great value to you in your career and of course in your life as well.

How should I choose a project?

Choose a project that you can complete in the time that you have. It is far better to do a thorough job with a small project, instead of taking on a big one and leaving it unfinished.

Here are some criteria you can use while choosing a project. They are given in the form of a checklist of questions you can ask yourself:

- Attraction of the topic: Do I have a basic interest in the topic? Am I going along just because others have chosen the topic?
- Usefulness: Will I be able to make a modest and yet a useful contribution to a given real-life situation?
- Specialist knowledge needed: Does the project need any specialist information and if so, can I equip myself with it before I begin work? Are there accessible experts whom I can consult?
- Work by other groups: Are there reports prepared by others who have been to the site earlier? Does the project sound interesting, judging from these reports?
- Measurability: Will I be able to measure the impact of my work?

- Accessibility and facilities: Is the project location easily accessible? If I am carrying out the project on a full-time basis, is there any accommodation available at or near the project site?
- Weekend projects: If it is going to be a weekend project, will I be able to do some work each week, adding up to a meaningful whole?
- Future work: Given the short time frame, it may not be possible for me to achieve all my objectives; is there scope to do things later on as follow-up measures?

At the end of each chapter in this book, suggestions for project work have been given. More ideas are listed below:

- Documenting environmental resources like a river, forest, grassland, hill, or mountain in a specific area
- The study of a polluted site in an urban, rural, industrial, or agricultural area
- The study of simple ecosystems: ponds, river, hill slopes, etc.

Project ideas classified by discipline of student:

- Science: Study of plants, insects, birds, etc., in a given area; tracing the origins of different plants, trees, vegetables, and fruits available in an area; study of any seed bank, preserving indigenous species; testing water quality in an area, resource mapping of a village, etc.
- Design, fine arts, crafts, etc.: Designing products with waste material; improving the design of solar energy products like solar lanterns and solar cookers
- Philosophy: Studies on environmental ethics
- Economics: Study of alternative economic models
- Architecture: Study buildings designed using ideas of ecological architecture
- Performing arts: Effective ways of communicating with people and mobilizing them to preserve our environment (street theatre performances or designing a brochure on a specific topic)

Should I seek a project within an organization? It is best to work as a volunteer or a project trainee with an organization that has been working in the area or on the site. It could be a voluntary organization (an NGO), a government department, or even a company that is funding and managing an environmental or social project.

How should I prepare myself for the project work?

- Define objectives: Start with a list and revise it as you go along.
- Formulate questions: List the questions you expect to answer through the project.
- Conduct a literature survey: Do a thorough study of the literature available on the topic (books, reports, news items, etc.).
- Keep notes: During the literature survey and throughout the project period, write down all relevant information. A good practice is to use cards of about 10 cm x 15 cm to write your notes. Each card should contain one main idea and should carry the name of the topic/sub-topic at the top right-hand corner. This will make it easy to shuffle and arrange the cards according to the topic. It will enormously simplify the writing process at the end. Remember to assign a card for each reference and write down the full bibliographic details right when

you access the item the first time. Otherwise you will be desperately searching for details while writing the report.

- Meet the concerned persons: If you are going to interview a person, go prepared with a set of questions. Take a tape recorder along, but take the person's permission before recording the interview.
- Obtain any permission needed for entering the project site.
- Letter from college/university: Always carry a letter of introduction from your institution and your identity card.
- Collect any equipment that you may need for the project.

How should I conduct the field study?

- Plan each visit or trip meticulously.
- Keep a diary and make detailed notes on cards at the end of each day.
- Always be punctual and keep any deadlines you have agreed to.
- Keep your guides (on site and at your institution) informed of your progress and seek their advice.
- At the end of the project, make sure that you return all borrowed material and leave the organization on a pleasant note.
- Take a certificate for your work from the organization. Your institution may ask for it and it will also be useful to you later on.
- Remember to give the organization a copy of your report later on.

How should the report be written?

- Time frame: Allow yourself enough time to write a good report. Ultimately, what you have done only becomes evident through your report.
- The general format of the report could be as follows (but follow any guidelines set by your institution or guide):
  - ♦ Executive summary
  - ♦ Contents
  - ♦ Acknowledgements
  - ♦ Preface
  - ♦ Introduction: the problem or ecosystem you are studying in brief and the objectives of study
  - ♦ Background: literature survey
  - ♦ Observations and results
  - ♦ Conclusions and recommendations
  - ♦ References
  - ♦ Photographs
- Language and style: Read the slim volume *Elements of Style* by Strunk and White (1972). It is a classic and contains many useful tips. Keep the language as simple as possible. If your language abilities are weak, get the draft corrected by a professional.
- Word processing: Do not depend too much on the software to correct the grammar and usage.
- Approval of the draft report: Get the draft approved by the organization and your guides.

- Revision and final version: Check for typographical errors.
- Printing and submission: Follow any guidelines given on the size, number of copies, etc.

How should I prepare for a viva-voce examination, if there is one?

- Prepare overhead transparencies or a computer presentation.
- Each slide should have not more than 10 lines of text. Do not read out what everyone can see. The points on a slide are only to trigger your memory.
- Prepare ahead for possible questions.

In which other ways can I use the project work and the report?

- A good project report can be used in job interviews and in applications for admission to courses.
- If you are looking for a career in environment, send an electronic copy of the report to organizations that have openings.
- Write an article on your work for the appropriate journal, magazine, newspaper, or website.

What does the Muse say?

The following is part of the poem *Little Gidding*, one of the *Four Quartets* by the Anglo-American critic, dramatist, and poet, T.S. Eliot (1888–1965):

We shall not cease from exploration
and the end of all our exploring
shall be to arrive where we started
and know the place for the first time.

MODULE 9

Conclusion

CHAPTER 22

Conclusion: Another World is Possible

Anyone who believes
exponential growth can go on forever
in a finite world is
either a madman or an economist.

Kenneth Boulding

If the world is saved,
it will be saved by people with changed minds,
people with a new vision.
It will not be saved by people
with old minds and new programs.
It will not be saved by people
with the old vision but a new program.

Daniel Quinn

Box 22.1 Too much of a success: Turning barren land into forest

We began this book with the crisis in Kalahandi, Orissa, a story of environmental degradation leading to economic deprivation and migration. We end the book with another story from the same region. This is a story of hope, of what is possible when a community gets together to save the environment. Yet, it is also a story of despair for it shows that too much of success even in environmental conservation can bring trouble in its wake!

In the story, 'The Man Who Planted Trees' (Module 1, Prologue: Ecosystems), you would have read the following passage:

> The Forest Department noticed the change and thought that a natural forest had come up by itself. One day, a forest ranger told Elzeard that he could not light a fire in his hut, because the natural forest had to be protected! Elzeard then moved further away, but continued his work.

Something similar happened in reality in the village of Kendupatti. In the mid-1980s, the people of this village started reviving the forests around them. They did not grow the forest to make money. They just felt the need to plant trees to improve the area.

The villagers' efforts succeeded beyond belief and, by 1992, there were half a million trees over 140 acres of government land. In the normal course they would have used the trees selectively for construction or other purposes without destroying what they had created. Their huge success, however, brought problems in its wake.

When the Forest Department discovered the trees, they promptly claimed them and appointed a guard! Obviously, the commercial value of the trees was a great attraction and the trees were on government land anyway. Other hitherto uninvolved people like the sarpanch of a neighbouring village wanted a piece of the cake.

The outraged villagers tried to protect the forest and keep the encroachers and the forest guards away. The dispute led to charges being filed against the villagers!

Who owns the forest? The villagers who grew it with their own efforts? Or the government, which owns the land? Can the Forest Department take control and sell the timber? And the greatest horror of all: Should the villagers be punished for creating a beautiful forest? These questions are as yet unanswered.

Source: Sainath (1996)

What does the Kendupatti story tell us?

The story shows that simple village folk have an interest in maintaining tree cover and can do wonders when they come together. It demonstrates also the hurdles that such a community faces in its quest for environmental regeneration.

This book has been full of disquieting stories about the state of the environment, peppered with some hopeful news. In Chapter 18 we saw several successful examples of community-based water harvesting. We end the book with more stories of hope, examples of what committed individuals have done in this country in the cause of environmental conservation.

Farming, the organic way

Thursday is visitors' day at Narayan Reddy's five-hectare farm near Bangalore. Many people, some from very far away, come to visit this farm. Reddy grows the same kinds of crops that other farmers in the area cultivate: coconuts, fruits, vegetables, mulberries, finger millet, and rice. Also, like many other farmers in the area, he keeps dairy cows and rears silkworms. Why then do people want to visit this farm?

Reddy practises organic farming and has made his farm almost self-sufficient. All the food for the family is grown on the farm. In fact, except for tea and white sugar, he does not get anything else from the 'outside.' He also grows coffee!

Many activities on the farm are interlinked. The cows give milk and dung. The dung goes into a biogas plant, which produces enough gas to take care of all the cooking. The biogas plant also produces slurry, which makes good fertilizer. This is used to grow Napier grass, which becomes fodder for the cows.

He has dispensed with chemical fertilizers and pesticides. In addition to the biogas slurry, he makes compost using farm and kitchen wastes. He uses earthworms to speed up and improve the composting process. He makes his own insecticide from ground neem seeds, cow urine, and other materials available on the farm.

When Reddy shifted to organic farming, his yields went down a little, but because his costs also went down, his profit remained more or less the same. With time, as the soil improved, and he started growing a number of different crops each season, his overall profit increased.

Reddy has combined the best of traditional methods with modern technology, with great success. He learns from others and experiments and innovates all the time. He has implemented many new ideas in organic farming.

Narayan Reddy has perhaps found the solution to our agricultural crisis. We can only hope that many farmers will learn from him and shift to organic farming.

Creating a forest

The story began in Kasargode, Kerala in the early 1970s. Abdul Karim spent his savings on a piece of wasteland on a rocky laterite hill. The barren region had hardly ever seen water and was barely habitable. His family and friends thought he was out of his mind. Worse was to come. Karim started boring deep holes into the hard rock to plant saplings.

Abdul Karim had been running a travel and placement service in Nileswar for Keralites eager to go to the Gulf. His wife came from the village of Puliyamkulam, about 20 km away. While visiting Puliyamkulam, he came by his hill.

'I would walk around the area and see barren hill sides,' he says. 'It was a heartache of a sight and the pull on me was strong. I suddenly realised that I had often dreamt of the Kaavu of India's collective memory. They were the sacred groves that every village had once upon a time. I had been told of them as a child. I think I had subconsciously yearned for one.'

On an impulse he bought two hectares of barren rock with a pathetic well and started planting trees. None of them survived the first year. Abdul Karim, however, would not give up. He started planting the saplings all over again. He kept trudging great distances just to fetch some precious water for his little plants.

Then one day, during the third year, a miracle happened. A defiant sapling burst through the scorched surface. Encouraged, he began planting even more. And slowly, Nature began to relent and the trees began to multiply.

When the trees started growing, the water level in the well rose. Abdul Karim had meanwhile bought more of the land, ending up with 25 hectares. He now began to plant the whole area in a frenzy. He chose a variety of plants plucked from the wild and let nature do the rest.

Birds began to arrive and disseminate all types of seeds. Weeds grew and amidst them rare herbs and medicinal plants—none chosen by Abdul Karim. Water levels in villages within a 10 km radius rose. The once barren hill was now a forest and a water sponge. He now lives in a house in the forest.

He has never weeded his acres, never lopped a tree, never swept the leaves, never hunted game, never selected a species and of course, never used a chemical of any kind.

'Deep inside everyone of us is a call to the wild,' he says. 'Much of the impatience, discontent or violence around us is due to a lack of opportunity to reconnect with where we came from. For sanity and generosity of spirit, we should be able to witness nature at its unceasing, rejuvenating work.'

It is said that he is still poor with a growing family. One hopes that he will get support to continue his work. All over India, there are many Abdul Karims, quietly planting trees.

Banishing plastic waste

If Ahmed Khan has his way, there will soon be no plastic waste left in Bangalore. It will all be disposed off—under the roads. Khan's company has developed Polyblend, a fine powder of recycled plastic with a few proprietary additives. This powder is mixed with the bitumen that is used to lay the roads.

Khan has already laid 40 km of road in 12 different areas of Bangalore. These have been tested for over 18 months including two monsoons. Polyblend appears to have strengthened the roads by enhancing the bitumen's bonding ability, which has made the roads last longer by rendering them more impervious to water. Khan has a contract to lay 800 km of road in the city.

56-year-old Ahmed Khan has been producing plastic sacks for 20 years. About eight years ago he realized that plastic waste was a real problem. He set out to create an opportunity out of the situation.

Khan developed Polyblend and sought the help of Prof. Justo of Bangalore University, who in turn got the students of R.V. College of Engineering, Bangalore, to experiment with various blends of bitumen and Polyblend. The experiments showed that Polyblend could increase road life by a factor of three, offsetting the extra cost.

Khan managed to get the support of the then Chief Minister of Karnataka. In 2002, Khan received the contract lay about 40 km of road from the Bangalore City Corporation using his method.

Against the price of Rs 0.40 per kg that rag pickers had been getting for plastic waste, Khan now offers Rs 6. Barring rigid plastics, he accepts any film waste. If the corporation continues with this method, Khan will soon be running short of waste.

Environmentalists are against the incineration of plastic, which release dioxins. Technologists point out that many of the recycling solutions, like recovering liquid fuels from waste plastic, yield less in energy terms than what was put into the process.

Khan has perhaps found the ideal and acceptable solution to the problem of plastic waste. If his idea spreads to other cities, we might still avoid being smothered by plastic waste.

Building for the people

There are 13 of them, each custom-designed for the location. In some, local artists have painted murals on the walls. Some communities let out the first floor space for private functions. In some there are gyms and in others night schools or a creche and so on. They are all public toilets in the slums of Pune! They have been designed and built by architects Pratima Joshi and Srinanda Sen of Shelter Associates.

The new approach to building public toilets came about when Ratnakar Gaikwad, the Commissioner of the Pune Municipal Corporation, realized that people had to be consulted and involved in the design and maintenance of toilets. Instead of giving the job to contractors, he asked NGOs to build the toilets. Shelter Associates, which had already built low-cost houses for the poor, was one of them.

Pratima and Srinanda began by setting up 'Baandhinis' or women's local groups in the project areas that would research, counsel, plan, build, administer, and maintain the community toilets. The Baandhinis discussed and narrated their needs and the architects drafted plans keeping costs and other factors in mind.

Many innovations resulted from the discussions: septic tanks built of ferro-cement in situ; a water connection for each toilet; a multi-purpose space on the first floor; absolute separation of men's and women's enclosures; hard-wearing gently-sloping stone slabs instead of ceramic floor tiles; airy and bright spaces; and, finally, the delightful innovation of the 'baby channel' within the ladies' section. Here children can safely squat within their mothers' view and the waste can be driven off with a little water. No more squatting in the streets!

The Baandhinis supervised quality, kept a vigil on materials, maintained accounts and rallied everyone into active participation. Each of the user families pays about Rs 20 per month to the Baandhini and this meets the salary of a caretaker, maintenance costs, and the electricity bills. The Baandhinis have travelled to Solapur and Sangli to spread the idea.

These public toilets may not be the ultimate solution to the sanitation problem, but right now they have made a great difference to the people, especially the women. The Baandhinis will hopefully continue to keep the toilets clean and maintain them properly.

These four cases are just a sample of the many inspiring stories of individual and group efforts in India to save the environment and improve the quality of life.

Where can I find more information?

The website *www.goodnewsindia.com* is the source of the stories described above. The site itself is the initiative of D.V. Sridharan, who was tired of hearing bad news all the time and began collecting and sharing stories of hope, describing the good things that happen in the country. Another site, *www.indiatogether.org*, has many interesting stories of Indian events.

With so many hopeful stories, why should we worry about India's environment?

While we can draw some comfort from these stories, they do not add up to any real change in the country. Part of the reason is the NIMBY syndrome. NIMBY stands for 'NOT IN MY BACKYARD' and describes the prevalent attitude of the majority that wants to enjoy modern comforts and consumer products, but is not willing to deal with the adverse effects of such a lifestyle. They are happy as long as the waste that they produce is not dumped in their backyard!

It is the NIMBY syndrome that is preventing us from solving many environmental problems. Individuals must think about their own lifestyles and ecological footprints.

What is my ecological footprint?

This is a question that you must ask yourself. As we saw in Chapter 1, your ecological footprint is the amount of productive land area required to sustain you on this Earth.

With current population levels, there are about 1.9 hectares of productive area per person in this world. However, the average ecological footprint is already 2.3 hectares. That is, we already need 1.5 Earths to live sustainably.

The citizens of the US have the largest ecological footprint at 9.57 hectares per person. If we all start consuming at that rate today, we will need 25 Earths. We survive because there are still countries like Bangladesh, which has a footprint of just 0.5 hectare and because we use fossil fuels, which represent 'ghost acreage' from the distant past. What will happen, however, when India and China reach population figures of 1.5 billion each?

The conclusion is:

If the Earth's environment and humanity have to survive, we cannot continue to exploit nature the way we have done over the past 100 years or more.

The people in the rich countries cannot continue with their wasteful, patently unsustainable consumption and those in the poor countries cannot continue to yearn for such lifestyles.

Humankind has to move towards a simpler life in harmony with nature.

It will indeed be very difficult for people in Western countries to give up their cars and consumer products. It will be equally difficult for the people of developing countries to stop dreaming of catching up with the West.

What is your ecological footprint? How much do you consume and where do these items of consumption come from? What is the impact of your consumption on other organisms and the environment? Why does your mind yearn for the consumer products and lifestyles advertised in the media?

A profound change is possible only if all of us change our mindset.

What does 'changing one's mindset' mean?

The great scientist Albert Einstein said, 'You cannot solve a problem with the same mindset which created it in the first place'. That is why token responses like buying recycled paper, using a CFC-free refrigerator, or driving an electric car are not going to save the environment.

What is it that we are trying to protect the environment from? Is it to protect it from 'them', the polluters and the destroyers? Perhaps we must first protect the environment from ourselves. It is our own lifestyle, greed, selfishness, and lack of awareness that is the starting point of all the problems. In short, as the cartoon character Pogo says, 'We have seen the enemy and it is us.'

We mistakenly see a difference between other beings and ourselves. When we truly begin to see ourselves as an indivisible part of the Universe, there would be a fundamental shift in our mindset.

What then can I do in my daily life?

You can begin with the following:

- Step out of the process of consumerism and preserve your individual freedom. Exercise your choice in how you live.
- Determine your genuine needs for living a rational, reasonable, and truly contented life and do not be misled by artificial needs imposed by society.
- Believe that the collective interests of all organisms in the world are above the interests of any particular entity, race, or group of beings. Consider the order of importance of interests as global, regional, national, community, family, and then self.
- Develop a sense of universal responsibility, caring for others and for the coming generations.

Mahatma Gandhi captured this thought in just a few words:

Be the change you wish to see in the world.

Begin changing yourself and you will then find that:

Another world is possible!

Appendices

Appendix A1

Bibliography

This list includes all the references cited in the text as well as sources from which many ideas have been borrowed in the course of writing this book. The author would like to thank all the authors whose painstaking work has furthered the cause of environment.

Agarwal, Anil and Sunita Narain, eds. (1997). *State of India's Environment-4, Fourth Citizens' Report : Dying Wisdom: Rise, Fall and Potential of India's Traditional Water Harvesting Systems*. New Delhi: Centre for Science and Environment.

Agarwal, Anil, Sunita Narain and Indira Khurana, eds. (2001). *Making Water Everybody's Business: Practice and Policy of Water Harvesting*. New Delhi: Centre for Science and Environment.

Agarwal, Anil, Sunita Narain and Srabani Sen, eds. (1999). *State of India's Environment-5, Fifth Citizens' Report, Part I: National Overview and Part II: Statistical Database*. New Delhi: Centre for Science and Environment.

Aggarwal, Partap (2004). 'Punjab Pilgrimage'. Private communication.

Alvares, Claude, ed. (1996). *The Organic Farming Sourcebook.* Goa: The Other India Press.

Annan, Kofi A. (2002). 'Toward a Sustainable Future', *Environment* 44 (7), pp. 10–15.

Anuradha, R.V. (2000). *Sharing the Benefits of Biodiversity: The Kani—TBGRI Deal in Kerala, India*. Pune: Kalpavriksh.

Appenzeller, Tim (2004). 'The Case of the Missing Carbon', *National Geographic* 205(2), pp. 88–117.

———. (2004). 'The End of Cheap Oil'. *National Geographic* 205(6), pp. 80–109.

Ayres, Ed (1999). *God's Last Offer: Negotiating for a Sustainable Future*. New York: Four Walls Eight Windows.

———. (2004). 'The Hidden Shame of the Global Industrial Economy', *World Watch* 17(1).

Banerjee, Ruben (2001). *The Orissa Tragedy: A Cyclone's Year of Calamity*. New Delhi: Books Today.

Barik Satyasundar, Ranjan Panda, Deepa Kozhisseri, and Mahesh L. (2004). 'The Job Queue', *Down To Earth* 13 (13), pp. 26–34.

Bavadam, Lyla (2004). 'In Troubled Waters', *Frontline* 21(1), pp. 112–13.

———. (2004a). 'Sardar Sarovar Project: Rehabilitation Realities', *Frontline* 29, pp. 43–44.

———.(2004b). 'Sardar Sarovar Project: Flood Of Fears', *Frontline* (August 13), pp. 94–95.

Beans, Bruce E. (2002). 'The Waste That Didn't Make Haste', *Washington Post,* 17 July.

Bell, Ruth Greenspan, Kuldip Mathur, Urvashi Narain, and David Simpson (2004). 'Clearing the Air: How Delhi Broke the Logjam on Air Quality Reforms', *Environment* 46 (3), 22–39.

Bhatia, Bela (2001). 'Jadugoda: Fighting an Invisible Enemy', in *The Hindu Survey of Environment 2001, The Hindu*, Chennai, pp. 129–35.

Bhatia, Gautam (1991). *Laurie Baker: Life, Work, Writings*. New Delhi: Viking Penguin.

Botkin, Daniel and Edward Keller (1995). *Environmental Science: Earth as a Living Planet*. New Yorl: John Wiley.

Brandon, C., K. Hommann, and N.M. Kishore (1995). 'The Cost of Inaction: Valuing the Economywide Cost of Environmental Degradation in India'. Paper presented at the UNU Conference on the Sustainable Future of the Global System, Tokyo, October 16–18.

Brown, Lester R., Janet Larsen, and Bernie Fischlowitz-Roberts (2002). *The Earth Policy Reader*. The Earth Policy Institute. Indian Edition (2003). Hyderabad: Orient Longman.

Bunsha, Dionne (2004). 'Developing Doubts', *Frontline* (June 18), pp. 44–45.

Carson, Rachel (1962). *Silent Spring,* Indian Edition. Goa: Other India Press.

Chiras, Daniel D. (2001). *Environmental Science: Creating a Sustainable Future* 6e. Sudbury, MA: Jones and Bartlett Publishers.

Chowdhury, Namrata (2004). 'Ship-breaking Industry: Need for Global Solutions'. *The Hindu Survey of the Environment 2004, The Hindu*, Chennai, pp. 125–29.

CIIR (1993). *Biodiversity: What is at Stake?*. London: Catholic Institute for International Relations.

Comte, Douglas L. (2004). 'A Year of Extremes: 2003's Global Weather', *Weatherwise,* (March/April).

Costanza, R. et al. 'The Value of the World's Ecosystem Services and Natural Capital', Nature (387).

Centre for Science and Environment (CSE) (2000). *Our Ecological Footprint: Think of Your City as an Ecosystem*. New Delhi: Centre for Science and Environment.

Cunningham, William P. and Barbara Woodworth Saigo (1995). *Environmental Science: A Global Concern* 3e. Boston: Wm. C. Brown Publishers.

Cunningham, William P. and Mary A. Cunningham (2004). *Principles of Environmental Science: Inquiry and Applications* Indian Reprint, New Delhi: Tata McGraw-Hill Pub. Co.

Das, Anjana and Vikram Dayal (1998). 'An Energy Shortage Closer to Home', *The Economic Times*, 15 January.

Das, Prafulla (2001). 'Orissa Turtles: Dance of Death'. *The Hindu Survey of the Environment 2001, The Hindu*, Chennai, pp. 149–54.

Dasgupta, Kumkum (2001). 'Poverty Amidst Plenty: The Punjabi Tale', *UNESCO Courier,* January (this can also be accessed from the website *http://www.unesco.org/courier/2001 01/uk/doss22.htm*).

de Villiers, Marq (1999). *Water*, Toronto: Stoddard,.

Divan, Shyam and Armin Rosencranz (2001). *Environmental Law and Policy in India* 2e. New Delhi: Oxford University Press.

DTE (2003a). 'Gulp: Bottled Water Has Pesticide Residues', *Down To Earth* 11(18), pp. 27–34.

DTE (2003b). 'Sacred Groves: For Fragments' Sake', *Down To Earth* 12(15), Extra Supplement.

DTE (2004). 'Industrial Water Use: Overused, Underrated', *Down To Earth*, pp. 61–80.

Easwaran, Eknath (1989). *The Compassionate Universe: The Power of the Individual to the Heal the Environment*. Reprinted in India in 2000 by Penguin India, New Delhi.

Ede, Piers Moore (2003). 'Çome Hell or High Water', *Alternatives Journal* 29 (1), pp. 8–9.

Enger, Eldon D. and Bradley F. Smith (2004). *Environmental Science: A Study of Interrelationships* 9e. New York: McGraw-Hill.

Fukuoka, Masanobu (1978). *The One-Straw Revolution: An Introduction to Natural Farming*. Reprinted in India in 1995 by Friends Rural Centre, Rasulia, M.P.

Gadgil, Madhav, H.N. Chandrasekhariah, and Anuradha Bhat (2001). 'Freshwater Fish: Out of Sight, Out of Mind', *The Hindu Survey of the Environment 2001, The Hindu*, Chennai, pp. 137–42.

Gibbs, Lois Marie (1982). *Love Canal: My Story*. New York: State University of New York Press.
———.(1995). *Dying from Dioxin*. Boston: South End Press.
Gifford, Eli and R. Michael Cook, eds (1992). *How Can One Sell the Air? Chief Seattle's Vision*. Summertown, Tennessee: The Book Publishing Company.
Giono, Jean (1995). *The Man Who Planted Trees,* Indian Edition. Friends of Elzeard Bouffier.
Goldsmith, Katherine (1997). 'A Gentle Warrior', *Resurgence* 181(March/April), p. 31.
Hazra, Monali Zeya (2004). 'Backyard Mess', *Down To Earth* 13 (6), pp. 38–41.
Herbert, Alan (2002). 'On Dark Streets with Solar', *Auroville Today,* December.
Hill, Julia Butterfly (2000). *The Legacy of Luna,* San Francisco: HarperCollins.
The Hindu (1983). 'Forests' Impact on Weather Patterns', July 18.
Jackson, Andrew R.W. and Julie M. Jackson (2000). *Environmental Science: The Natural Environment and Human Impact* 2e. Harlow, Essex: Pearson Education.
James, Barry (2003). 'Water: A Flood of Promises, A Trickle of Progress', *The New UNESCO Courier*, pp. 38–61.
Jamwal, Nidhi (2003). 'e-Waste', *Down To Earth* 12 (12), pp. 50–51.
———. (2004). 'Small Farms Can be Profitable', *Down To Earth* 12 (23), pp. 50–52.
———. (2004a). *The Telegraph*, 'Dharavi Dreaming', February 22.
Jamwal, Nidhi and D.B. Manisha (2003). 'The Dark Zone', *Down To Earth* 11 (22), pp. 27–41.
Johnsingh, A.J.T. and Deki Yonten (2004). 'Beautiful Bhutan', *Frontline* 21 (18), pp. 65–72.
Joshi, Sopan (2001). 'Children of Endosulfan', *Down To Earth* 9 (19). pp. 28–35.
———. (2004). 'Inevitable Tragedy', *Down To Earth* 13 (4). pp. 13–22.
Kalyanwala, Shveta (1997). *The Chilika Lake Adventure*. New Delhi: Centre for Science and Environment.
Katakam, Anupama (2001). 'On the Scrap Heap', *Frontline* 18 (15), pp. 89–92.
Koenig, Kevin (2004). 'Çhevron-Texaco on Trial', *World Watch* 17 (1), pp. 10–19.
Kothari, Ashish (1995). *Conserving Life: Implications of the Biodiversity Convention for India*. New Delhi: Kalpavriksh.
Krishna Kumar, R. (2004). 'Resistance in Kerala', *Frontline* 21 (3), pp. 38–40.
Lapierre, Dominique and Javier Moro (2001). *It Was Five Past Midnight in Bhopal,* Indian Edition. New Delhi: Full Circle Publishing.
Leakey, Richard and Roger Lewin (1996). *The Sixth Extinction: Patterns of Life and the Future of Humankind*. New York: Anchor Books, Random House.
Lubchenco, J. (2001). 'Understanding Human Impact on Earth's Ecosystems', in Peter H. Raven and Linda R. Berg, *Environment.* Fort Worth, USA: Harcourt College Publishers, p. 65.
Lynas, Mark (2004). 'Warning from a Warming World', *Geographical* (May), pp. 51–54.
Mabogunje, Akin L. (2002). 'Poverty and Environmental Degradation: Challenges within the Global Economy', *Environment* 44 (1), pp. 8–18.
Mahapatra, Richard (2002a). 'Consigned to Flames: Central and State Authorities Fiddle While Jharia Burns', *Down To Earth* 11 (13), pp. 25–29.
———. (2002b). 'Phulmai's Walk', *Down To Earth* 11 (14), pp. 25–34.
———. (2003). 'The Multi-billion-dollar Fuelwood Trade is the Last Resort for India's Poor', InfoChange News & Features, June (from the website *http://www.infochangeindia.org/features112.jsp*).
Mahapatra, Richard and Ranjan Panda (2001). 'The Myth of Kalahandi: A Resource-rich Region Reels under a Government-induced Drought', *Down To Earth* 9 (21).

Manu, K. and Sara Jolly (2000). *Pelicans and People: The Two-tier Village of Kokkare Bellur*. New Delhi: Kalpavriksh.

Mason, Colin (2003). *The 2030 Spike: Countdown to Global Catastrophe*. London: Earthscan.

Mathew, Sebastian (2004). 'Jambudwip: In the Eye of a Storm', *The Hindu Survey of the Environment 2004, The Hindu*, Chennai.

Menon, Meena (2002). 'Tawa Matsya Sangh: Fishing for their Lives', *The Hindu Survey of Environment 2002*. Chennai : The Hindu, pp. 99–103.

———. (2003). *The Hindu*. 'Development or Displacement?', November, 9.

Menon, Parvathi (2003). 'A Conflict on the Waves', *Frontline* 20 (6), pp. 65–73.

Middleton, Nick (1999). *The Global Casino: An Introduction to Environmental Issues* 2e. London: Arnold.

Miller, G. Tyler (2004). *Living in the Environment: Principles, Connections, and Solutions* 13e. Pacific Grove, CA Brooks/Cole–Thomson Learning,.

Mishra, Vinod, Robert D. Retherford, and Kirk R.Smith (2002). *Indoor Air Pollution: The Quiet Killer*. Hawaii: East West Center.

Ministry of Environment and Foprests (MoEF) (2004). *Draft National Environment Policy 2004,* Government of India, New Delhi.

Mollison, Bill (1988). *Permaculture: A Designer's Manual*. Tyalgum, Australia: Tagari Publications.

Morrow, Rosemary (1993). *Earth User's Guide to Permaculture*. East Roseville, Australia: Kangaroo Press.

Narain, Sunita (2003). 'Killing the High-End Killer', *Down To Earth* 12 (4), p. 5.

Nebel, Bernard J. and Richard T. Wright (1998). *Environmental Science: The Way the World Works* 6e. New Jersey: Prentice-Hall International.

Padmanabhan, Geeta (2003). *The Hindu*. 'Solar Power at Sundown', January, 23.

Pal, H. Bhisham (1982). *Sands Bloom* New Delhi: Publications Division.

Pattnaik, A.K. (2003). 'Chilika: Blue Lagoon Once More', *The Hindu Survey of Environment 2003*. Chennai: The Hindu, pp. 147–53.

Pauly, Daniel (2004). 'Empty Nets', *Alternatives Journal* 30 (2), pp. 8–10.

Pauly, Daniel and Reg Watson (2003). 'Counting the Last Fish', *Scientific American*, (July), pp. 42–47.

Polycarp, Clifford and Shams Kazi (2004). 'Managing the Woods', *Down To Earth* 13 (11), pp. 44–45.

Postel, Sandra (1992). *The Last Oasis: Facing Water Scarcity*. World Watch Environmental Alert Series. London: Earthscan.

———. (1999). *Pillar of Sand: Can the Irrigation Miracle Last?*. New York: W.W. Norton.

Price, Tom (2003). 'High Tide in Tuvalu', *Sierra,* (July/August), pp. 35–37.

Ministry of Information and Broadcasting. *India 2004: A Reference Annual*. New Delhi: Publications Division.

Raghunathan, Meena and Mamata Pandya (1999). *The Green Reader: An Introduction to Environmental Concerns and Issues*. Ahmedabad: Centre for Environmental Education.

Rauber, Paul (2003). 'The Melting Point', *Sierra,* (July/August), pp. 28–32.

Raven, Peter H. and Linda R. Berg (2004). *Environment* 4e. Hoboken, NJ: John Wiley & Sons.

Robbins, Ocean and Sol Solomon (1994). *Choices for Our Future: A Generation Rising for Life on Earth*. Summertown, TN: Book Publishing Co.

Rodes, Barabara K. and Rice Odell (1992). *A Dictionary of Environmental Quotations*. Baltimore: Johns Hopkins University Press.

Rodgers, W. A. and H.S. Panwar (1988). *Biogeographical Classification of India*. Dehradun: Wildlife Insititute of India.

Rogers, John J.W. and P. Geoffrey Feiss (1998). *People and Earth: Basic Issues on the Sustainability of Resources and Environment*. Cambridge: Cambridge University Press.

Rosegrant, Mark W., Ximing Lai, and Sarah A.Cline (2003). 'Will the World Run Dry? Global Water and Food Security', *Environment* 45 (7), pp. 24–36.

Rosset Peter, Joseph Collins, and Frances Moore Lappé (2000). 'Lessons from the Green Revolution', (accessed on the website *http://www.foodfirst.org/media/opeds/2000/4-greenrev.html*).

Sagreiya, K.P. (1994). *Forests and Forestry*. New Delhi: National Book Trust.

Sainath, P. (1996). *Everybody Loves a Good Drought: Stories from India's Poorest Districts*. New Delhi: Penguin India.

———. (2004). 'Seeds of Suicide – I', (accessed on the web site *http://www.indiatogether.org/2004/jul/psa-seeds1.htm*).

Sambandan, V.S. (2002). 'The Plastics Ban: Powerful Interests at Play', *The Hindu Survey of Environment 2002*. Chennai: The Hindu, pp. 121–24.

Sebastian, Sunny (2003). 'Rajasthan: Mine Over matter', *The Hindu Survey of the Environment 2003*. Chennai: The Hindu, pp. 117–21.

Sekhsaria, Pankaj (2002). 'Logging Off, For Now', *Frontline* 19 (1), pp. 65–67.

———. (2003). 'Andaman's Last Chance: An Opportunity Lost', *The Hindu Survey of the Environment 2003*. Chennai: The Hindu, pp. 129–31.

Sethi, Nitin (2004). 'The Bandipur-Brazil Corridor', *Down To Earth* 13 (11), pp. 49–52.

Sethi, Nitin and Vikas Parashar (2004). 'Officially Bankrupt', *Down To Earth* 13 (5), pp. 22–35.

Sharma, Kalpana (2000). *Rediscovering Dharavi*. New Delhi: Penguin India.

Shashikumar, V.K. (2004). *Tehelka*. 'The Sunderbans: Killer Assault', October, 23, pp. 18–24.

Strunk, William and E.B. White (1972). *The Elements of Style*. New York: Macmillan Publishing.

Subba Rao K.V. and R.Shankar (2003). *Story of the Oceans*. Bangalore: Geological Society of India.

Subramanium, Meena (2004). *The Hindu, Young World Supplement*. 'Mercury in our Backyard', December, 4.

Suzuki, David (1994). *Time to Change: Essays*. Toronto: Stoddart.

Swami, Praveen, Parvathi Menon, and Lyla Bavadam (2001). 'The Aftershocks', *Frontline* 18 (4), pp. 4–20.

Tompkins, Peter and Christopher Bird (1991). *Secrets of the Soil,* London: Viking Arkana.

Tripathi, Purnima S. (2004). 'Tragedy at Tehri', *Frontline,* (August 27), pp. 33–39.

Tyler, Patrick E. (2004). 'For an Environment of Peace', *Frontline* 21 (22), pp. 46–48.

United Nations Environment Programme (UNEP) (2001). *India: State of the Environment 2001*. Thailand: United Nations Environment Programme, Regional Resource Centre for Asia and the Pacific.

UNEP (2003). *Global Environment Outlook 3*. Geneva: UNEP Global Resource Information Division, Division of Early Warning and Assessment.

UNEP, IUCN, and WWF (1991). *Caring for the Earth: A Strategy for Sustainable Living,* Nairobi: UNEP.

United Nations Human Settlements Programme (UN-HABITAT) (2003). *Water and Sanitation in the World's Cities: Local Action for Global Goals*. London : UN-HABITAT and Earthscan.

Varshney, Vibha (2004). *Down To Earth* 12 (20), pp. 27–34.

Venkatesan, V. and Praveen Swami (2001). 'The Killer Earthquake', *Frontline* 18 (3), pp. 4–16.

Verma, Prabhanjan (2002). 'Fishy Deaths', *Down To Earth* 11 (4), pp. 7–8.

Vijayalakshmi, E. (2003). 'Calling the Shots: Cola Major Gets a Taste of Panchayat Power in Kerala', *Down To Earth,* (December 15), pp. 15–19.

Vyas, Sharad (2004). *Tehelka*, 'Get Ready for E-waste', November, 27, p. 5.

Wackernagel, Mathis and William E. Rees (1996). *Our Ecological Footprint: Reducing Human Impact on the Earth*. Gabriola Island, Canada: New Society Publishers.

World Commission on Environment and Development (WCED) (1987). *Our Common Future*. Oxford: Oxford University Press.

Winblad, Uno and Wen Kilama (1985). *Sanitation Without Water*. London: Macmillan.

Wright, Richard T. and Bernard J. Nebel (2004). *Environmental Science: Towards a Sustainable Future* 8e., Indian Reprint. New Delhi: Prentice-Hall of India.

Yadav, Kushal P.S. (2003). 'Dubiously Acquitted: Evidence Ignored as Endosulfan is Declared Not Guilty', *Down To Earth* 11 (23), pp. 7–8.

Yadav, Kushal P.S. and S S. Jeevan (2002). 'Endosulfan Conspiracy', *Down To Earth* 11(4), pp. 25–34.

Youth, Howard (2003). *Winged Messengers: The Decline of Birds*. World Watch Paper 165. Washington D.C.: World Watch Institute.

Zytaruk, Melinda (2003). 'Life After Oil', *Alternatives Journal* 29 (4), pp. 2–25.

Appendix A2

Resources: Web links and periodicals

Web links

Through the Internet, you can access the World Wide Web, which today contains more than six million pages on every conceivable topic. These pages are stored in thousands of computers across the world.

There are hundreds of sites on the subject of the environment. This annotated list—collected from various websites (such as, *www.soilandhealth.org*, *www.journeytoforever.org*, and *www.scidev.net*, etc.), Internet discussion groups, USENET, and various other sources—is only a sample of what you can find on the Web.

They are classified into various categories for your convenience. Many of these sites provide links to other related sites.

Happy surfing!

United Nations

United Nations Environment Programme (UNEP)
www.unep.org

UNEP-World Conservation Monitoring
www.unep-wcmc.org

Convention on Biological Diversity
www.biodiv.org

Equator Initiative: Projects that alleviate poverty through biodiversity conservation
www.undp.org/equatorinitiative

Global Environment Facility
www.gefweb.org

World Meteorological Organization
www.wmo.ch

Joint United Nations Programme on HIV/AIDS
www.unaids.org

World Health Organization
www.who.int

Food and Agriculture Organization
www.fao.org

Intergovernmental Panel on Climate Change (IPCC)
www.ipcc.ch

Biodiversity issues

Developments in Biodiversity
www.iisd.org

Global Biodiversity Outlook
www.biodiv.org/gbo/gbo-pdf.asp

Red List of Threatened Species
www.redlist.org

Plants for a Future
www.pfaf.org

Biodiversity Hotspots: Twenty-five areas of the world that contain the largest number of species under the greatest threat.
www.biodiversityhotspots.org

Global Biodiversity Information Facility
www.gbif.net

Campaigns for Biodiversity and Community
www.foei.org/publications/corporates/clashestext.html

Water

Water Supply and Sanitation Collaborative Council
www.wsscc.org

World Water Council
www.worldwatercouncil.org

World Water Forum
www.worldwaterforum.org

Energy

Oil Depletion
www.lifeaftertheoilcrash.net

Energy related documentation, especially about the fossil fuel crisis
www.dieoff.org

The Alliance to Save Energy
www.ase.org

American Council for an Energy-Efficient Economy
http://solstice.crest.org/efficiency/aceee/index.html

Centre for Renewable Energy and Sustainable Technology
www.solstice.crest.org

Energy Conservation Center of Japan
www.eccj.or.jp

International Solar Energy Society
www.ises.org

Climate change

Biodiversity and Climate Change: Key Issues and Opportunities Emerging from the Kyoto Protocol
www.wri.org/ffi/climate/kyotfrst.htm

Forest Carbon and Local Livelihoods: Assessment of Opportunities and Policy Recommendations
www.cifor.cgiar.org/publications/pdf_files/OccPapers/OP-037.pdf

Carbon Sinks in the Post-Kyoto World
www.weathervane.rff.org/features/feature050.html

Transportation in Developing Countries: An overview of greenhouse gas reduction strategies
www.pewclimate.org/global-warming-in-depth/all_reports/transportation_overview/index.cfm

Climate Action Network (CAN): The Climate Action Network is a global network of over 287 non-governmental organizations (NGOs).
www.climatenetwork.org

ChangingClimate
www.changingclimate.org

Greenhouse Gas Online
www.ghgonline.org

Information portals/Web magazines

Eco-zine: A conservation list from Bangalore, India, operated by Gopakumar and hosted off yahoogroups. To subscribe, send a mail to *eco-zine-owner@yahoogroups.com*

Environmental News Network
www.enn.com

International Environmental Law Research Centre
www.ielrc.org

International Development Research Center of Canada: It hosts a significant number of important books on development paradigms from its digital library called Booktique.
web.idrc.ca

Worldbank: Various documents and reports
www.worldbank.org

Soil and Health Library: It has a veritable collection of very good online books on topics ranging from radical agriculture and homesteading to health and hygiene.
www.soilandhealth.org

Journey to Forever: It is an NGO dealing with biofuels and has quite a few online books on farming and biofuels.
www.journeytoforever.org

Mindfully.org: The goal of mindfully.org is to provide information that people on limited budgets who would not obtain this information otherwise. Mindfully.org is to be used as a non-profit research tool. It has over 3500 files and that number is increasing daily.
www.mindfully.org

Humanity Libraries: It contains more than 5000 online books on development and many other topics.
www.payson.tulane.edu/libraries

Virtual Library on International Development
www.acdi-cida.gc.ca/virtual.nsf/pages/index_e.htm

India-specific links

Centre for Ecological Sciences at the Indian Institute of Science, Bangalore: It is the gateway to a lot of information on the research at one of India's foremost ecology research centers.
http://wgbis.ces.iisc.ernet.in

Wildlife Institute of India: An autonomous institute for research and training in conservation, ecology, and other wildlife-related disciplines.
www.wii.gov.in

Environmental Information Service (ENVIS) India: An initiative to network multiple governmental and non-governmental organizations to disseminate information about the environment. Links to various ENVIS nodes and centres are of tremendous interest as these include some of the best organizations working in the areas of environmental research and education in India. Also notable is the contact information for various governmental agencies.
www.envis.nic.in

Kalpavriksh: An active group on environmental education, research and action. It has brought out several publications.
www.kalpavriksh.org

The Wildlife Trust of India (WTI): It is an NGO working on conservation and animal rehabilitation.
www.wildlifetrustofindia.org

Wildlife Protection Society of India (WPSI): It is one of the leading conservation NGOs in India. It is notable for its work on Wildlife Law enforcement, the creation of a Wildlife Crime Database, and work on conservation of Olive Ridley Sea turtles in Orissa.
www.wpsi-india.org/wpsi/index.php

The Energy and Resources Institute (TERI)
www.teriin.org

Centre for Environment Education: A organization supported by the Ministry of Environment and Forests, Government of India. Its main aim is to provide environmental information to the public.
www.ceeindia.org

Prayogpariwar: It is a spontaneously organized network of knowledge ventures. Pioneered by Shripad Dabholkar, Prayogpariwar has provided a new way of demystifying science, which can bring about ecologically viable economic development through participatory learning.
www.prayogpariwar.net

Periodicals

Down to Earth: A fortnightly published by the Centre for Science and Environment, established by the Late Anil Agarwal.
www.downtoearth.org.in

Sanctuary Magazine; Sanctuary Cub (for children): *The Ecologist Asia* (Indian edition of *The Ecologist*, UK); All three periodicals are published by the noted environmentalist Bittu Sahgal.
www.sanctuaryasia.com

Appendix A3
Units of measure

Some Common Prefixes

Prefix and Symbol	Meaning	Example
Giga (G)	billion	1 gigaton = 1,000,000,000 tons
Mega (M)	million	1 megawatt (MW) = 1,000,000 watts
kilo	thousand	1 kilogram (kg) = 1,000 grams
centi	hundredth	1 centimetre (cm) = 0.01 metre
milli	thousandth	1 millilitre (ml) = 0.001 litre
micro	millionth	1 micrometre = 0.000001 metre
nano	billionth	1 nanometre = 0.000000001 metre
pico	trillionth	1 picocurie = 0.000000000001 curie

Numbers

1 million = 1,000,000 (10 lakh or lac)
1 billion = 1,000 million (100 crores)

Length

1 metre = 39.37 inches
1 inch = 2.54 cm
1 mile = 1.6 km

Area

1 hectare (ha) = 10,000 sq m
= 2.471 acres
1 sq km = 1,000,000 sq m
= 100 hectares

Volume

1 litre = 1000 cu. cm
1 U.S. gallon = 3.785 litres

Mass

1 metric ton (tonne) = 1000 kg
1 pound = 453.6 grams

Appendix A4
Options for you: Environmental organizations and careers

You can join an environmental organization as a volunteer and gain experience. Some of them offer short-term fellowships and regular positions. If you are enthusiastic and enterprising, you can even start your own organization.

Selected international and Indian organizations are listed below. The websites of many of these organizations provide very detailed information on areas of their interest.

International organizations

Campaign for Nuclear Disarmament *www.cnduk.org*

Climate Action Network *www.climatenetwork.org*

Conservation International (CI): A US-based group set up in 1987 to work on habitat conservation and community participation in Bolivia, Costa Rica and Mexico *www.conservation.org*

Earth Island Institute *www.earthisland.org*

Earth Policy Institute

Earthwatch *www.earthwatch.org*

Food First *www.foodfirst.org*

Friends of the Earth: Formed in the US in 1970, it is primarily a campaigning body that concentrates on public education and political lobbying. Its independent research and data gathering work on environmental matters is highly respected by governments and other NGOs *www.foe.org*

Genetic Resources Action International (GRAIN) *www.grain.org*

Global Commons Institute (GCI): A UK-based organization that was founded in 1990 after the Second World Climate Conference *www.gci.org.uk/*

Greenpeace: An international environment protest group, formed in 1971 out of an attempt by US and Canadian citizens to halt nuclear weapons testing in the Aleutian Islands. Greenpeace is best known for its non-violent direct action campaigns against environmental abuses, notably—nuclear weapons testing, the slaughter of whales, and the dumping of toxic wastes into the ocean *www.greenpeace.org*

International Coral Reef Action Network *www.icran.org*

International Ocean Institute: An organization that deals with ocean and environment issues including coastal areas *www.ioinst.org*

Natural Resources Defense Council *www.nrdc.org*

One World: An online civil society network, which campaigns for equitable and sustainable distribution of wealth amongst the world's population. *www.oneworld.net*

Rainforest Action Network *www.ran.org*

Rainforest Alliance *www.rainforest-alliance.org*

Sierra Club: A US-based NGO and thinktank dealing with development and environmental issues *www.sierra.org*

Toxics Link: An NGO working on toxic and waste substances *www.toxicslink.org*

Union of Concerned Scientists *www.ucsusa.org*

World Conservation Union (IUCN): The oldest, best-known, and most politically-savvy of all the international conservation groups. Founded in 1948 its members include 75 states, 750 NGOs, and 10,000 scientists *www.iucn.org*

World Resources Institute (WRI): A US-based organization working on environmental research and policy *climate.wri.org*

World Wide Fund for Nature (WWF) *www.panda.org*

Worldwatch Institute *www.worldwatch.org*

Indian organizations

Many of the international organizations listed above have Indian branches. In addition, there are many Indian NGOs working to save the environment. The selected ones are listed here.

Beauty without Cruelty (India) *www.bwcindia.org*

Centre for Environment Education *www.ceeindia.org*

Centre for Science and Environment *www.cse.org*

C.P.R. Environmental Education Centre *www.cpreec.org*

Foundation for the Revitalisation of Local Health Traditions (FRLHT) *www.frlht.org*

Kalpavriksh *www.kalpavriksh.org*

People for Ethical Treatment of Animals (PeTA) *www.petaindia.com*

The Energy and Resources Institute (Previously called Tata Energy Research Institute) *www.teri.org*

Forming an organization

NGOs are voluntary groups, often set up by committed individuals, who want to make a positive change in society. Normally, an NGO focuses on a particular field that is of concern to the founder—

women's development, eradication of child labour, homes for street children and orphans, education of the poor and underprivileged, welfare of the handicapped, etc. In recent years, many NGOs in India have begun focusing on environmental conservation.

An NGO may lack official backing and large funds, but that can be made up by dedication, moral authority, and efficient and effective use of limited funds. Some hints for starting an environmental organization:

- Identify a group of likeminded youngsters who will be ready to commit their time and effort to the cause of the environment.
- Start small, begin with a simple neighbourhood activities and modest aims.
- Gain credibility by showing results.
- Learn from other NGOs by working as a volunteer.
- Register your organization as a society or trust.
- Attend workshops on fund raising, accounting, managing NGOs, etc.
- Apply for funds under state and central government schemes; check with the Ministry of Environment and Forests (MoEF), Department of Science and Technology.
- Raise funds from foreign sources; you will need clearance from the Union Ministry of Home Affairs under the Foreign Contribution (Regulation) Act.

Environment-related careers

Here is a sample list of careers related to environmental issues:

- Air quality engineer
- Conservation biologist
- Consultant for Environmental Impact Assessment
- Ecological scientist
- Energy auditor
- Environmental auditor in a financial institution
- Environmental health scientist
- Environmental journalist
- Environmental lawyer
- Environmental scientist or engineer in a company or a Pollution Control Board
- Environmental toxicologist
- Forest ecologist
- Government regulator
- Noise-control specialist
- Salaried staff member with Indian and global NGOs
- Solid waste manager
- Water quality technologist

Appendix A5

Environment timeline and calendar

Environment Timeline

The most important events in the world from the point of view of the environment, beginning with the 1960s:

1962–72

- Rachel Carson publishes *Silent Spring*
- Thalidomide cases
- The *Torrey Canyon* oil spill on the northern coast of France
- Acid rain kills thousands of fish in the Swedish lakes

1972

- United Nations Conference on the Human Environment, Stockholm, Sweden
- United Nations Environment Programme (UNEP) established
- The Club of Rome publishes *Limits to Growth*
- UNESCO Convention Concerning the Protection of the World Cultural and Natural Heritage
- London Dumping Convention

1973

- Convention on International Trade in Endangered Species of Wild Fauna and Flora (CITES)
- Drought in the Sahel, Africa, kills millions of people
- First oil crisis

1975

- Great Barrier Reef Marine Park declared in Australia

1976

- Dioxin released in an industrial accident at a pesticide plant in Seveso, Italy
- Major earthquake in Tangshan in eastern China
- Major earthquake in Guatemala

1977

- Toxic chemicals leak into homes in Love Canal, USA
- United Nations Conference on Desertification, Nairobi, Kenya
- Green Belt Movement established in Kenya

1978

- Floods in West Bengal kill 1300 people and destroy 1.3 million homes

1979

- Major accident at the Three Mile Island nuclear power plant, USA
- First World Climate Conference, Geneva, Switzerland
- 640 km oil slick in the Gulf of Mexico from a blow-out
- Convention on the Conservation of Migratory Species of Wild Animals (CMS)

1980

- World Climate Programme established
- *World Conservation Strategy* launched by IUCN, UNEP, and WWF
- Start of the International Decade for Drinking Water and Sanitation
- Brandt Commission publishes *North-South: a Programme for Survival*

1982

- United Nations Convention on the Law of the Sea (UNCLOS)
- United Nations General Assembly adopts the World Charter for Nature

1983

- Monsoon storms in Thailand claim 10,000 victims

1984

- Famine in Ethiopia caused by exceptional and long-lasting drought
- Chemical accident in Bhopal kills thousands and maims many more
- World Industry Conference on Environmental Management
- Typhoon in the Philippines kills 1300 and leaves 1.12 million homeless

1985

- Vienna Convention for the Protection of the Ozone Layer
- Size of Ozone Hole measured for the first time
- International Conference on the Assessment of the Role of Carbon Dioxide and other Greenhouse Gases, Villach, Austria

1986

- World's worst nuclear disaster occurs at Chernobyl, USSR, spreading radioactive fall-out over large areas of Europe
- International Whaling Commission imposes a moratorium on commercial whaling
- Fire in Basel, Switzerland, releases toxic chemicals into the Rhine, killing fish as far north as the Netherlands

1987

- Montreal Protocol on Substances that Deplete the Ozone Layer
- *Our Common Future* (the Brundtland report) publicizes the idea of sustainable development

1988

- UN resolution recognizes climate change as a 'common concern of mankind'
- Hurricane Gilbert kills 350, leaves 750,000 homeless, and causes damage of US$ 10 billion in the Caribbean, Mexico, and USA

1989

- *Exxon Valdez* releases 50 million litres of crude oil into Prince William Sound, Alaska
- Basel Convention on the Transboundary Movements of Hazardous Wastes and their Disposal
- Intergovernmental Panel on Climate Change (IPCC) established

1990

- First IPCC Assessment Report warns of impending global warming
- Second World Climate Conference, Geneva, Switzerland

- Global Climate Observing System created

1991
- Millions of litres of crude oil spilled and burned during the Gulf War
- Global Environment Facility (GEF) established
- *Caring for the Earth* published by IUCN, UNEP and WWF

1992
- UN Conference on Environment and Development (the Earth Summit) in Rio de Janeiro, Brazil
- Convention on Biological Diversity
- UN Framework Convention on Climate Change

1993
- Chemical Weapons Convention
- World Conference on Human Rights, Vienna, Austria

1994
- First Meeting of the Factor 10 Club, Carnoules, France
- UN Convention to Combat Desertification
- International Conference on Population and Development, Cairo, Egypt
- Global Conference on the Sustainable Development of Small Island Developing States, Bridgetown, Barbados
- Burst pipeline spills thousands of tonnes of crude oil on the tundra on the Kori Peninsula, Russian Federation

1995
- Year of the Sea Turtle
- World Summit for Social Development, Copenhagen, Denmark
- Fourth World Conference on Women, Beijing, China
- Second IPCC Assessment released, acts as major incentive for the Kyoto Protocol
- World Business Council for Sustainable Development created

1996
- UN Conference on Human Settlements (Habitat II), Istanbul, Turkey
- World Food Summit, Rome, Italy
- ISO 14000 created for an Environmental Management System in industry
- Comprehensive Nuclear Test Ban Treaty

1997
- Kyoto Protocol adopted
- Rio +5 Summit reviews implementation of *Agenda 21*

1998
- Warmest year of the millennium
- Extensive forest fires in Amazonia and Indonesia
- Rotterdam Convention on the Prior Informed Consent Procedure for Certain Hazardous Chemicals and Pesticides in International Trade

1999
- Launch of *Global Compact* on labour standards, human rights, and environmental protection
- World population reaches six billion

2000
- Size of Ozone Hole reaches record high, affecting tip of South America
- Cartagena Protocol on Biosafety adopted
- Millennium Summit, New York, USA
- World Water Forum, The Hague, The Netherlands

2001
- IPCC publishes its Third Assessment, with increased estimates of global warming
- Stockholm Convention on Persistent Organic Pollutants (POPs)

2002
- World Summit on Sustainable Development, Johannesburg, South Africa

2003
- 3rd World Water Forum, Kyoto, Japan
- New figures for deforestation in Amazonia are the highest in six years
- Serious floods in Mexico and Venezuela
- International Coral Reef Symposium in Okinawa, Japan,
- Earth Observation Summit, Washington, D.C., USA
- Heatwaves in Europe
- The Cartagena Protocol on safety and trade in genetically modified organisms and their derivatives comes into force.

- Planet's list of protected areas tops 100,000 mark
- Size of Ozone Hole over the Antarctic matches the record set three years previously
- World Climate Change Conference, Moscow, Russia
- Sustained drought in Argentina, Bolivia, Paraguay and Uruguay.
- Break-up of the Arctic's largest ice-shelf—the Ward Hunt
- Big forest fires in California, USA
- Catastrophic earthquake in the city of Bam in southeastern Iran

2004

- Floods in Haiti and the Dominican Republic
- Russia ratifies Kyoto Protocol
- Tsunami, triggered by a massive undersea earthquake off Sumatra in Indonesia, leaves over 300,000 dead and many more homeless in India, Sri Lanka, and South-East Asia.

Source: UNEP (2002). *Global Environment Outlook 3: Past, Present and Future Perspectives*. Also the website *www.unep.org/geo/yearbook/002.htm.*

Environment Calendar

Some of the important days and weeks you can observe during the year:

Date	Day/Week	Reference
February 2	World Wetlands Day	Box 4.6
March 8	International Women's Day	Chapter 17
March 21	World Forestry Day	Chapter 9
March 22	World Water Day	Box 7.5
April 5	National Maritime Day	Box 4.7
April 22	Earth Day	Box 2.4
May 22	International Day of Biological Diversity	Box 6.8
June 17	World Day to Combat Desertification and Drought	Chapter 10
July 1–7	Vanamahotsava Week	Chapter 3
July 11	World Population Day	Chapter 15
September 16	International Day for the Preservation of the Ozone Layer	Chapter 19
September 22	Car Free Day	Box 11.7
October (First Monday)	World Habitat Day	Chapter 15
October (Second Wednesday)	International Day for Natural Disaster Reduction	Chapter 14
October 16	World Food Day	Chapter 10
October 29	National Day of Disaster Reduction	Chapter 14
November 14	National Children's Day	Chapter 17
November 21	World Fisheries Day	Box 10.7
November 21–27	National Land Resources Conservation Week	Chapter 10
December 1	World AIDS Day	Box 17.7
December 3	Global Day of Action Against Corporate Crime (Bhopal Gas Tragedy Day)	Box 16.8
December 11	International Mountain Day	Chapter 3
December 14	National Energy Conservation Day	Chapter 8
December 23	Farmer's Day	Chapter 10

Appendix A6

Links to important policies, conventions, and related documentation

This appendix presents a list of important documents and policy statements that are relevant to Environmental Studies. It is intended for those readers who wish to delve deeper into the subject. This is by no means an exhaustive list and the readers are encouraged to spread their net wider.

Conventions, declarations, agreements...

- Convention on Biological Diversity (UN) (CBD), Rio, 1992: To conserve the biological diversity, the sustainable use of its components, and the fair and equitable sharing of the benefits arising out of the utilization of genetic resources, taking into account all rights over those resources and to technologies.
  www.biodiv.org/doc/legal/cbd-en.pdf
- Framework Convention on Climate Change (UNFCCC), New York, 1992: To achieve stabilization of greenhouse gas concentrations in the atmosphere at a level that would prevent dangerous anthropogenic interference with the climate system. Such a level should be achieved within a time-frame sufficient to allow ecosystems to adapt naturally to climate change, to ensure that food production is not threatened, and to enable economic development to proceed in a sustainable manner.
  unfccc.int/resource/conv/conv.html
- Convention on Persistent Organic Pollutants (The POP Stockholm Convention), Stockholm, 2001: To protect human health and the environment from persistent organic pollutants.
  www.pops.int/documents/convtext/convtext_en.pdf
- Convention on the Prior Informed Consent Procedure for Certain Hazardous Chemicals and Pesticides in International Trade (UNEP/FAO) (The PIC Rotterdam Convention), Rotterdam, 1998: To promote shared responsibility and co-operative efforts among Parties in International Trade of certain hazardous chemicals, in order to protect human health and the environment from potential harm and to contribute to their environmentally sound use.
  www.pic.int/en/ViewPage.asp?id=104#Preamble
- Convention on Access to Environmental Information, Public Participation in Environmental Decision-making and Access to Justice (The Aarhus Convention), Aarhus, 1998: To contribute to the protection of the right of every person of present and future generations to live in an environment adequate to his or her health and well-being, each Party shall guarantee the rights of access to information, public participation in decision-making, and access to justice in environmental matters in accordance with the provisions of this Convention.
  www.unece.org/env/pp/documents/cep43e.pdf
- Convention to Combat Desertification in Countries Experiencing Serious Drought and/or Desertification, 1994, 1998: To combat desertification and mitigate the effects of drought in countries experiencing serious drought and/or desertification, particularly in Africa, through effective action at all levels.
  www.unccd.int/convention/text/convention.php
- International Tropical Timber Agreement (ITTA), Geneva, 1994, 1996, 1997: To promote and apply comparable and appropriate guidelines and criteria for the management, conservation, and sustainable development of all types of timber producing forests.
  www.itto.or.jp
- Convention on Transboundary Prevention of, Preparedness for and Response to Industrial Accidents—Effects of Industrial Accidents (UN/ECE) (The Accident Convention), Helsinki, 1992
  www.unece.org
- Convention on the Protection and Use of Transboundary Watercourses and International Lakes (The UN/ECE Water Convention), Helsinki, 1992: To protect and manage in an ecologically sound manner, transboundary surface waters and

groundwaters. To prevent, control, and reduce water pollution from point and nonpoint sources.
www.unece.org/env/water/text/water_convention/text11toc.htm

- Convention on the Control of Transboundary Movements of Hazardous Wastes and their Disposal (Basel Convention), Basel, 1989: To ensure that the management of hazardous wastes and other wastes including their transboundary movement and disposal is consistent with the protection of human health and the environment whatever the place of disposal.
www.basel.int/text/con-e.htm
- Convention for the Protection of the Ozone Layer (UNEP) (Vienna Convention), Vienna, 1985: To protect human health and the environment against adverse effects resulting or likely to result from human activities which modify or are likely to modify the ozone layer.
www.unep.org/ozone/vienna.shtml
- Protocol to the Convention for the Protection of the Ozone Layer on Substances that Deplete the Ozone Layer (MONTREAL Protocol), as amended, Montreal, 1987: To protect the ozone layer by taking precautionary measures to equitably control total global emissions of substances that deplete it, with the ultimate objective of their elimination on the basis of developments in scientific knowledge, taking into account technical and economic considerations and bearing in mind the developmental needs of developing countries.
www.unep.org/ozone/montreal.shtml
- Convention on Long-range Transboundary Air Pollution (CLRTAP) (UN-ECE), Geneva, 1979: To protect man and his environment against air pollution and to limit and, as far as possible, gradually reduce and prevent air pollution including long-range transboundary air pollution.
www.unece.org/env/lrtap/lrtap_h1.htm
- Protocol to the Convention on Long-range Transboundary Air Pollution on Persistent Organic Pollutants (POPs), Aarhus, 1998: To control, reduce or eliminate discharges, emissions, and losses of persistent organic pollutants.
www.unece.org/env/lrtap/pops_h1.htm
- Convention on the Conservation of Migratory Species of Wild Animals (CMS) (UNEP), Bonn, 1979: To conserve migratory species and take action to this end, paying special attention to migratory species the conservation status of which is unfavourable, and taking individually or in co-operation appropriate and necessary steps to conserve such species and their habitat.
www.wcmc.org.uk/cms
- Convention for the Protection of the Mediterranean Sea against Pollution (Barcelona Convention), Barcelona, 1976: To prevent, abate, and combat pollution of the Mediterranean Sea area and to protect and enhance the marine environment in that area.
www.unep.ch/seas/main/med/medconvi.html
- Convention on International Trade in Endangered Species of Wild Fauna and Flora (CITES), 1973
www.cites.org/eng/disc/text.shtml
- Convention on the Physical Protection of Nuclear Material, 1980
www.unodc.org/unodc/terrorism_convention_nuclear_material.html
- UN Conference on Environment and Development, 1992
www.un.org/geninfo/bp/enviro.html
    - ♦ Rio Declaration on Environment and Development, Principles of Forest Management
    www.iisd.org/educate/learn/rio_decl.doc
    - ♦ *Agenda 21*: Global Programme of Action for Sustainable Development
    www.unep.org/Documents/Default.asp?DocumentID=52&ArticleID=49
    - ♦ Convention on Biological Diversity
    www.biodiv.org/convention/articles.asp
- United Nations Convention on the Law of the Sea of 10 December 1982 (UNCLOS)
www.un.org/Depts/los/convention_agreements/texts/unclos/UNCLOS-TOC.htm
- United Nations Agreement on Straddling Fish Stocks and Highly Migratory Fish Stocks (UNFA)
www.un.org/Depts/los/convention_agreements/convention_overview_fish_stocks.htm
- The FAO Code of Conduct for Responsible Fisheries
www.fao.org/fi/agreem/codecond/codecon.asp

- A Compendium of Fisheries Agreements
  www.oceanlaw.net/texts/index.htm
- World Summit on Sustainable Development (WSSD Johannesburg) Reports
  www.uneptie.org/outreach/wssd/joburgreport.htm
- UN Water Development Report: Water for people, water for life
  www. unesdoc.unesco.org/images/0012/001295/129556e.pdf
- Report of the World Commission on Dams
  www.dams.org
- Citizens' Guide to the World Commission on Dams
  www.irn.org/wcd/wcdguide.pdf
- World Commission on Forests and Sustainable Development
  www.iisd.org/wcfsd/
- Independent World Commission on the Oceans
  www.waterland.net/iwco/
- Living Planet Report, October 2004
  www.panda.org/downloads/general/lpr2004.pdf
- Global Environment Outlook 3
  www.unep.org/geo/geo3/english/pdf.htm
- World Scientists' Warning to Humanity, 1992
  www.ucsusa.org/ucs/about/page.cfm?pageID=1009
- Scientists' Call for Action (1997)
  www.ucsusa.org/global_environment/ archive/page.cfm?pageID=530
- Millennium Development Goals
  www.undp.org/mdg
- World Charter for Nature
  www.un.org/documents/ga/res/37/a37r007.htm

Appendix A7

Environmental studies: UGC core module syllabus (Six-month compulsory course for undergraduates)

The core module will be integrated with teaching programmes of all undergraduate courses. The syllabus includes classroom teaching and fieldwork. The syllabus is divided into eight units covering 50 lectures. The first seven units comprise 45 lectures to enhance the knowledge of environment. The eighth unit is based on field activities and provides students firsthand information on local environmental aspects.

The duration of the course will be 50 lectures.

Syllabus

(The number of lecture hours for each module is given in brackets.)

Module 1: Multidisciplinary nature of environmental studies (2)

Definition, scope and importance, need for public awareness

Module 2: Eco-systems (6)

Concept of an ecosystem, structure and function of an ecosystem, producers, consumers and decomposers, energy flow in an ecosystem, ecological succession, food chains, food webs and ecological pyramids

Introduction, types, characteristic features, structure and function of: forests, grassland, desert, and aquatic (ponds, streams, lakes, rivers, oceans, estuaries) ecosystems

Module 3: Biodiversity and its conservation (8)

Introduction, definitions: genetic, species and ecosystem biodiversity, biogeographical classification of India, value of biodiversity, consumptive use, productive use, social, ethical, aesthetic and option values, biodiversity at global, national and local levels, India as a mega-diversity nation, hotspots of biodiversity, threats to biodiversity: habitat loss, poaching of wildlife, man-wild life conflicts, endangered and endemic species of India Conservation of biodiversity: in-situ and ex-situ conservation

Module 4: Renewable and non-renewable natural resources (8)

Forest resources: use and over-exploitation, deforestation, case studies, timber extraction, mining, dams and their effects on forests and tribal people

Water resources: use and over-utilization of surface and ground water, floods, drought, conflicts over water, dams: benefits and problems

Mineral resources: use and exploitation, environmental effects of extracting and using mineral resources, case studies

Food resources: world food problems, changes caused by agriculture and over-grazing, effects of modern agriculture, fertilizer-pesticide problems, water logging, salinity, case studies

Energy resources: growing energy needs, renewable and non-renewable energy sources, use of alternate energy sources, case studies

Land resources: land as a resource, land degradation, man-induced landslides, soil erosion and desertification

Role of an individual in conservation of natural resources

Equitable use of resources for sustainable lifestyles

Module 5: Environmental pollution (8)

Cause, effects and control measures of: Air pollution, water pollution, soil pollution, marine pollution, noise pollution, nuclear hazards

Solid waste management, urban and industrial wastes: causes, effects and control measures

Disaster management: floods, earthquakes, cyclones, landslides

Role of an individual in prevention of pollution, pollution case studies

Module 6: Social issues and the environment (7)

From unsustainable to sustainable development, urban problems related to energy, water conservation, rainwater harvesting, watershed management Resettlement and rehabilitation of people: problems, concerns, case studies Environmental ethics: issues and possible solutions

Climate change, global warming, acid rain, ozone layer depletion, nuclear accidents and holocaust, case studies, wasteland reclamation, consumerism and waste products

Environmental law: Environment Protection Act, Air (Prevention and Control of Pollution) Act, Water (Prevention and Control of Pollution) Act, Wildlife Protection Act, Forest Conservation Act

Issues involved in enforcement of environmental legislation, public awareness

Module 7: Population and the environment (6)

Population growth, variation among nations, population explosion, family welfare programme, environment and human health, human rights, value education, HIV/AIDS, women and child welfare, role of information technology in environment and human health, case studies

Module 8: Field work (5)

Visit to a local area to document environmental assets: river, forest, grassland, hill, mountain

Visit to a local polluted site: urban, rural, industrial, agricultural

Study of common plants, insects, birds

Study of simple ecosystems: ponds, river, hill slopes

Exam pattern

Part A: Short-answer Questions	25 marks
Part B: Essay Questions	50 marks
Part C: Field Work	25 marks

Appendix A8

Glossary

Abiotic conditions The non-living components of ecosystems.

Abyssal zone The cold and dark zone at the bottom of the ocean.

Acid rain Rain, mist, or snow formed when atmospheric water droplets combine with a range of man-made chemical air pollutants.

Agent Orange A compound herbicide, used extensively as a defoliant by the US Army in the Vietnam War.

AIDS (Acquired Immune Deficiency Syndrome) A condition in which the body's immune system is weakened and therefore less able to fight certain infections and diseases. It is caused by the Human Immunodeficiency Virus (HIV).

algal bloom A population explosion of some pigmented marine algae seen as an explosion of colour on the ocean—orange, red, or brown.

Aquaculture The artificial production of fish in ponds or underwater cages.

Aquatic life zone The non-terrestrial part of the biosphere including wetlands, lakes, rivers, estuaries, inter-tidal zone, coastal ocean, and open ocean.

Background extinction The gradual disappearance of species due to changes in local environmental conditions.

Bathyal zone The dimly lit middle level zone in the ocean, roughly between 200 m and 1500 m in depth.

Benthos Bottom-dwelling organisms adapted to living on the floor of a water body.

Biodiversity The numbers, variety, and variability of living organisms and ecosystems. It covers diversity within species, between species, as well as the variation among ecosystems. It is concerned also with their complex ecological interrelationships.

Biodynamic farming A type of organic farming that exploits bio- and solar rhythms. It is based on the ideas of Rudolf Steiner.

Biogeochemical cycle A cycle (with biological, geological, and chemical interactions) through which matter moves through ecosystems, powered directly or indirectly by solar energy. The water and carbon cycles are examples.

Bioinformatics A field of study that combines biology with computer science and thus uses the power of information technology to study and conserve biodiversity.

Biointensive farming A type of organic farming involving intensive garden cultivation using deep-dug beds.

Biological extinction The complete disappearance of a species. It is an irreversible loss with not a single member of the extinct species being found on Earth.

Biome A group of similar or related ecosystems with a distinct climate and life forms adapted to the climate. A biome is more extensive and complex than an ecosystem. It is the next level of ecological organization above a community and an ecosystem.

Biomedical waste Waste that originates mainly from hospitals and clinics and includes blood, diseased organs, poisonous medicines, etc.

Biosphere That portion of the planet and its environment, which can support life. The biosphere includes most of the hydrosphere, parts of the lower atmosphere and the upper lithosphere.

Biota The living component of an ecosystem, also called the biotic community. The biotic community includes the plants, animals, and micro-organisms.

Biotechnology Technology that manipulates the genes in an organism to change its characteristics.

Bog Water-logged soil, which may or may not have trees.

Carbon cycle Cyclic movement of carbon in various forms from the environment to organisms and back to the environment.

Chlorofluorocarbon (CFC) A type of chemical that is used as a refrigerant or aerosol propellant. When it breaks apart in the atmosphere and releases chlorine atoms, it causes ozone depletion.

Coastal zone The area that extends from the high tide mark on land to the edge of the continental shelf, which is the submerged part of the continent.

Common effluent treatment plant A plant that treats waste collected from several units in an industrial area.

Community The assemblage of all the interacting populations of different species existing in a geographical area. It is a complex interacting network of plants, animals, and micro-organisms.

Coniferous forest A type of forest with an abundance of coniferous trees like spruce, fir, pine, and hemlock.

Consumer An organism that feeds on producers or other organisms, also called a heterotroph.

Continental shelf The submerged part of the continent at the edge of the coastal zone, where there is a sharp increase in the depth of water. It marks the beginning of the open ocean.

Coral reef A formation by huge colonies of tiny organisms called polyps that secrete a stony substance around themselves for protection. When the corals die, their empty outer skeletons form layers and cause the reef to grow.

Crude birth rate The number of live births per 1000 people in a population in a given year.

Crude death rate The number of deaths per 1000 people in a population in a given year.

Cyclone Violent tropical storm in which strong winds move in a circle. It occurs mainly in the Indian Ocean.

DDT (Dichloro-diphenol-trichloroethane) An insecticide that protects crops and human beings from insects. Being harmful to organisms, it is now banned in many countries.

Decibel (db) A logarithmic scale for measuring the intensity of sound.

Decomposer An organism that gets its nourishment from dead organic material.

Desertification Land degradation in arid and semi-arid areas caused by human activities and climatic changes.

Detrivore A consumer that feeds on detritus, which refers mainly to fallen leaves, parts of dead trees, and faecal wastes of animals.

Dioxin A chemical compound that belongs to a group that occurs as unwanted contaminants in a number of industrial processes and products. Dioxins are highly toxic to humans and animals and are carcinogens.

Ecological architecture Architecture that seeks to minimize the ecological footprint of the house, building, or complex that is being designed and constructed.

Ecological extinction The state of a species when so few members are left that the species can no longer play its normal ecological role in the community.

Ecological footprint A measure of the ecological impact of an entity, expressed as the extent of land needed to completely sustain the entity.

Ecological niche All the physical, chemical, and biological factors that a species needs in order to live and reproduce. The niche is characterized by the particular food habits, shelter-seeking methods, ways of nesting and reproduction, etc., of the species.

Ecological pyramid A graphical representation of the change that occurs as we move from one trophic level to the next in a food chain. Pyramid of energy and pyramid of biomass are examples.

Ecological sanitation (EcoSan) A sustainable closed-loop sanitation system that uses dry composting toilets.

Ecological succession The orderly process of transition from one biotic community to another in a given area.

Ecology The science that studies the relationships between living things and their environment. It is often considered to be a discipline of biology.

Ecosystem A defined area in which a community (with its populations of species) lives with interactions taking place among the organisms and between the community and its non-living physical environment.

Ecosystem diversity The variety of habitats found in an area, that is, the variety of forests, deserts, grasslands, aquatic ecosystems, etc., that occur in the area.

Ecosystem service The ecological service provided by an ecosystem, such as the maintenance of the biogeochemical cycles, modification of climate, waste removal and detoxification, and control of pests and diseases.

Ecotone The transitional zone between adjoining ecosystems.

Edge effect The unique nature of biodiversity in an ecotone. An ecotone contains additional species not found in the adjoining ecosystems. Or, the ecotone may have greater population densities of certain species than in the adjoining communities.

Electromagnetic fields Electromagnetic radiation emitted by sources like television receivers, computer monitors, overhead electric lines, and mobile phones.

Endangered species The state of a species when the number of survivors is so small that it could soon become extinct over all or most of its habitat. Unless it is protected, it will move into the critically endangered category, before it becomes extinct.

Energy flow Flow of energy in a food chain from one organism to the next in a sequence.

Environment The natural world in which people, animals, and plants live.

Environmental ethics Moral principles that try to define our responsibility towards the environment.

Environmental health Those aspects of human health that are determined by physical, chemical, biological, social, and psychosocial factors in the environment.

Environmental Impact Assessment (EIA) An assessment of the environmental impact of any development project.

Environmental science The systematic and scientific study of our environment and our role in it.

Environmental studies The branch of study concerned with environmental issues. It has a broader canvas than environmental science and includes the social aspects of the environment.

Euphotic zone The upper part of the open ocean where there is enough light for the phytoplankton to carry out photosynthesis.

Eutrophication The enrichment of a standing water body by nutrients such as phosphorus and nitrogen.

E-waste Electronic waste that results from discarded devices like computers, televisions, telephones, and music systems.

Exclusive Economic Zone (EEZ) The area of the ocean up to a distance of 200 nautical miles from a country's shoreline over which the country has the exclusive right to exploit the resources.

Exponential growth The growth of a quantity with time in such a way that the growth curve is relatively flat in the beginning, but becomes steeper and steeper with time.

Ex-situ conservation Conservation that attempts to preserve and protect the species in a place away from their natural habitat.

Extractive reserve Protected forests, in which local communities are allowed to harvest products like fruits, nuts, rubber, oil, fibres, and medicines in ways that do not harm the forest.

First Law of Thermodynamics Energy can neither be created nor destroyed. It can only be transformed from one form to another.

Fluorosis An ailment caused by the excess intake of fluoride.

Food chain A sequence of organisms, in which each is the food for the next.

Food web A complex network of interconnected food chains.

Fossil fuel Remains of organisms that lived 200–500 million years ago that were converted by heat and pressure into coal, oil, and natural gas.

Fuel cell An electrochemical unit that burns hydrogen to produce electricity.

Genetic diversity The variety in the genetic makeup among individuals within a species.

Global warming The occurrence of higher temperatures on Earth due to an abnormal increase in the concentration of greenhouse gases.

Grassland Regions where the average annual precipitation is high enough for grass and a few trees to grow.

Green economics A model of economics that seeks to conserve the environment and at the same time increase employment and income levels. It measures the success of the economy by the improvement in the well-being of all people and the development of healthy, vibrant, diverse communities, rather than by the total monetary value of economic activity.

Green livelihood Productive and sustainable livelihoods based on natural resources.

Green Revolution The rapid increase in world food production, especially in the developing countries, during the second half of the twentieth century, primarily through the use of lab-engineered high-yielding varieties of seeds (HYVs).

Greenhouse gas A gas like carbon dioxide that surrounds the Earth and prevents some of the Sun's heat from being reflected back out again.

Habitat The area where a species is biologically adapted to live. It is marked by the physical and biological features of its environment, such as the vegetation, climatic conditions, presence of water and moisture, and soil type.

Habitat fragmentation The process by which continuous areas of species habitat are reduced in extent or divided into a patchwork of isolated fragments due to human impact.

Headwaters The streams that form the source of a river.

Heap-leach mining A gold mining process, in which streams of cyanide are poured over huge piles of low-grade ore to extract the metal.

HIV (Human Immunodeficiency Virus) The virus that causes AIDS.

Hubbert Curve A curve proposed by the geophysicist M. King Hubbert that describes the pattern of oil availability in a field over time.

Hurricane Violent storm with very strong winds experienced mainly in the western Atlantic Ocean.

Indicator species Species that are very sensitive indicators of environmental problems. They give us early warning of problems that could potentially affect other species.

In-situ conservation Conservation that tries to protect species where they are, that is, in their natural habitat.

Intertidal zone The area of shoreline between the low and high tides. It is the transition between the land and the ocean.

Joint forest management (JFM) A model of forest management in which the local communities are involved in the planning of the conservation programme.

Judicial activism The pro-active role of the judiciary in responding to the complaints of citizens or public interest litigation (PIL) and in assuming the role of policy maker, educator, administrator, and, in general, friend of the environment.

Keystone species Species that play roles affecting many other organisms in an ecosystem. They determine the ability of a large number of other species to survive.

Landfill An area, usually located just outside the city, on which municipal waste is dumped.

Local extinction The state of a species when it is no longer found in the area it once inhabited. It is, however, present elsewhere in the world.

Malthusian Theory of Population Theory on growth of human population enunciated by the British economist and demographer Thomas Robert Malthus. It predicted that human population would outgrow the capacity of land to produce food and that famine, plagues, natural disasters, and wars would then control the population.

Mangrove A unique salt-tolerant tree with interlacing roots that grows in shallow marine sediments.

Marsh Wetland with few trees.

Mass extinction A global, catastrophic extinction of species, with more than 65 per cent of all species becoming extinct over some millions of years. It is characterized by a rate of disappearance significantly higher than background extinction.

Maximum sustainable yield (MSY) The yield of a species of fish that we could harvest annually leaving enough breeding stock for the population to renew itself. It is the amount of a fish species that we can catch every year indefinitely.

Natural farming No-tillage farming, pioneered by Masanobu Fukuoka of Japan.

Nautical mile The unit of distance used in the ocean. One nautical mile equals 1.85 km.

Nekton Relatively strong-swimming aquatic organisms such as fish and turtles.

New economics See green economics

Old-growth forest Forests that have not been seriously disturbed by human activities or natural disasters for several hundred years or more.

Opportunistic infection Infection that attacks the body when its immune system has broken down, say, by the HIV.

Organic farming A method of farming that does not use chemical fertilizers and chemical pesticides. It is a return to the traditional methods like crop rotation, use of animal and green manures, and some forms of biological control of pests.

Ozone Depleting Potential (ODP) A number that refers to the amount of ozone depletion caused by a substance.

Ozone layer A layer of ozone that exists in the upper atmosphere, or stratosphere, between 10 and 50 km above the Earth.

PCBs (Polychlorinated biphenyls) A group of toxic chemical compounds that are very stable, have good insulation qualities, are fire resistant, and have low electrical conductivity. They can also pass through food chains.

Permaculture A way of designing sustainable human settlements through an approach to land use that weaves together microclimate, annual and perennial plants, animals, soils, water management, and human needs into intricately connected, productive communities.

Persistent organic pollutants (POPs) A group of persistent, toxic chemicals that can accumulate in organisms and can contaminate sites far removed from their source. They cause reproductive, immunological, and neurological problems in marine organisms and possibly in humans.

Photochemical smog A form of outdoor air pollution formed by chemical reactions between sunlight, unburnt hydrocarbons, ozone, and other pollutants.

Photosynthesis The process used by producers to convert inorganic material into organic matter.

Photovoltaic cell A device that converts solar energy directly into electricity.

Phytoplankton Photosynthetic producers of the ocean that form the basis of the ocean's food web.

Plankton Free-floating micro-organisms that cannot swim easily and are buffeted about by the waves and currents.

Polyp Tiny organisms that form corals.

Population The members of a species living and interacting within a specific geographical region. The term refers only to those members of a certain species that live within a given area.

Precipitation All the forms in which water comes down on Earth, for example, rain, snow, and hail.

Primary air pollutant A harmful chemical that is released directly from a source into the atmosphere.

Producer An organism such as a green plant that can produce food from simple inorganic substances. Also called autotroph.

Public interest litigation (PIL) Petitions filed in courts by individual citizens, groups, and voluntary organ-izations against pollution and degradation of the environment.

Pyramid of biomass A graphical representation of the reduction in biomass of organisms as we move from one trophic level to the next.

Pyramid of energy A graphical representation of the reduction in usable energy of organisms as we move from one trophic level to the next.

Rainforest A type of forest found in the hot and humid regions near the Equator These regions have abundant rainfall and little variation in temperature over the year.

Renewable energy Type of energy (like solar energy) that is replaced by natural processes and can be used forever.

Reverse osmosis The purification of water by forcing it through a semi-permeable membrane.

Richter scale A measure of the severity of an earthquake. It is a measure of the amount of energy released, which is indicated by vibrations in a seismograph.

Sacred grove A forest that is protected by the local community through social traditions and taboos that incorporate spiritual and ecological values.

Savanna Tropical grasslands with widely scattered clumps of low trees. They are marked by low rainfall and prolonged dry periods.

Second Law of Thermodynamics Whenever we transform energy from one form into another, a part of it is lost as heat. That is, all the energy is not available to do work. Some of it just disperses into the environment as heat.

Secondary air pollutant Harmful chemical that is produced from chemical reactions involving primary pollutants.

Second-growth forest A forest that results from secondary ecological succession that takes place when forests are cleared and then left undisturbed for long periods of time.

Seismograph An instrument that measures the amount of energy released in an earthquake by sensing the vibrations.

Sentinel species See Indicator species

Ship-breaking The process of breaking up decommissioned ships for recycling the parts to the extent possible.

Smog A form of outdoor pollution that is localized in urban areas, where it reduces visibility. The term was originally used to describe a combination of smoke, fog, and chemical pollutants that poisoned the air in industrialized cities.

Social forestry The planting of trees, often with the involvement of local communities, in unused and fallow land, degraded government forest areas, in and around agricultural fields, along railway lines, roadsides, river and canal banks, in village common land, government wasteland, and panchayat land.

Species A species is a set of organisms that resemble one another in appearance and behaviour. The organisms in a species are potentially capable of reproducing naturally among themselves. The term includes all members of a certain kind, even if they exist in different populations in widely separated areas.

Species diversity The number of plant and animal species present in a community or an ecosystem.

Sustainable development Development that meets the needs of the present without compromising the ability of future generations to meet their own needs.

Swamp A wetland dominated by trees and shrubs.

Tailings Waste products that come out of the initial processing of the uranium ore.

Temperate forest A type of forest with seasonal variations in climate—freezing in winter and warm and humid in summer.

Threatened species A species that is still found in reasonable size in its natural habitat, but in declining numbers. Unless conservation measures are taken, it is likely to move into the next category, that is, the endangered list.

Trophic level The specific feeding stage of an organism in the ecosystem.

Tsunami An extremely large wave in the ocean caused by an earthquake.

Tundra The forests in the Arctic. They occur in the extreme northern latitudes, where the snow melts seasonally.

Typhoon Violent storm with very strong winds experienced mainly in the western Atlantic Ocean.

Water cycle The cycle of water that includes evaporation, precipitation, and flow to the ocean.

Water harvesting The process of catching the rain when and where it falls.

Wetlands Land surfaces covered or saturated with water for a part or whole of the year.

Zooplankton The primary consumers that feed on phytoplankton.

Zooxanthellae The tiny single-celled algae that live inside the tissues of the polyps that form the corals. They produce food and oxygen through photosynthesis and give colour to the corals.

Index